AQA

GCSE SCIENCE

Editors: Vic Pruden and Keith Hirst

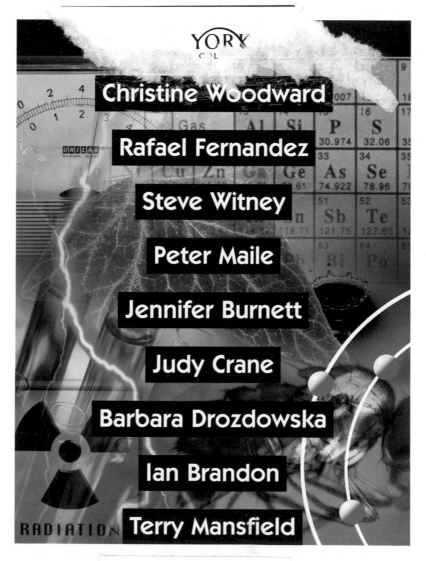

Christine Woodward

Rafael Fernandez

Steve Witney

Peter Maile

Jennifer Burnett

Judy Crane

Barbara Drozdowska

Ian Brandon

Terry Mansfield

ton

GROUP

Photo acknowledgements

The publishers would like to thank the following individuals, institutions and companies for permission to reproduce photographs in this book. Every effort has been made to trace ownership of copyright. The publishers would be happy to make arrangements with any copyright holder whom it has not been possible to contact:

Action Plus (34 top, 296, 366); Alcan Smelting & Power UK (184); Andrew Lambert (146, 262, 273 bottom); Associated Press (376 top); Bruce Coleman Ltd (102 all, 112 fennec fox, 120 middle, 292, 298 all); Corbis (84, 170, 206, 245, 343 top, 347, 373 top right, 381 both, 382 Niels Bohr); Hodder & Stoughton (100, 386); Holt Studios (73 both, 74 both, 75, 83, 96, 98 bottom); Life File (103 right, 120 right, 261, 267 bottom, 269, 273 top); Hulton Archive (128, 382 top); Museum of London (60); Natural History Museum (104 top, 338 bottom left, 339 top); PA Photos (323); RSPCA Photo Library (98 top left); Science Museum (248); Science Photo Library (1, 3, 11, 19, 23, 31 both, 34 bottom, 43 both, 57, 58, 59, 76, 82, 86, 92, 94, 95, 97, 98 top right, 103 left, 104 middle, 112 arctic fox, cacti, 113 both, 114, 116, 120 left, 127, 142, 173, 189, 199 all, 204, 213 all, 216, 249, 250 both, 251, 252, 263, 267 top, 287, 316, 331, 336, 338 top right & bottom right, 339 left, 341, 342, 343 bottom, 345, 346 both, 348, 372 all, 373 top left, middle & bottom, 375, 376 bottom left, 377 bottom); Robin Marshall (382 Ernest Marsden); Victoria and Albert Museum (190); Wellcome Trust (376 bottom right, 377 top).

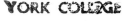
Orders: please contact Bookpoint Ltd, 130 Milton Park, Abingdon, Oxon OX14 4SB. Telephone: (44) 01235 827720, Fax: (44) 01235 400454. Lines are open from 9.00–6.00, Monday to Saturday, with a 24 hour message answering service. Email address: orders@bookpoint.co.uk

British Library Cataloguing in Publication Data
A catalogue record for this title is available from The British Library

ISBN 0 340 80247 2

First published 2000
This edition published 2001

Impression number   10   9   8   7   6   5   4   3   2   1
Year   2006   2005   2004   2003   2002   2001

Cover illustration by Sarah Jones, Debut Art
Typeset by J&L Composition Ltd, Filey, North Yorkshire.
Printed in Italy for Hodder & Stoughton Educational, a division of Hodder Headline Ltd, 338 Euston Road, London NW1 3BH.

Contents

About this book

The contents of this book are designed to cover all aspects of the knowledge and understanding needed if you are following any of the AQA GCSE specifications (courses) in:

- Science: Double Award (Co-ordinated)

- Science: Single Award (Co-ordinated)

- Science: Double Award (Modular)

- Science: Single Award (Modular)

The knowledge and understanding required by the Double Award specifications makes up the core content for each of the related separate sciences. This means that you can use this textbook to help you understand the core content if you are following any of the AQA GCSE specifications in:

- Biology

- Chemistry

- Physics

- Biology (Human)

How the content links with the courses

The contents are divided into 18 **chapters**. Chapters 1–6 deal with the biological aspects, chapters 7–12 deal with the chemical aspects and chapters 13–18 deal with the physical aspects. Each chapter is divided into a number of **sections**. The heading to each section clearly identifies which GCSE course the content addresses. The specification matching grids, in Tables 1.1, 1.2 and 1.3, show how the sections of the book relate to each of the courses.

Specification Matching Grids

Table 1.1   Life and Living Processes

Content		AQA specification references			
		Co-ordinated		Modular	
Chapter	Section	DA	SA	DA	SA
1 Cell activity	1.1 Animal cells	10.1	10.1	01	13
	1.2 Plant cells	10.1		02	
	1.3 Diffusion	10.2		01	
	1.4 Osmosis and active transport	10.2		02	
	1.5 Cell division	10.3	10.2	04	14
2 Humans as organisms	2.1 Nutrition	10.4	10.3	01	13
	2.2 Composition of the blood	10.5	10.4	01	13
	2.3 The circulatory system	10.5		01	
	2.4 Breathing	10.6		01	
	2.5 Respiration	10.7		01	
3 Response, co-ordination and health	3.1 Nervous system	10.8	10.5	02	13
	3.2 Hormones and the menstrual cycle	10.9	10.6	02/04	13/14
	3.3 Hormones and diabetes	10.9		02	
	3.4 Homeostasis	10.10	10.7	02	13
	3.5 Fighting disease	10.11	10.8	01	13
	3.6 Drugs	10.12	10.9	02	13
4 Green plants as organisms	4.1 Plant nutrition	10.13		02	
	4.2 Plant hormones	10.14		02	
	4.3 Transport and water relations	10.15		02	
5 Variation, inheritance and evolution	5.1 Variation	10.16	10.10	04	14
	5.2 Genes	10.17	10.11	04	14
	5.3 DNA	10.17		04	
	5.4 Controlling inheritance	10.18	10.12	04	14
	5.5 Evolution	10.19	10.13	04	14
6 Living things in their environment	6.1 Adaptation and competition	10.20	10.14	03	14
	6.2 Human impact on the environment	10.21	10.15	03	14
	6.3 Energy and nutrient transfer	10.22		03	
	6.4 Nutrient cycles	10.23		03	

Table 1.2 Materials and their Properties

Chapter	Section		AQA specification references			
			Co-ordinated		Modular	
			DA	SA	DA	SA
7 Getting together	7.1	Atoms	11.1/ 12.23	11.1/ 12.15	08/12	16/18
	7.2	Bonding	11.2		08	
	7.3	Quanitative chemistry	11.8		07	
8 Representing reactions	8.1	Representing chemical symbols, formulae and reactions	11.7	Intro	07/08	15/16
	8.2	Types of chemical reactions	Intro	Intro	07/08	15/16
	8.3	Exothermic and endothermic reactions	Intro		07/08	
9 The Atmosphere	9.1	Changes to the Atmosphere	11.9		06	
	9.2	Useful products from the air	11.6		07	
10 The Earth	10.1	The rock record	11.10		06	
	10.2	Useful products from metal ores	11.4	11.4	05	15
	10.3	Useful products from rocks	11.5	11.4	06	15
	10.4	Useful products from crude oil	11.3	11.2	06	15
11 Patterns of behaviour	11.1	The development of the periodic table	11.11	11.3	08	16
	11.2	Patterns in the periodic table	11.11	11.3	08	16
	11.3	Metals and the periodic table	11.11	11.3	08	15
	11.4	Patterns in the transition elements	11.12		05	
	11.5	Patterns in the reactions of metal halides (halogens)	11.12		08	
	11.6	Patterns in making metal compounds	11.12	11.4	05	15
12 Chemistry in action	12.1	Energy transfers in chemical reactions	11.16		07	
	12.2	Reversible reactions	11.15		07	
	12.3	Rates of reaction	11.13	11.5	07	16
	12.4	Reactions involving enzymes	11.14	11.6	07	16

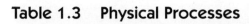

Table 1.3   Physical Processes

Chapter	Section		AQA specification references			
			Co-ordinated		Modular	
			DA	SA	DA	SA
13 Electricity and magnetism	13.1	Electric charge	12.5		10	
	13.2	Circuits	12.1	12.1/ 12.2	10	17
	13.3	Energy and power in a circuit	12.2		10	
	13.4	Mains electricity	12.3	12.3	10	17
	13.5	Paying for electricity	12.4	12.4	09	17
	13.6	Electromagnetic forces	12.20		09	
	13.7	Electromagnetic induction	12.21	12.13	10	17
14 Forces and motion	14.1	Speed, velocity and acceleration	12.6		11	
	14.2	Force and acceleration	12.7		11	
	14.3	Frictional forces and non-uniform motion	12.8		11	
15 Waves	15.1	Characteristics of waves	12.9	12.5	12	18
	15.2	The wave equation	12.9		12	
	15.3	The electromagnetic spectrum	12.10	12.6	12	18
	15.4	Sound and ultrasound	12.11	12.7	12	18
	15.5	Seismic Waves	12.12		12	
	15.6	Tectonics	12.13		12	
16 The Earth and beyond	16.1	The solar system	12.14	12.8	11	18
	16.2	The wider Universe	12.15	12.9	11	18
17 Using energy and doing work	17.1	Thermal energy transfers	12.16	12.10	09	17
	17.2	Efficiency	12.17	12.11	09	17
	17.3	Energy resources	12.18	12.12	09	17
	17.4	Work, power and energy	12.19		09/11	
18 Radioactivity	18.1	Types, properties and uses of radioactivity	12.22	12.14	12	18
	18.2	Atomic structure and radioactivity	12.23	12.15	12	18
	18.3	Half life	12.22		12	
	18.4	Nuclear Fission	12.23		12	

What is in each chapter?

At the beginning of each chapter is a list of **Key Terms** used in that chapter. When used for the first time in the text these key terms are emboldened. Some key terms are coloured. These are the extra terms that you need to know if you are to be entered for the Higher tier papers in the final examination. All the key terms together with their meanings are also found in the **Glossary** on page 392.

The contents of each chapter are divided into several **sections**. Each section concentrates on one topic. A symbol at the start of each section shows clearly whether the topic is part of the single award or double award specification and which particular module, or modules, the topic addresses.

Some of the content has a yellow tinted background. This is the extra content you need to know for the Higher tier papers in the final examination.

Throughout the book are a number of **Did you know?** boxes. The information in these boxes will not have to be learnt, but is provided to give you further background and extra interest to the topic.

At the end of each section is a **Summary** and a set of **Topic Questions**. The summary provides a brief analysis of the important points covered in the section.

The topic questions are included to help you understand what you have read in the section. Do not worry if you have to go back to read the section again when you try to answer the questions. Reading the section again to answer the topic questions will help you to learn the work. Because the topic questions have been designed to produce answers that you could use as a set of revision notes it is recommended that you write down the questions as well as their answers. The questions written on a yellow background are the more demanding questions expected to be answered by a grade B/A/A★ student.

At the end of each chapter are some **GCSE questions** taken from past AQA (SEG) or AQA (NEAB) examination papers.

Answering the GCSE questions will help to give you an idea of what is wanted when you sit your written papers at the end of the course. Again do not worry if you have to go back to read the section again when you try to answer the questions. The questions written on a yellow background are the more demanding questions expected to be answered by a grade B/A/A★ student.

The answers to the topic and GCSE questions are not bound into this book, but will be available to your teachers as part of an online resource from the publishers.

Ideas and evidence in Science

 You will find that many sections contain information which is marked with a bell and a vertical stripe in the margin. This is material to support the 'Ideas and Evidence in Science' part of your course. It will provide you with information about:

- how scientific ideas were developed and presented,
- how scientific arguments can arise from different ways of interpreting the evidence,
- ways in which scientific ideas may be affected by the contexts in which it takes place (for example, social, historical, moral and spiritual) and how these contexts may affect whether or not ideas are accepted,
- the problems science has in dealing with industrial, social and environmental questions, including the kinds of questions science can and cannot answer, uncertainties in scientific knowledge, and the ethical issues involved.

Each of the 'Ideas and evidence' contexts needed for whatever course you are following is included in this book. A guide to these contexts, where they fit in your Double or Single Award course and whether they are needed for Core or Higher tier is given in Tables 1.4, 1.5 and 1.6.

Table 1.4   Contexts for the delivery of 'Ideas and evidence' in Life and Living Processes

Section	DA	SA	Core/ HT	Context
3.2	✓	✓	core	Benefits and problems caused by the use of hormones to control fertility
3.5	✓	✓	core	How living conditions and life style are related to the spread of disease
3.6	✓	✓	core	Why the link between smoking tobacco and lung cancer gradually became accepted
5.1	✓	✓	core	Why Mendel proposed the idea of separately inherited factors and why this discovery was not immediately recognised
5.4	✓	✓	core	The economic, ethical and social issues raised by the development of cloning and genetic engineering
5.5	✓	✓	core	How fossil evidence supports the theory of evolution
5.5	✓	✓	core	How over-use of antibiotics can lead to the evolution of resistant bacteria
5.5	✓	✓	core	Why Darwin's theory of natural selection was only gradually accepted
6.2	✓	✗	core	How the managing of food production for human needs is a compromise between competing priorities
6.2	✓	✓	core	Some of the major environmental issues facing society
6.3	✓	✗	core	Problems involved with the large scale production of food

Ideas and Evidence in Science

Table 1.5 Contexts for the delivery of 'Ideas and evidence' in Materials and their Properties

Section	DA	SA	Core/HT	Context
7.1	✓	✓	core	How the idea of the atom became generally accepted after Dalton reintroduced the idea about 200 years ago
9.2	✓	✗	HT	How economic factors affect the conditions under which the Haber process is carried out
9.3	✓	✓	core	How benefits from the use of nitrogenous fertilisers need to be balanced with the potential contamination of water supplies
10.4	✓	✓	core	How the burning of hydrocarbon fuels affects the environment
10.4	✓	✓	core	How the disposal of plastics affect the environment
11.1	✓	✓	core	How early attempts to classify elements systematically led to the development of the modern periodic table
11.1	✓	✓	core	Why the periodic table gradually became accepted as an important summary of the structure of atoms
12.4	✓	✓	core	How the use of microbes and enzymes to bring about chemical reactions has advantages and disadvantages

Table 1.6 Contexts for the delivery of 'Ideas and evidence' in Physical Processes

Section	DA	SA	Core/HT	Context
15.3	✓	✓	core	The dangers or possible dangers of exposure to different types of electromagnetic radiation and measures that can be taken to reduce such exposure
15.6	✓	✗	core	Why the accurate prediction of earthquakes and volcanic activity is difficult
15.6	✓	✗	core	Why Wegener's theory of Continental Drift took a long time to be accepted
16.2	✓	✓	core	How scientists have tried to discover whether there is life elsewhere in the Universe
16.2	✓	✓	HT	Why the theories of the origin of the Universe have to account for the 'red-shift'
17.3	✓	✓	core	The advantages and disadvantages of using different energy sources to generate electricity
17.3	✓	✓	HT	How different energy sources compare financially and economically in the generation of electricity
18.3	✓	✗	core	How the Rutherford and Marsden scattering experiment led to the replacement of the 'plum-pudding' model of the atom with the present model of the atom

Some hints about doing well in the final written examinations

Some frequently used command words and what they mean

Before you can answer a question, you need to know what is expected. Question-writers use command words or phrases that inform you of the style of answer they expect you to give. A list of the most frequently used command words and phrases is given below. Question-writers assume that you have learned the meanings of the words or phrases.

Calculate or **work out** .... means that a calculation is needed together with a numerical answer.

Compare .... means that a description is needed of the similarities and/or differences in the information that has been provided.

Complete .... means that spaces in a diagram, a table or a sentence or sentences need to be completed.

Describe .... means that the important points about the particular topic must be provided.

Draw a bar chart .... – if the axes are already labelled and scales have been given then the values given must be plotted as bars.
– if the axes are labelled but no scales have been given then scales need to be added and the values given need to be plotted as bars.

Draw a graph .... – if the axes are already labelled and scales have been given then the values given need to be plotted as points and a line (or curve) appropriate to the points plotted must be drawn.
– if the axes are labelled but no scales have been given then the scales need to be added, the values given must be plotted as points and a line (or curve) appropriate to the points plotted must be drawn.

Explain how or **Explain why** .... means that scientific theory must be used to show an understanding of how or why something happens.

Give a reason or **How** or **Why** ... means that the answer requires a cause for something happening based on scientific theory.

Give or **Name** or **State** or .... means that a short answer with no supporting
Write down scientific theory is needed.

List .... means that a number of short answers are needed, each one being written one after the other.

Predict or **Suggest** .... means that the answer is based on a *consideration* of various pieces of information and suggesting, without supporting theory, what is likely to happen.

Sketch a graph .... means that a line (or curve) is to be drawn to show a trend or pattern without the need to plot a series of points.

Exam hints

Use the information	means that the answer must be based on the information provided in the question.
Use your understanding of ... to	this is the science topic around which the answer needs to be built.
What is meant by	means that the answer is likely to be a definition.

Some more hints

Obviously if you want to do well you need to have learned and understood as much as you can. However here are some hints about answering questions.

- Do not rush – no marks are awarded for finishing first. A paper worth 100 marks is designed to allow you about 90 minutes to finish it. This means that you have nearly one minute of time to think and write down 1-marks worth of answer.

- Read each question carefully at least twice before you write down your answer. If you need to do rough working to sort out your thoughts use the gaps in the margins – but make sure you put a line through this rough working.

- Look at the number of marks awarded for each part of each question. Each mark is given for a different piece of information:
  - *1 mark* means that one piece of information is needed.
  - *2 marks* mean that two pieces of information are needed etc.

- Lots of questions ask you to give a reason for something or to explain something. Such answers are usually worth 2 or more marks. Your answers to these should include a 'because' part.

- Do not throw away marks. Marks are often given for:
  - units such as joules, °C, ohms etc. Learn all the units and what they measure.
  - the names and symbols of chemical elements – so learn them
  - equations, such as 'potential difference = current × resistance' – so learn the equations and how to use them in calculations. Remember that all equations need an '=' sign in them.

- If you are writing an answer that needs several sentences, make sure that each sentence is saying something new and is not just rewording an earlier sentence.

- Try to avoid using the words 'it', 'they' or 'them' in an answer. The marker may find it difficult to understand what you mean.

- Take care when you are drawing graphs. Make sure all the points are correctly plotted. When you draw in the line for your points use a pencil with a fine point and try to draw the complete line in one go.

- If you have learned the work you should finish the paper in good time. Go through the paper again and check what you have written – it could save you throwing away some marks for silly mistakes.

Chapter 1
Cell activity

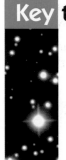

Key terms

aerobic respiration · active transport · allele · alveolus · asexual reproduction · cell · cellulose · cell membrane · cell wall · chlorophyll · chloroplast · chromosome · concentration gradient · cytoplasm · diffusion · enzyme · fertilisation · gamete · gene · meiosis · mitochondria · mitosis · nucleus · organ · organism · organ system · osmosis · partially permeable membrane · photosynthesis · sexual reproduction · root hair cells · stomata · tissue · vacuole

1.1	
Co-ordinated	Modular
DA 10.1	DA 1
SA 10.1	SA 13

Animal cells

A study of **organ systems** in humans shows that they are made up of **cells**, **tissues** and **organs**. To show how they are all linked, consider the digestive system (see section 2.1).

- An **organism** is made up of a number of organ systems.

- An organ system is a number of organs linked so that they work together to perform a particular function. In the case of the digestive system, the stomach is just one of the organs involved.

- An organ is made of a number of different tissues, each with a particular role. Different tissues are combined to form the stomach.
  - It has an inner lining of tissue that secretes **enzymes** and mucus.
  - It has muscle tissues that churn the stomach contents by their contraction and relaxation.
  - It has nervous tissue and a supply of blood (a connective tissue).

- Each tissue is made of a group of cells that all have the same structure and are working together to carry out a particular function. In the stomach there are four different groups of cells each working together to make the stomach perform all its functions.

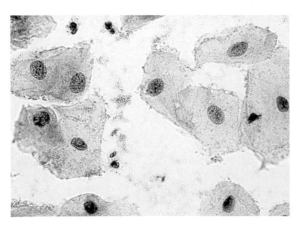

All living things are made up of cells

Figure 1.1

Figure 1.1 provides a summary of the levels of organisation within organisms.

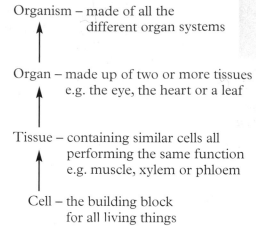

Organism – made of all the
different organ systems

↑

Organ – made up of two or more tissues
e.g. the eye, the heart or a leaf

↑

Tissue – containing similar cells all
performing the same function
e.g. muscle, xylem or phloem

↑

Cell – the building block
for all living things

All living things are made of cells – they are the 'building bricks' of life. There are a huge variety of specially adapted cells carrying out specific functions; for example humans have nerve cells, red blood cells, white blood cells, and skin cells.

Despite the huge range of different cells, most animal cells contain the same parts (see Figure 1.2).

The important parts of a cell are:

- the **nucleus**.
  - it contains the **chromosomes** that carry the **genes** (see sections 1.5 and 5.2). It is the genes which control the characteristics of the cell.
  - it is responsible for controlling all the chemical activities going on in the cell.
  - it controls cell division.

- the **cell membrane** – this is a thin barrier between the cell contents and the outside of the cell. It controls the movement of substances into and out of the cell. This includes the entry of useful chemicals such as water, oxygen and glucose, and the removal of waste chemicals such as carbon dioxide.

Because the cell membrane controls which chemicals pass into and out of the cell, it is described as being a **partially permeable membrane**.

- the **cytoplasm** – this is the substance that fills the space within the cell membrane. All the chemical reactions take place in the cytoplasm.

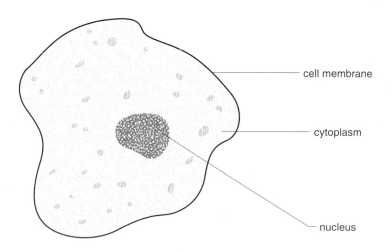

cell membrane

cytoplasm

nucleus

Figure 1.2
A typical animal cell

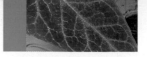

The chemical reactions are controlled by **enzymes**.

- mitochondria. These are found in the cytoplasm and can be described as the 'power station' of the cell, because they are the site of **aerobic respiration** (see section 2.5). Although tiny, their inner membrane is highly folded, providing a very large surface area where the chemical reactions that release the energy needed by the cell take place. Cells that require a lot of energy such as muscle cells and sperm cells have a large number of mitochondria.

Figure 1.3
A mitochondrion – the site of aerobic respiration in the cell

Summary

◆ **Organ systems** are made of **organs**.

◆ Each system carries out a particular function or range of functions.

◆ **Organs** are made of **tissues**.

◆ **Tissues** are collections of **cells** working together to carry out a particular function.

◆ Most cells have a **nucleus** which controls the activities of the cell, **cytoplasm** in which chemical reactions take place and a **cell membrane** which controls the passage of substances in and out of the cell.

◆ The chemical reactions in a cell are controlled by **enzymes**.

◆ Energy from respiration is released by the mitochondria in the cytoplasm.

Topic Questions

1   Put the following in order of size, smallest first:
    cell   nucleus   organ   organ system   tissue.

2   Complete the gaps in the following table.

Name of part	Function
nucleus	
cytoplasm	
	this controls the passage of substances moving into and out of the cell

3   a) What controls the chemical reactions that take place in a cell?
    b) What are mitochondria?

1.2	**Plant cells**

Co-ordinated	Modular
DA 10.1	DA 2
SA n/a	SA n/a

All plant cells, like animal cells have a nucleus, a cell membrane, cytoplasm and mitochondria. In addition, the following features are found only in plant cells.

- The **cell wall**. This is important in a plant cell because it is a rigid layer, made mostly of **cellulose**, which helps to strengthen the cell. Cellulose allows water and other substances to move freely in and out of the cell. The presence of water in the plant cell also helps to give the cell shape and to support the plant.

- **Chloroplasts** – the site of **photosynthesis**. Chloroplasts are small and disc shaped, containing molecules of the green pigment **chlorophyll** which absorbs light energy.

- Large **vacuoles**. These contain cell sap, which is mostly water and dissolved substances like sugars. Vacuoles and their cell sap also help the plant cell keep its shape and give support to the young plant.

Figure 1.4
A typical plant cell

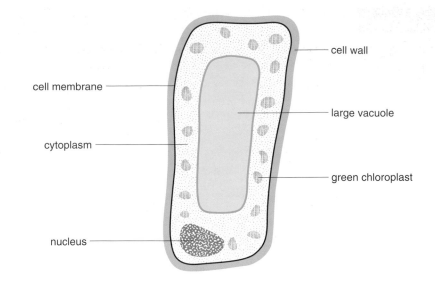

Figure 1.5 lists those parts found in animal and plant cells.

Figure 1.5

Part	Plant cell	Animal Cell
a nucleus	✓	✓
a cell membrane	✓	✓
cytoplasm	✓	✓
mitochondria	✓	✓
a cell wall	✓	✗
a vacuole	✓	✗
chloroplasts	✓	✗

Summary

◆ Plant cells like animal cells have a nucleus, a cell membrane, cytoplasm and mitochondria. In addition most plant cells have **chloroplasts** which absorb light energy to make food, a large **vacuole** which is filled with cell sap, and a **cell wall** which strengthens the cell.

Topic Questions

1 a) What is the name of the green substance found in chloroplasts?
   b) Complete the gaps in the following table.

Name of part	Function
chloroplast	
cell wall	
	filled with cell sap to give support

2 a) List the four parts that both animal and plant cells have.
   b) What three extra parts do plant cells have?

1.3	
Co-ordinated	**Modular**
DA 10.2	DA 1
SA n/a	SA n/a

Diffusion

Substances enter and leave cells through the cell membrane and cell wall by a number of processes, one of which is **diffusion**.

Diffusion is the movement of particles from a region where there is a high concentration of the particles to where there is a lower concentration of the particles. For example, if a girl sprayed herself with perfume, the perfume particles would move very quickly away from the girl and soon become evenly spread throughout the room. Diffusion is the process that allows the smell of the perfume to spread around the room.

Diffusion in liquids and gases

In a gas, particles move around freely and, like the girl's perfume, will spread out completely within their container. Liquid particles move about less freely within the volume of the liquid.

If a dye is put carefully into a beaker of water, the colour can be seen spreading slowly through the water even if the water is not stirred.

Figure 1.6
Demonstrating diffusion in a liquid

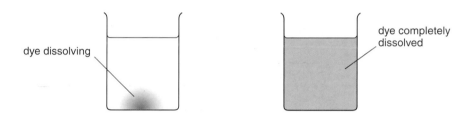

dye dissolving

dye completely dissolved

Similarly, bromine, a dark brown gas, will quickly spread throughout a jar of air until the whole of the inside is a uniform colour. This shows that the air and bromine are completely mixed.

Figure 1.7
Diffusion of bromine vapour

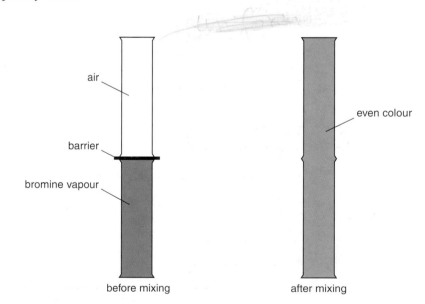

air

barrier

bromine vapour

even colour

before mixing

after mixing

Diffusion in gases is much faster than diffusion in liquids because the particles in a gas move much faster than those in a liquid.

5

Diffusion in living things

Oxygen and carbon dioxide can move quickly into and out of cells by diffusion. This is because the gases are made up of small molecules. Bigger molecules, such as those of glucose, can also pass through the cell membrane but they take a little longer to do so.

The rate at which substances pass through depends on their concentration (i.e. the number of molecules of each substance) on either side of the membrane – if there is a big difference in concentration (a large **concentration gradient**), the movement is rapid.

Diffusion does not use any energy. It is the way in which oxygen leaves the alveoli (see section 2.4) and enters the red blood cells and carbon dioxide leaves the blood and goes into the alveoli during gas exchange in the lungs. It is also the way small molecules leave the small intestine and enter the bloodstream during digestion.

Figure 1.8
Diffusion of gases between an alveolus and a blood capillary in the lung

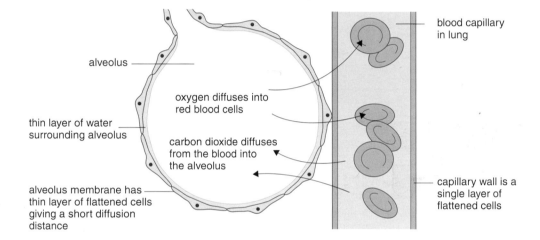

blood capillary in lung

alveolus

oxygen diffuses into red blood cells

thin layer of water surrounding alveolus

carbon dioxide diffuses from the blood into the alveolus

alveolus membrane has thin layer of flattened cells giving a short diffusion distance

capillary wall is a single layer of flattened cells

In plants, water diffuses into the **root hair cells**. During **photosynthesis** carbon dioxide diffuses into the leaf through the **stomata** and oxygen diffuses out through the stomata.

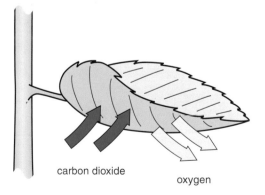

carbon dioxide

oxygen

Figure 1.9
The diffusion of gases into and out of the lower surface of a leaf

Summary

◆ **Diffusion** is the movement of particles from a region where they are at a higher concentration to a region where they are at a lower concentration.

◆ During diffusion the particles move down a concentration gradient.

◆ The greater the difference in concentration the faster the rate of diffusion.

Topic Questions

1 What is diffusion?

2 a) Complete the following table.

% concentration of oxygen		In which direction will oxygen move?	Why?
Region A	Region B		
20	5		
5	30		
10	0		

b) In which pair of concentrations will diffusion take place most rapidly? Give a reason.

Co-ordinated	Modular
DA 10.2	DA 2
SA n/a	SA n/a

1.4 Osmosis and active transport

Osmosis

Osmosis is a special kind of diffusion involving the movement of water molecules. Water molecules are small and can pass easily through a partially permeable membrane that will prevent the movement of larger molecules.

This special kind of diffusion only occurs when a partially permeable membrane separates two solutions. So osmosis is the movement of water through a partially permeable membrane from a region of high water concentration to a region of lower water concentration. The water is said to move down a concentration gradient.

partially permeable membrane

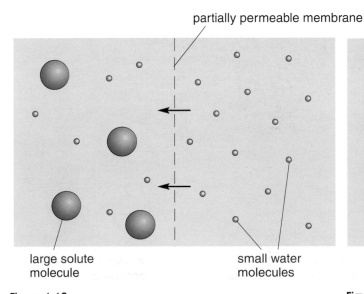

large solute molecule

small water molecules

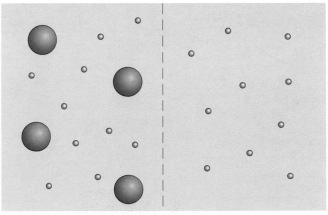

Figure 1.10
A concentration gradient across a partially permeable membrane

Figure 1.11
The water molecules have moved through the partially permeable membrane until there is the same concentration on each side

7

If a potato chip is put into a beaker containing water (Figure 1.12), then after a while the chip weighs more than it did at first and the chip has become firmer. Water has moved into the chip by osmosis.

But when a similar chip is put into a beaker of sugar solution, after a while the chip has lost weight and is very soft because water has moved out of the chip into the sugar solution by osmosis.

Dialysis tubing can be used in a similar way (see Figure 1.13). Dialysis tubing is a partially permeable membrane.

Osmosis is a very important process. It enables plants to take in water through their roots because there will usually be a lower concentration of water inside the cells of the root hairs than outside in the soil.

As water enters the root hair cell, the contents of the cell become more dilute, so water moves by osmosis to the next cell, which in turn becomes more dilute and so water passes on to the next cell. In this way water reaches the xylem and then travels up the stem by the transpiration pull (see section 4.3). The root hair cell keeps losing water to the next cell and so keeps gaining water by osmosis because the concentration gradient is maintained.

Figure 1.12
Water moves from a region of high water concentration to a region of lower water concentration by osmosis

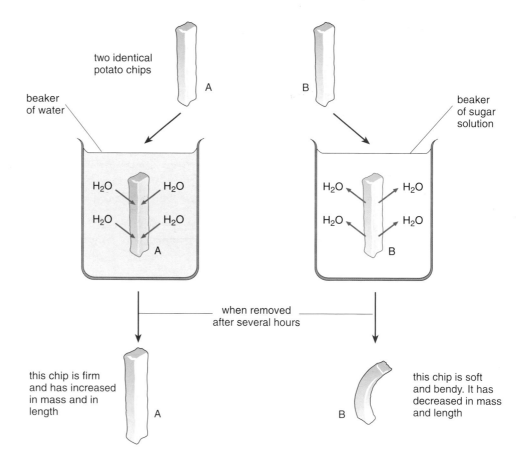

Figure 1.13
The results of an osmosis experiment using dialysis tubing

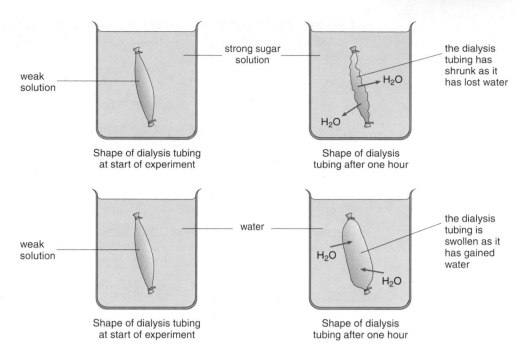

Shape of dialysis tubing at start of experiment

weak solution

strong sugar solution

Shape of dialysis tubing after one hour

the dialysis tubing has shrunk as it has lost water

H_2O

H_2O

weak solution

water

Shape of dialysis tubing at start of experiment

Shape of dialysis tubing after one hour

the dialysis tubing is swollen as it has gained water

H_2O

H_2O

Active transport

Sometimes it is necessary to move substances against a concentration gradient. In these situations there will be a greater concentration inside the cell than outside. If the cell needs to take in some of these substances then energy transfer is needed. The cell must transfer energy to take in more of the substance from outside the cell. The taking in of a substance against a concentration gradient is called active transport. It is called active transport because there is an energy transfer involved.

Active transport:

- allows a plant to take in mineral ions from the soil.
- explains why many mitochondria are found in the root hair cells.

Figure 1.14
Water moves by osmosis from a region of high water concentration to a region of low water concentration

HIGH WATER CONCENTRATION

LOW WATER CONCENTRATION

Concentration gradient

Summary

◆ **Osmosis** is the diffusion of water through a **partially permeable membrane** from a region of high water concentration to a region of lower water concentration.

◆ The partially permeable membrane allows the passage of water molecules but not solute molecules.

◆ **Active transport** is the absorption of substance against a concentration gradient. This absorption requires the energy from respiration.

9

Topic Questions

1 a) What is osmosis?
  b) What is a partially permeable membrane?
  c) What part of a cell is the partially permeable membrane?

2 Put these sugar solutions in order of **increasing** concentration.

Solution	Volume of water (cm³)	Amount of sugar dissolved (g)
A	100	8
B	100	12
C	50	3
D	50	5
E	200	10
F	200	18

3 Complete the following table.

% concentration of sugar		In which direction will water move?	Why?
Region A	Region B		
10	15		
50	15		
12	12		

4 In what way is active transport different from either osmosis or diffusion?

5 a) What does the cell need to supply if active transport is to occur?
  b) Which living process supplies this need?

6 Why are the cells in root hairs well supplied with mitochondria?

1.5   Cell Division

Co-ordinated	Modular
DA 10.3	DA 4
SA 10.2	SA 14

Mitosis

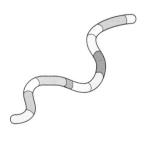

The nucleus of a cell contains **chromosomes**. Each chromosome contains large numbers of **genes** that control characteristics such as eye colour. In body cells chromosomes are normally found in pairs. Therefore in each cell there will be two genes for each characteristic. Many genes, including the gene for eye colour, has different forms called **alleles**. One allele might be responsible for blue eyes and the other allele responsible for brown eyes.

Body cells divide to produce additional cells during growth or to produce replacement cells. When body cells divide, each cell receives an identical copy of the genetic information of the parent cell.

Figure 1.15
A chromosome

The division of body cells is by a process called **mitosis**. During mitosis a copy of each chromosome is made. When the cell divides each new body cell contains identical genetic information.

Some organisms are able to reproduce through mitosis. For example, strawberry plants grow runners which separate into new individuals. The new plants are genetically identical to the parent plant. This is called **asexual reproduction**.

Meiosis

Sexual reproduction involves two parents who each produce sex cells that must be joined together at **fertilisation** (**fusion**) to develop into the new individual.

Meiosis is cell division that leads to the formation of the **gametes**. In meiosis a copy of each chromosome is made. The cell divides twice, so that each cell produced by meiosis has half the number of chromosomes.

In humans meiosis takes place in the testes and ovaries. Each individual has 23 pairs of chromosomes, one of each pair coming from the mother and the other from the father. At the start of meiosis these chromosomes form pairs so that during cell division the gametes formed only have one chromosome from each pair. This process is completely random so all gametes will be different because they will each have an assortment of mother and father chromosomes. Another important point about mciosis is that it keeps the chromosome number the same in all generations with 23 pairs remade when the two sets of 23 single chromosomes in the gametes join at fertilisation.

The important points about meiosis are:

- The gametes formed have half the number of chromosomes of the parent cell. This means that when fertilisation occurs, the number in the new cell will be the same as in the original parent cell.

- The gametes are all genetically different to each other and so the offspring formed by joining two different gametes together will have a unique set of genetic information.

Figure 1.16
A human egg surrounded by sperm. Notice the huge difference in size between the two types of cells

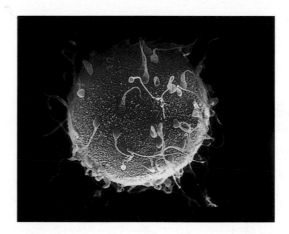

Figure 1.17
*A summary of
sexual reproduction*

Male

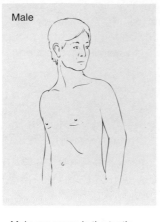

Female

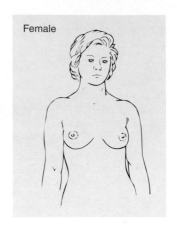

Normal body cells have 46
chromosomes
arranged as 23 pairs

Male sex organ is the testis

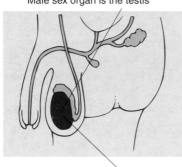

Female sex organ is the ovary

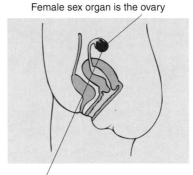

Meiosis takes place in the sex
organs to form sex cells with
23 single chromosomes.

Meiosis is reduction division –
the number of chromosomes
is reduced from 23 pairs to 23
singles to form the gametes.

Remember
MEiosis = REduction

Male sex cell is the **sperm**.
This is formed in the testis

nucleus has
23 chromosomes

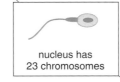

nucleus has
23 chromosomes

Female sex cell is the **ovum**.
This is formed in the ovary

Sex cells are called gametes

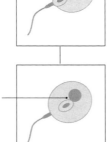

Fertilisation is the joining
together of the sperm
nucleus with the ovum
nucleus to form a new cell

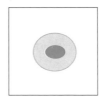

23 pairs of chromosomes

This new cell will have 23 pairs
of chromosomes and will divide
by normal cell division (mitosis)
to form a ball of cells which
develops into the fetus

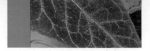

Summary

- A cell nucleus contains **chromosomes**.

- Each chromosome carries a large number of **genes** which control the characteristics of the body.

- Many genes have different forms called **alleles**.

- In body cells the chromosomes are found in pairs.

- During **mitosis** a copy of each chromosome in a body cell is made. Cell division takes place and each new body cell contains identical genetic information.

- During **meiosis** the cells in the reproductive organs divide to form **gametes**. In this cell division copies of the chromosomes are made, the cell divides twice to form four gametes each with half the number of chromosomes of the parent cell.

- During **fertilisation** gametes join to form a single body cell with the complete set of paired chromosomes.

Topic Questions

1 Put these in increasing order of size:
   cell   chromosome   gene   nucleus

2 What are alleles?

3 In a body cell chromosomes are normally found:
   A on their own
   B in pairs
   C in groups of three
   D in groups of four.

4 What happens to a body cell during mitosis?

5 What happens to the cells in the reproductive organs during meiosis?

Examination questions

1 *Amoebae* are single-celled animals. The students took some *Amoebae* from the pond. They looked at one under a microscope.

a) Use the words from the box to name the parts of this *Amoeba*. You may use each word once or not at all. *(3 marks)*

| cell membrane cell wall chloroplast cytoplasm |
| nucleus vacuole |

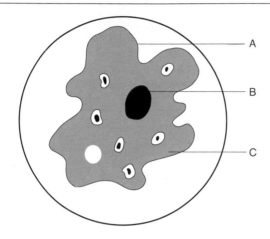

b) Draw lines to link each part of the *Amoeba* cell with its function.

Part | Function

A | controls the characteristics of the cell

B | allows substances to move in and out of the cell

C | where chemical reactions take place

(3 marks)

c) In which of these parts would you find genes? *(1 mark)*

2 These diagrams represent types of cells.

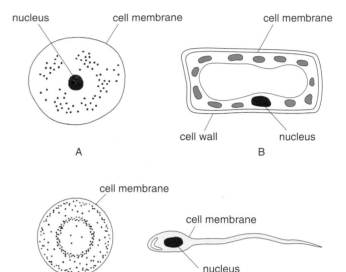

Which diagram A, B, C or D shows a plant cell? *(1 mark)*

3 The diagram shows an experiment as it was set up (X) and its appearance one hour later (Y).

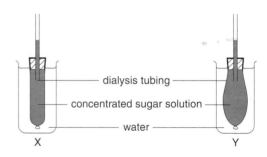

Which is the best explanation for the difference?

A Sugar has moved through the dialysis tubing by osmosis

B Water has been carried into the sugar solution by active transport

C Water has diffused through the dialysis tubing into the sugar solution

D Water has moved from a region of lower water concentration to a region of higher water concentration *(1 mark)*

4 The diagrams are experiments set up by a student to study the movement of particles. Test tube 1 shows the start of the experiment. Test tube 2 shows the same tube some time later.

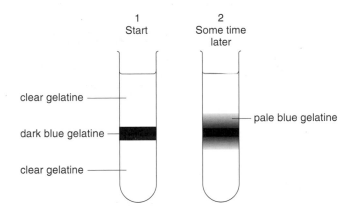

The movement of the blue colour is due to

A active transport
B diffusion
C osmosis
D transpiration

(1 mark)

5 The diagram shows what happens to a plant after 6 hours in a strong salt solution. Why has the plant wilted?

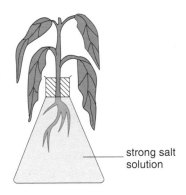

A Salt has entered the plant by diffusion
B Salt has left the plant by osmosis
C Water has entered the plant by diffusion
D Water has left the plant by osmosis

(1 mark)

6 Two chips, P and Q, both 50 mm long, were cut from a potato. P was put into a dish containing water, Q into a dish containing sugar solution.

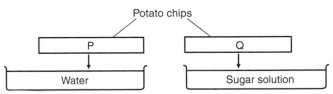

Which row of the table shows the *likely* lengths after one hour?

	Length of chip in mm	
	P	**Q**
A	48	54
B	48	50
C	50	54
D	54	48

7 Some students set up the equipment below to investigate osmosis.

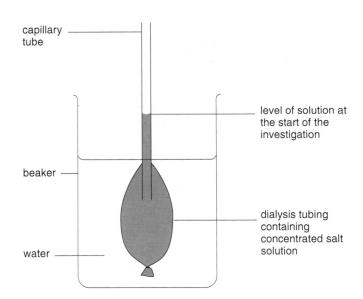

a) What is osmosis? *(3 marks)*
b) i) What will happen to the water level in the capillary tube during the investigation because of osmosis? *(1 mark)*

ii) Use your knowledge of osmosis to explain why this happens. *(2 marks)*

8 In the cell shown in the diagram as a box, one chromosome pair has alleles **Aa**. The other chromosome pair has alleles **Bb**. The cell undergoes meiosis.

a) Copy and complete the diagram of the four gametes to show the independent assortment, or reassortment, of genetic material during meiosis. *(2 marks)*

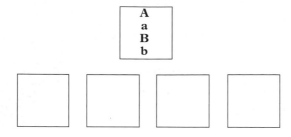

b) If the cell undergoes mitosis instead of meiosis, draw the two daughter cells which result to show the chromosomes in each. *(2 marks)*

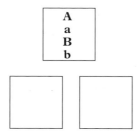

c) State the number of chromosomes in:

   i)   a normal human cell *(1 mark)*
   ii) a human gamete *(1 mark)*
   iii) the daughter cell from mitosis of a human cell. *(1 mark)*

9 The drawing shows part of a root hair cell.

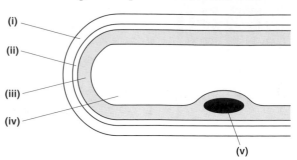

Use words from the list to label the parts of the root hair cell.

**cell membrane   cell wall   cytoplasm
nucleus   vacuole**

10 a) The diagram shows four ways in which molecules may move into and out of a cell. The dots show the concentration of molecules.

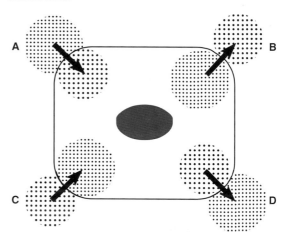

The cell is respiring aerobically.
Which arrow, **A**, **B**, **C** or **D**, represents:

   i)   movement of oxygen molecules
   ii) movement of carbon dioxide molecules?

(2 marks)

b) Name the process by which these gases move into and out of the cell.

(1 mark)

c) Which arrow, **A**, **B**, **C** or **D**, represents the active uptake of sugar molecules by the cell? Explain the reason for your answer

(2 marks)

Chapter 2
Humans as organisms

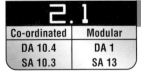

Key terms

absorption • aerobic respiration • alveoli • amino acids • amylase • anaerobic respiration • anus • aorta • artery • atria • bile • breathing • bronchi • bronchioles • cilia • capillaries • catalysts • denatured • deoxygenated blood • diaphragm • digestion • emulsifying • enzyme • exhale • faeces • fatty acids • gall bladder • gaseous exchange • glycerol • glycogen • gullet • haemoglobin • heart • inhale • insoluble • lactic acid • large intestine • lipase • lipids • lungs • liver • mitochondria • mucus • nutrition • oesophagus • oxygenated blood • oxygen debt • oxyhaemoglobin • pancreas • pH • plasma • platelets • proteases • pulmonary artery • pulmonary vein • red blood cells • respiration • respire • rib muscles • saliva • salivary glands • small intestine • soluble • stomach • substrate • thorax • tissue fluid • trachea • urea • vein • vena cava • ventilation • ventricles • villi • white blood cells

2.1 Nutrition

Co-ordinated	Modular
DA 10.4	DA 1
SA 10.3	SA 13

Unlike plants, that make their own food by photosynthesis, most animals take **insoluble** food in and then break it down into small, **soluble** molecules that can be absorbed into the blood. This process is called **digestion**. In mammals and many other animals, once the food has been broken down it is **absorbed** into the bloodstream. Digestion is made up of three stages.

Figure 2.1
The three stages of digestion

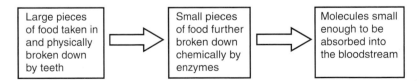

What happens during digestion?

The mouth is at the start of the process of digestion. In the mouth, the food is broken down into small pieces and mixed with a special chemical called **saliva**. The teeth break the food down into small enough pieces for swallowing and to allow the **enzyme amylase** to reach a large surface area of the food as quickly as possible. Amylase is made by the **salivary glands** which produce the saliva. The saliva enters the mouth as food is chewed, making the food moist enough to swallow. The amylase in the saliva starts to break the large starch molecules into smaller sugar molecules.

Figure 2.2
The human digestive system

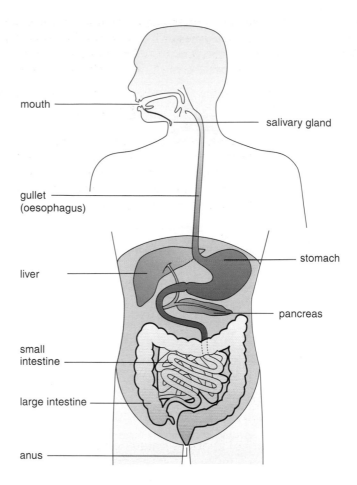

mouth

salivary gland

gullet (oesophagus)

liver

stomach

pancreas

small intestine

large intestine

anus

The mouth is the opening of a long tube. This long tube varies in width as it passes through the body to an opening at the other end called the **anus**. Food does not really enter the body until it passes through the walls of this tube and enters the bloodstream. Some of the food material that we eat, in particular fibre, never enters the body but passes out of the body as **faeces** through the anus.

The **gullet (oesophagus)** is the part of the tube which carries the food from the mouth to the stomach. It is a muscular tube and by repeated contractions, the food is pushed down towards the stomach. This movement through the digestive system is called peristalsis.

Figure 2.3
Peristalsis

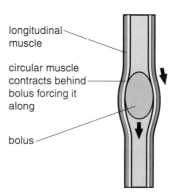

longitudinal muscle

circular muscle contracts behind bolus forcing it along

bolus

The **stomach** is a muscular bag. It has three layers of muscles that work in different directions so that when they contract, it means that the contents of the stomach are squeezed and moved in all directions. As a result, the contents are very well mixed with the gastric juices made in the stomach.

Gastric juices are made in the walls of the stomach and include dilute hydrochloric acid and enzymes called **proteases**. Hydrochloric acid helps to destroy bacteria; it also activates the protease and provides the right pH for the enzyme to work. The proteases are enzymes that break down proteins into **amino acids**.

The contents of the stomach are squirted into the **small intestine**. The small intestine is a long tube, coiled to take up less space.

The contents of the **gall bladder** – the **bile** – and the pancreatic enzymes are emptied into the first part of the small intestine through a small tube from the **pancreas**. Bile does not contain any enzymes but is very important because it breaks down (**emulsifies**) the fats into very small droplets giving them a larger surface area so that the fat digesting enzymes (**lipases**) can work more quickly and efficiently.

Figure 2.4
Bile breaking a fat molecule down into small droplets

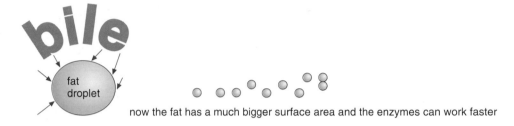

Glands in the wall of the small intestine produce amylase, protease and lipase enzymes that complete digestion. Proteins are digested into amino acids. The remaining starches are digested into glucose, fats are digested into **fatty acids** and **glycerol**.

Another important feature of the small intestine is that it is here that the small digested molecules are absorbed into the blood. The walls of the small intestine are lined with finger-shaped projections called **villi**.

As can be seen in Figure 2.5, the villi provide an efficient absorption surface because they have:

- a large surface area
- a moist surface
- a plentiful blood supply
- a thin membrance.

These features ensure that molecules can be absorbed rapidly into the bloodstream.

Figure 2.5
a) A cross section of the small intestine. b) Villi lining the small intestine

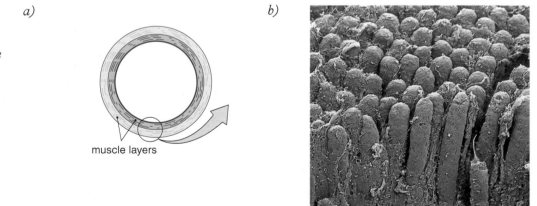

By the time the contents of the small intestine have been moved along to the **large intestine**, the small soluble molecules have been absorbed. The large intestine contains the material that cannot be digested and absorbed by our gut.

This is mostly the fibre in our diet. The fibre is, however, important in providing the bulk of the contents of the intestine, something for the muscles to push against, and helping to move the food through from the mouth to anus.

There is still one very important function to be done before the unusable contents of the large intestine can be removed as faeces. The water, which has made the 'food' soft and easy to push through the digestive system, is now reabsorbed through the lining of the large intestine back into the body. When the faeces are released we are said to defaecate.

Figure 2.6
Summary of digestion

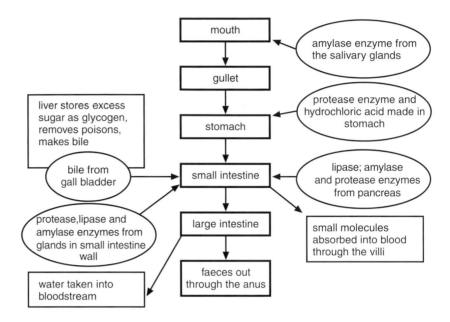

The liver

The **liver** is the largest organ in a mammal and only two of the many important jobs that it does will be considered here. Bile is made in the liver and stored in the gall bladder.

Bile is vital because it contains chemicals that alter the **pH** of the stomach contents as it enters the small intestine. Enzymes in the small intestine operate best at a higher pH. Remember that the stomach contains hydrochloric acid, so the bile salts must be alkaline.

Bile also changes the fats into smaller droplets to provide a larger surface area for the enzymes to work on (Figure 2.4).

Did you know?

One of the other important functions of the liver is the storage of glucose as **glycogen**. Glucose is very important in the process of respiration, so it is essential for the body to be able to obtain it quickly. However glucose alters the osmotic effect of a solution, so if the glucose was stored in the blood, the concentration in the blood would fluctuate constantly. To prevent this, but to be able to call on glucose reserves quickly, the body stores the 'spare' glucose as glycogen in the liver (and muscles).

One way that too much alcohol can damage an individual's health is because alcohol is a poison to the body. The liver cells are damaged as they break down the poisons from our blood.

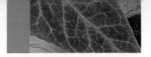

Digestion by enzymes

The teeth have already started to break down the food by the time it reaches the stomach but the pieces of food are still too big for the body to absorb. The next stage of digestion is that done by enzymes. Enzymes are **catalysts** which speed up the breakdown of large molecules into smaller molecules. There are three main enzymes involved in the digestion of starch, protein and fats.

Enzymes are chemicals with very specific tasks. The enzyme amylase can only break down the large starch molecules to smaller sugar molecules. Amylase cannot break down any other food molecule and no other enzyme can break down starch. The same is true for all the enzymes: they can only perform one job.

Protease enzymes break down proteins into amino acids, and lipases break fats down into fatty acids and glycerol (Figure 2.7). The molecule which reacts with an enzyme is called its **substrate**.

The specific nature of enzymes is explained in section 12.4. It is related to the particular shapes of the enzyme molecules. Overheating an enzyme disturbs the shape so that the substrate no longer fits with the enzyme. Such enzymes are described as having been **denatured** and will not work properly.

Figure 2.7
Enzymes break down starch, protein and fat during digestion

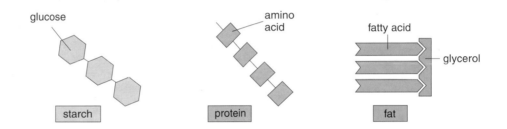

Figure 2.8

Enzyme	Substrate	Product
amylase	starch	sugars
proteases	protein	amino acids
lipase	lipids	fatty acids + glycerol

The sugars, amino acids, fatty acids and glycerol are all small enough to be absorbed through the gut wall into the bloodstream and then to travel to the cells that need them. So the enzymes have completed the task of digestion.

Did you know?

The body has a special way of preventing proteases from digesting its own body proteins. The proteases are not activated (switched on) until there is food present to digest.

Pepsin, the protease made in the stomach, has an optimum pH of 2, but most enzymes work best at pH 7.

Did you know?

The importance of pH to the digestion of starch
Starch makes up much of the diets of people throughout the world. Starch is an example of a group of foods called carbohydrates. The digestion of carbohydrates, such as starch, starts in the mouth through the action of amylase. This causes some of the large starch molecules to be broken down into smaller sugar molecules. However, the action of amylase ceases in the acidic conditions in the stomach and does not continue until the undigested starch reaches the alkali conditions in the small intestine. Here, the amylase produced by the pancreas causes the breakdown of the remaining starch into sugar. It is in the small intestine that another enzyme, maltase, causes the breakdown of the sugar molecules into even smaller molecules of glucose. It is this glucose that is important in respiration.

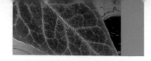

Summary

◆ **Digestion** is the process of breaking down large **insoluble** food molecules (such as starch, protein and fats) into smaller **soluble** molecules.

◆ **Absorption** is the movement of the small soluble molecules from the digestive system into the bloodstream.

◆ The absorption of the soluble substances takes place through the wall of the **small intestine**.

◆ The walls of the small intestine are lined with **villi** which are adapted to be efficient absorbers.

◆ Water is absorbed into the bloodstream in the **large intestine**.

◆ **Faeces** are indigestible food which leaves the body through the **anus**.

◆ **Enzymes** are biological **catalysts** that speed up the breakdown of the large molecules.

◆ **Amylase** (an enzyme produced in the **salivary glands, pancreas** and **small intestine**) speeds up the breakdown of starch into sugars.

◆ **Protease** enzymes (produced in the **stomach,** pancreas and small intestine) speed up the breakdown of protein into **amino acids**.

◆ **Lipase** enzymes (produced in the pancreas and small intestine) speed up the breakdown of **lipids** (fats and oils) into **fatty acids** and **glycerol**.

◆ The stomach produces hydrochloric acid which kills bacteria and produces the acidic conditions in which the enzymes in the stomach work most effectively.

◆ The **liver** produces **bile** which neutralises the acid added in the stomach. It produces the alkaline conditions in which the enzymes in the small intestine work most effectively.

◆ Bile breaks up (**emulsifies**) large fat drops into small droplets thereby increasing the surface area for enzymes to act on.

Topic questions

1 What is digestion?

2 Where are the breakdown products of digestion absorbed into the bloodstream?

3 How are villi adapted to ensure that the absorption of digested food is as rapid as possible?

4 Through which part of the digestive system is water reabsorbed into the bloodstream?

5 Copy and complete the following table.

6 Give two reasons why hydrochloric acid is produced in the stomach.

7 a) Which organ produces bile?
  b) Explain why bile is important in the process of digestion.

Enzyme	Where produced	Substrate	Products
amylase			
proteases			
lipase			

2.2	
Co-ordinated	Modular
DA 10.5	DA 1
SA 10.4	SA 13

Composition of the blood

Plasma

Plasma is the main transporting fluid in blood. It is a straw-coloured liquid but because it carries so many red cells, blood is red.

Plasma is an important transport medium. It carries

- **red blood cells, white blood cells** and **platelets**
- antitoxins and antibodies (see section 3.5)
- the soluble end products of digestion (amino acids, glucose, fatty acids and glycerol) from the small intestine to other organs
- carbon dioxide from the organs to the lungs
- **urea** from the liver to the kidneys (see section 3.4)
- mineral salts
- hormones (see section 3.2)
- heat, which is distributed around the body in the blood. This helps to regulate body temperature.

Red blood cells

Red blood cells are very numerous. Their function is to transport oxygen to the organs.

Figure 2.9
Red blood cells as seen in an electron micrograph

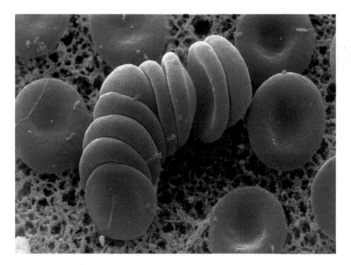

Mature red blood cells do not have a nucleus. This causes them to have a (biconcave) double-dish shape.

This shape is important as it gives

- more packing space for haemoglobin for transporting oxygen
- a large surface area for oxygen to diffuse into the red blood cell
- a short distance for the oxygen to diffuse through to combine with the haemoglobin to form oxyhaemoglobin.

The diameter of a red blood cell is similar to the diameter of a small capillary. But as the cell is surrounded by a flexible membrane, its shape can distort and it can squeeze through the smallest capillaries.

Red blood cells and oxygen transport

The only place in the body that the red blood cells can collect oxygen is the alveoli of the lungs. The oxygen dissolves in the moist lining of the alveoli and diffuses along the concentration gradient into the blood capillaries (see section 1.3). The oxygen then passes by diffusion into the red blood cells and combines with haemoglobin to form oxyhaemoglobin. The blood is now oxygenated and returns in the pulmonary vein to the heart. The heart pumps the blood around the body via the arteries. All tissues need oxygen, especially respiring muscles, and in areas such as these where there is a shortage of oxygen, the oxyhaemoglobin quickly breaks down giving free oxygen that diffuses into the cells where it will be used for respiration.

White blood cells

Some white blood cells fight against infection by ingesting and destroying bacteria (see section 3.5). White blood cells contain a nucleus.

Platelets

Platelets are cell fragments. They play a very important role in the formation of blood clots (see section 3.5). Platelets do not have a nucleus.

Summary

- The fluid part of the blood is called **plasma**.

- Plasma contains **white blood cells, platelets** and **red blood cells**.

- Plasma transports:
  - carbon dioxide from the organs to the lungs
  - soluble end-products of digestion from the small intestine to other organs
  - **urea** from the liver to the kidneys.

- **White blood cells** have a nucleus and are part of the body's defence system.

- **Red blood cells** transport oxygen from the lungs to the organs.

- **Platelets** are cell fragments that help blood to clot.

- **Red blood cells** have no nucleus and contain haemoglobin.

- In the lungs oxygen joins with haemoglobin to form oxyhaemoglobin.

- In organs, where the concentration of oxygen is low, oxyhaemoglobin releases oxygen and becomes haemoglobin.

Topic questions

1 What is plasma?

2 Which blood cells contain a nucleus?

3 a) Why do red blood cells have a large surface area?
   b) Explain why it is important that red blood cells have a flexible membrane.

4 a) Which cells take oxygen from the lungs to the organs of the body?
   b) Which part of the blood takes carbon dioxide from the organs to the lungs?

5 What is the function of white blood cells?

6 a) What are platelets?
   b) What is their function?

7 a) What is the name of the substance that gives red blood cells their colour?
   b) Where and when is oxyhaemoglobin formed?
   c) What happens to oxyhaemoglobin in organs which respire actively (e.g. the liver)?

2.3

Co-ordinated	Modular
DA 10.5	DA 1
SA n/a	SA n/a

Figure 2.10 shows the important parts of the circulatory system.

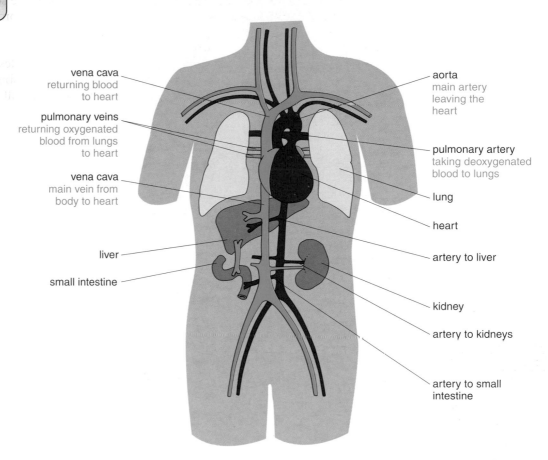

Figure 2.10
The main arteries and veins and the organs they supply

The heart

The **heart** is a four-chambered, muscular, double pump. One part pumps blood to and from the lungs, and the other part pumps blood around the body and back to the heart.

The two top chambers are the **atria** (left atrium and right atrium) – these receive the blood at low pressure from veins. The lower chambers are the pumping chambers called **ventricles**. There is a division between the two sides of the heart called the septum.

The heart is a double pump. The right ventricle pumps **deoxygenated blood** to the lungs (through the **pulmonary artery**) where it absorbs oxygen and returns to the heart via the **pulmonary vein**. The left ventricle pumps **oxygenated blood** through the **aorta** and around the body to the organs and tissues. Blood returns to the heart via the **vena cava**.

Figure 2.12 shows a human heart from the outside. The first artery to come from the aorta is the coronary artery. This supplies oxygenated blood to the heart muscles. The heart muscles are contracting all the time and so need a constant supply of oxygen and nutrients. If the coronary artery gets blocked, either by a fatty deposit or a blood clot, the shortage of oxygen to the muscles causes severe cramp-like pain – a 'heart attack'.

Oxygen is essential for living cells for respiration, and because organs carry out vital functions, each organ must receive a supply of oxygenated blood.

Did you know?

Babies born with a 'hole in the heart' have a gap in the septum. This sometimes grows together naturally but if it is large it may need surgical repair.

Figure 2.11
Blood flow through the heart

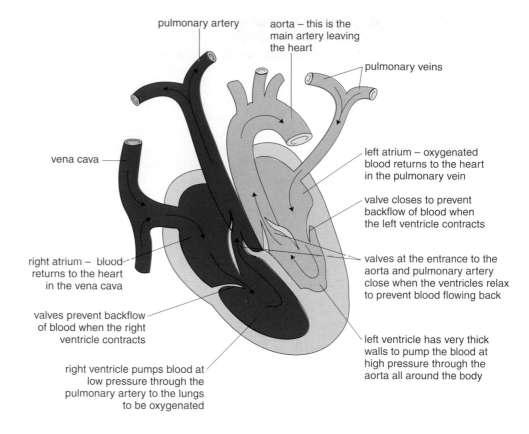

pulmonary artery

aorta – this is the main artery leaving the heart

pulmonary veins

vena cava

left atrium – oxygenated blood returns to the heart in the pulmonary vein

valve closes to prevent backflow of blood when the left ventricle contracts

right atrium – blood returns to the heart in the vena cava

valves at the entrance to the aorta and pulmonary artery close when the ventricles relax to prevent blood flowing back

valves prevent backflow of blood when the right ventricle contracts

left ventricle has very thick walls to pump the blood at high pressure through the aorta all around the body

right ventricle pumps blood at low pressure through the pulmonary artery to the lungs to be oxygenated

Figure 2.12
External view of the human heart showing the coronary arteries

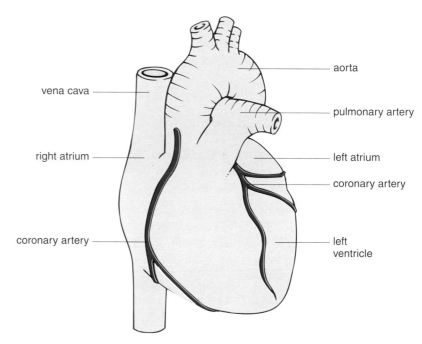

vena cava

aorta

pulmonary artery

right atrium

left atrium

coronary artery

coronary artery

left ventricle

Did you know?

- Ventricles pump out 60 cm^3 of blood in each beat.
- The human heart has a mass of 0.4% of body mass.
- The heart and blood vessels contain about 5 dm^3 blood in an adult male.

You can work out your blood volume using the rough rule of 1 dm^3 for each 10 kg body mass.

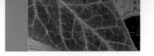

Structure of the blood vessels

Arteries

Arteries are vessels that carry blood away from the heart. The blood is under pressure and so artery walls must be strong.

> Remember AA = Arteries Away

Each heart contraction sends a volume of blood into the artery. The artery walls stretch to receive this and then contract to keep the blood moving. This is felt as a pulse. The lining of the artery in a healthy person is smooth and the lumen (the bore) is small.

Figure 2.13 ▶
Transverse section through an artery

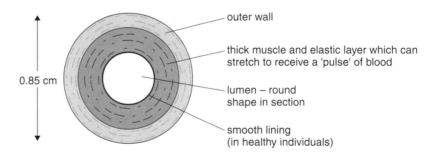

outer wall

thick muscle and elastic layer which can stretch to receive a 'pulse' of blood

lumen – round shape in section

smooth lining (in healthy individuals)

0.85 cm

Arteries subdivide into arterioles and these are important in controlling the supply of blood to the **capillary** networks.

Figure 2.14 ▶
Transverse section through a capillary

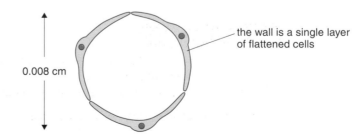

the wall is a single layer of flattened cells

0.008 cm

The capillary networks

Although **capillaries** are the smallest of the blood vessels, they are very important as they run close to all cells. Capillaries have walls made of a single layer of flat cells and they have a small lumen.

Plasma can pass out of capillaries to form the **tissue fluid**. Tissue fluid is a watery solution containing all the dissolved substances from the plasma that bathes the cells.

The cells take up substances (such as oxygen and glucose) from the tissue fluid and pass waste products (such as carbon dioxide and urea) into the tissue fluid along concentration gradients.

Figure 2.15 ▶
Exchange between the blood and the tissue cells at the capillary network

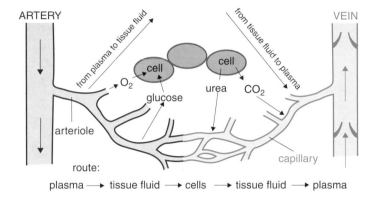

Veins

Veins receive blood from the capillary network and so carry blood at low pressure. The movement of the blood is helped by contraction and tone of the skeletal muscles and by pocket valves that prevent back-flow.

Veins join to the vena cava and return blood to the heart.

Some veins run close to the surface of the skin such as those which can be seen on the forearm and back of the hand.

Figure 2.16
Longitudinal section (a) and transverse section (b) through a vein

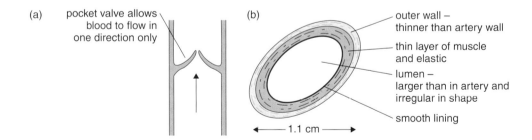

Summary of the transport functions of the blood

Transport of		
Soluble products of digestion	from villi	to all cells
Dissolved mineral salts and ions	from villi	to all cells
Oxygen in red blood cells (as oxyhaemoglobin)	from alveoli in lungs	to all respiring cells
Carbon dioxide (dissolved in plasma)	from respiring cells	to the alveoli of the lungs
Urea in solution in the plasma	from cells	to kidneys
Hormones in solution in the plasma	from endocrine glands (see section 3.2)	to target organs
Heat energy	from active muscles and the liver	to the skin capillaries, nasal capillaries and alveolar capillaries

Summary

- The circulatory system consists of the **heart, arteries, veins** and **capillaries**.

- The heart is a four-chambered muscular pump that pumps blood under pressure around the body.

- There are two separate circulatory systems:
  - one linking the heart and the lungs,
  - the other linking the heart with all the other organs.

- Blood flows from the heart through arteries to the organs.

- Blood returns from the organs to the heart through veins.

- Arteries have thick walls containing muscle and elastic fibres.

- Veins have thin walls and often have valves to prevent the back-flow of blood.

- Capillaries are very narrow blood vessels with very thin walls.

- Substances needed by the cells pass out of the blood through the walls of the capillaries.

- Substances produced by the cells pass into the blood through the walls of the capillaries.

Topic questions

1 In which direction does blood flow in:
   a) an artery
   b) a vein?

2 Describe the walls of:
   a) arteries
   b) veins
   c) capillaries.

3 The thickest muscular wall in the heart is that of the left ventricle. Why is this wall so thick?

4 The heart can be considered to be a double pump. Explain why?

5 a) What is tissue fluid?
   b) Why is tissue fluid important?

6 The veins and the heart contain valves. Why?

2.4	
Co-ordinated	Modular
DA 10.6	DA 1
SA n/a	SA n/a

Breathing

Breathing, gaseous exchange and respiration are not the same thing.

- Breathing is the process that moves air in and out of the lungs, also known as **ventilation**.

- Gaseous exchange is the exchange of oxygen and carbon dioxide at the lung surface.

- Respiration is the chemical process that takes place in each living cell which transfers energy from food molecules.

Breathing is important as it enables us to obtain oxygen and get rid of carbon dioxide. There is a difference between the inhaled (breathed in) air and the exhaled (breathed out) air. Figure 2.17 shows this difference.

Figure 2.17

	% composition inhaled air	% composition exhaled air
oxygen	20.9	16.3
carbon dioxide	0.04	4.1
nitrogen	79.0	79.5

Oxygen is removed from the air which is breathed into our lungs. The oxygen passes into the bloodstream for use in respiration. The oxygen is replaced by carbon dioxide produced by respiration. The exchange of oxygen and carbon dioxide in the lungs is called gaseous exchange.

Figure 2.18 ▼

The structure of the breathing system in humans

The breathing system

Our gas exchange organs are called the **lungs**. They are part of the breathing system, which is shown in Figure 2.18.

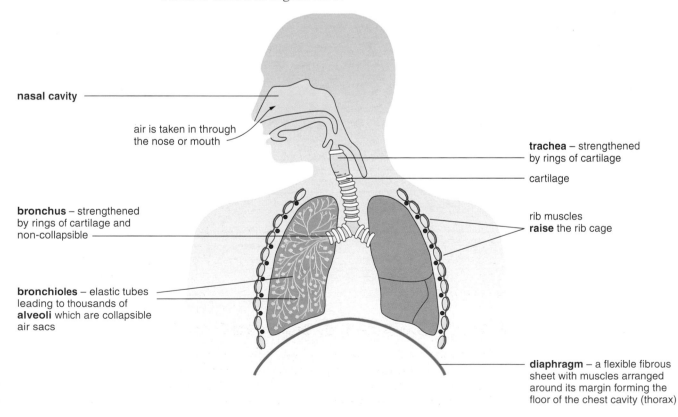

nasal cavity

air is taken in through the nose or mouth

trachea – strengthened by rings of cartilage

cartilage

bronchus – strengthened by rings of cartilage and non-collapsible

rib muscles **raise** the rib cage

bronchioles – elastic tubes leading to thousands of **alveoli** which are collapsible air sacs

diaphragm – a flexible fibrous sheet with muscles arranged around its margin forming the floor of the chest cavity (thorax)

We have two lungs inside the chest or **thorax**. The thorax is bounded and protected by the **ribs** and between the ribs are the **rib muscles**. Below the lungs is a dome-shaped sheet of smooth muscle which separates the thorax from the abdomen. This is called the **diaphragm**. The lungs are made airtight by two very thin sheets of tissue. A lubricating fluid helps the lungs to move smoothly inside the thorax. The lungs appear to be spongy but under the microscope they can be seen to consist of cavities (**alveoli**) and tubes through which air flows into and out of them.

Figure 2.19 ▶
Blood supply to the alveoli

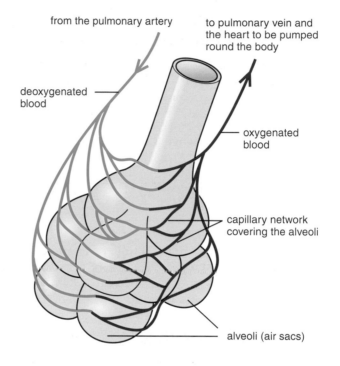

from the pulmonary artery

to pulmonary vein and the heart to be pumped round the body

deoxygenated blood

oxygenated blood

capillary network covering the alveoli

alveoli (air sacs)

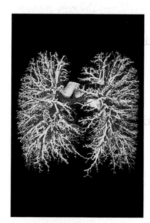

Figure 2.20 ▲
The respiratory tree

Figure 2.21 ▲
Micrograph of lung tissue showing the alveoli and their associated blood capillaries

The route of the air into the lungs

1 The air enters the body through the nose.

2 From the nose the air passes to the throat where there are small hair-like **cilia** and **mucus**.

3 From the throat the air goes to the **trachea** or windpipe. To stop you breathing and swallowing at the same time, there is a valve here called the epiglottis. It prevents food going into the windpipe and causing choking. The trachea must be kept open at all times. To stop it collapsing it has almost complete rings of a tough but flexible material called cartilage in its walls. These are shown in Figure 2.18.

4 Inside the chest cavity the trachea branches to form two smaller tubes, the **bronchi**. Inside the lungs the bronchi branch into smaller and smaller tubes called **bronchioles**. Figure 2.20 shows the bronchi and bronchioles after the rest of the lung tissue has been dissolved away. The bronchioles end in special bags called alveoli. The alveoli have very thin walls, which are surrounded by blood capillaries. Figures 2.19 and 2.20 show the alveoli with the blood capillaries surrounding them. It is at the surface of the alveoli that the exchange of oxygen and carbon dioxide takes place.

How are the gases exchanged?

- The alveoli have very thin walls as shown in Figure 2.21.
- The walls of the alveoli are moist and the oxygen dissolves in this moisture.
- The dissolved oxygen diffuses through the wall into the blood capillary.
- The carbon dioxide diffuses from the blood in the opposite direction.

- To make sure that there is always fresh air with lots of oxygen inside the lungs, air must be changed every few seconds. This is done during breathing.

The alveoli are well adapted for gaseous exchange. They have:

- a large surface area
- a moist surface
- a very thin membrane
- a rich capillary network.

Breathing

This has two parts:

1 Inhaling – when the rib cage is raised and the diaphragm contracts and flattens.

This increases the volume of the chest cavity and lowers the pressure so air rushes into the lungs.

2 Exhaling – when the rib cage is lowered and the diaphragm raised.

The volume of the thorax is reduced, the pressure inside is raised and the air is forced out. Figure 2.22 shows how this happens.

Breathing takes place automatically. Breathing is controlled in the brain where there are special sense cells that can detect any changes in the amount of carbon dioxide in the blood. If the level of carbon dioxide begins to rise, the brain makes us breathe more quickly and more deeply to increase the amount of oxygen and reduce the amount of carbon dioxide in the body. Carbon dioxide can be very poisonous if it is allowed to build up in the body.

Summary

- During the process of **gaseous exchange** oxygen is removed from the air and diffuses into the bloodstream and carbon dioxide diffuses out of the bloodstream into the air.

- The **alveoli** are well adapted for **gaseous exchange** with their very large, moist surfaces and their rich supply of blood capillaries.

- The action of the ribcage and diaphragm causes air to pass into and out of the lungs. The movement of air is called **breathing** or **ventilation**.

- Inhaling causes the volume of the chest cavity to increase, so lowering the pressure causing air to enter the lungs.

- Exhaling causes the volume of the chest cavity to decrease, so increasing the pressure causing air to be forced out of the lungs.

Topic questions

1 What happens to the ribcage and the diaphragm:
   a) when you breathe in
   b) when you breathe out?

2 During breathing what gas:
   a) diffuses from the air into the bloodstream
   b) diffuses from the bloodstream into the air?

3 Describe as fully as possible what causes the air to pass into your lungs when you breathe in.

4 In what ways is the structure of the alveoli adapted to ensure the rapid diffusion of gases?

Figure 2.22
The breathing action

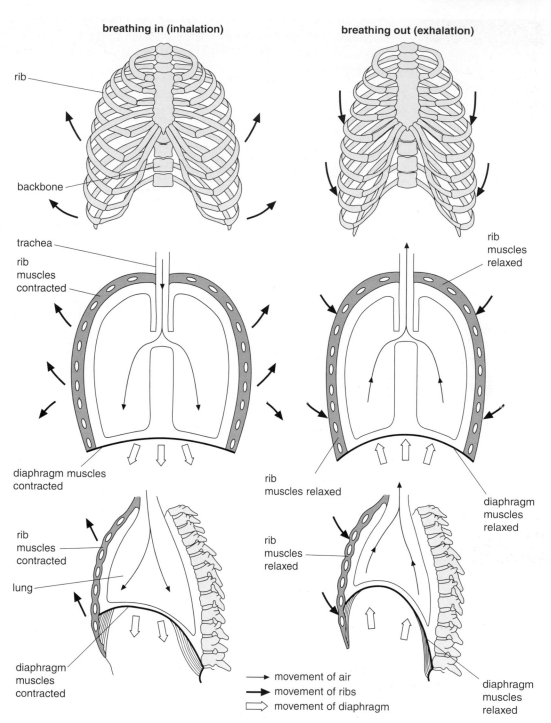

breathing in (inhalation)

rib

backbone

trachea

rib muscles contracted

diaphragm muscles contracted

rib muscles contracted

lung

diaphragm muscles contracted

breathing out (exhalation)

rib muscles relaxed

rib muscles relaxed

diaphragm muscles relaxed

rib muscles relaxed

diaphragm muscles relaxed

→ movement of air

→ movement of ribs

⇨ movement of diaphragm

2.5	
Co-ordinated	Modular
DA 10.7	DA 1
SA n/a	SA n/a

Respiration

Respiration is a little like burning fossil fuels – oxygen is used to enable the transfer of energy from the fossil fuels in the form of heat. At the same time, water and carbon dioxide are released.

A similar chemical reaction is happening in every living cell. To release energy, cells use the oxygen that is breathed in and the glucose from the food that is eaten. Water and carbon dioxide are given off as waste products.

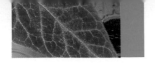

Which cells respire the most? To answer that question it is necessary to think about the cells which need the most energy. Muscle cells and sperm cells are examples. Cells in the roots of plants where minerals are taken in by active transport also need a lot of energy so they respire more than other plant cells (see section 1.4).

Figure 2.23▶
Intake and release of products from a cell

Substances taken in: Substances released:

oxygen ⟶   ⟶ carbon dioxide

glucose ⟶   ⟶ water

Figure 2.24 ▲
Increased heart rate and breathing rate ensure that more oxygen and glucose are delivered to the muscles for running

This process transfers a vast amount of energy. The word equation for respiration is:

$$\text{oxygen} + \text{glucose} \rightarrow \text{carbon dioxide} + \text{water} \; [+\text{energy}]$$

The energy transferred during respiration is used:

- to help cells build up large molecules from smaller ones, e.g. proteins from amino acids
- to help muscles contract
- to help maintain a steady body temperature in cold environments
- in the active transport of materials across boundaries.

Where does the energy come from?

Air is breathed into the lungs and oxygen diffuses into the bloodstream.

Haemoglobin in red blood cells combines with oxygen to make oxyhaemoglobin.

Food is digested and glucose is absorbed into the bloodstream.

The heart pumps blood to the tissues.

The blood contains oxyhaemoglobin in the red blood cells and glucose in the blood plasma.

Figure 2.25 ▼
Micrograph of a mitochondrion

What comes out?

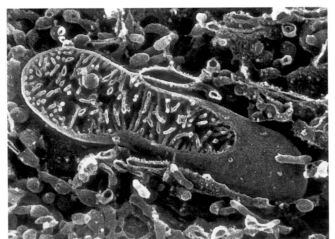

Muscles respire producing heat, carbon dioxide and water as waste products.

The skin releases excess heat.

The heart pumps the blood containing carbon dioxide, dissolved in the blood plasma and in the red cells, from the muscles to the lungs.

Carbon dioxide is breathed out from the lungs (heat energy and water vapour are also released from the lungs).

Aerobic respiration

Remember, it is called aerobic respiration because oxygen is being used.

Plants and animals need to respire all the time so that they can stay alive. We even need energy when we sleep so that our heart can pump, we can breathe and our brain can function.

Aerobic respiration takes place in the mitochondria.

Anaerobic respiration

When there is insufficient oxygen available to meet the oxygen requirements of cells they can respire without oxygen!

If one arm is raised and the other kept by your side and both fists are clenched regularly for several minutes the raised arm will start to ache quite badly.

But why does the raised arm ache so much? The arm muscles had to respire to release energy, but the raised arm did not get enough oxygen and had to respire without oxygen – **anaerobic respiration**.

Anaerobic respiration in humans

Compared to aerobic respiration, very little energy is transferred during anaerobic respiration because the breakdown of glucose is incomplete.

$$\text{glucose} \rightarrow \textbf{lactic acid} \; [+ \text{ energy}]$$

The lactic acid makes the muscle hurt – it is a mild poison that causes muscle fatigue.

A sprinter can run very fast for a short period of time. The muscles respire anaerobically for a short period of time, but the lactic acid has to be removed and oxygen is used to do this. It is said that the muscles have built up an **oxygen debt** because the cells are 'owed' oxygen. That is why sprinters breathe deeper and faster than normal after the race has finished.

Much less energy is transferred by anaerobic respiration since glucose is only partially broken down.

Rapid breathing enables more oxygen to reach the lactic acid and oxidise it to carbon dioxide and water, or convert lactic acid back to glucose.

Summary

- **Aerobic respiration** can be summarised as

  glucose + oxygen ⟶

  carbon dioxide + water [+ energy]

- The energy transferred during respiration is used to:
  - build up larger molecules from smaller ones
  - enable muscles to contract
  - maintain a steady body temperature in cold surroundings
  - enable active transport to take place.

- Aerobic respiration in cells takes place in the mitochondria.

- During vigorous exercise muscles may get short of oxygen. They then obtain energy from glucose by **anaerobic respiration**.

- **Lactic acid** is the waste product of anaerobic respiration. Lactic acid is a poison.

- Less energy is transferred during anaerobic respiration compared to aerobic respiration. This is because during anaerobic respiration the glucose is only partially broken down into lactic acid.

- Anaerobic respiration results in an **oxygen debt**. The oxygen is used to oxidise the lactic acid into carbon dioxide and water.

Topic questions

1 Arrange the following into the equation for aerobic respiration:

glucose	→	carbon dioxide	water	+

+		oxygen	[+ energy]

2 Which substance is produced during aerobic respiration by animal and plant cells?

   A Alcohol   **B** Carbon dioxide
   C Lactic acid   **D** Nitrogen

3 Cells which line the small intestine need a lot of energy to absorb digested food. Which of the following must be present in large numbers in these cells?

   A Chromosomes   **B** Fat molecules
   C Mitochondria   **D** Villi

4 Explain why it is not possible for a person to run and respire anaerobically for a long time.

5 What is meant by an oxygen debt?

Examination questions

1 Blood contains plasma, platelets, red cells and white cells. Each has one or more important functions.

Copy the table below and draw a line from each part to its function.

One part has two functions. Draw lines from this part to both functions.

Name of part of blood

red cell

platelet

plasma

white cell

Function of part of blood

fights bacteria

carries dissolved hormones

carries dissolved urea

transports oxygen around the body

helps blood to clot

(5 marks)

2 a) The sentences are about breathing. Choose words from the list in the box to complete the sentences that follow. Each word may be used once or not at all.

| decreases diaphragm in increases lungs |
| out rib ventilation |

Air is drawn into the _____ as the thorax _____ in volume.
This change in volume is caused by the _____ muscles contracting and moving the rib cage _____ at the same time as the _____ is moved down by its muscles.

(5 marks)

3 The diagram shows a human heart.

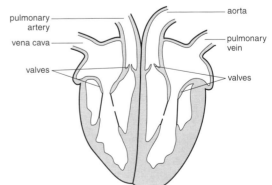

a) Complete these sentences. Use the information in the diagram to help.

Blood from the body enters the right side of the heart through the blood vessel called the _____.

The blood is pumped through the pulmonary _____ to the lungs.

The blood returns to the heart through the blood vessel called the _____.

The blood is pumped to the body through the blood vessel called the _____. *(4 marks)*

b) What do valves do in the heart? *(1 mark)*

c) Describe the differences between arteries and veins. *(4 marks)*

d) Complete this sentence.

In the lungs the blood loses _____ gas and picks up _____ gas. *(2 marks)*

4 a) Copy and complete the table to give one site where digestive substances are made.

Digestive substance	One site of production
bile	
amylase	
lipase	
protease	

(4 marks)

b) Describe **two** ways that the mouth can break down starchy foods. *(2 marks)*

c) Describe how the small intestine is adapted for absorbing food. *(5 marks)*

d) Describe how the liver helps to digest fats. *(2 marks)*

5 The diagram shows four parts of blood.

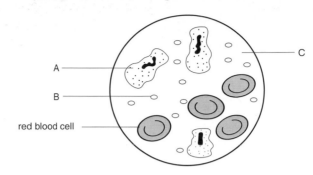

a) Complete the table to give the name and function of the parts labelled A, B and C.

Letter	Name	Function
A	....	....
B	....	....
C	....	....

(6 marks)

b) Red blood cells contain haemoglobin. Explain how this enables red blood cells to pick up oxygen from the alveoli and release it to cells in other parts of the body. *(4 marks)*

Chapter 3
Response, co-ordination and health

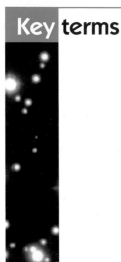

Key terms	addiction · ADH · antibodies · antitoxins · bacteria · bladder · brain · cancer · ciliary muscles · constrict · cornea · diabetes · dialysis · dilate · effectors · endocrine · environment · excretion · eye · fertility drugs · filtration · focus · FSH · gland · glucagon · glycogen · homeostasis · hormones · image · insulin · iris · kidney · lens · LH · motor neurone · mucus · negative feedback · nerves · nerve impulse · neurone · nicotine · oestrogen · optic nerve · oral contraceptives · pancreas · pituitary gland · pupil · reabsorption · receptors · reflex action · reflex arc · relay (connector) neurone · renal vein · renal artery · response · retina · sclera · sensory neurone · skin · stimulus · suspensory ligaments · synapse · target organ · toxin · urea · urine · vaccines · virus

3.1	Co-ordinated	Modular
	DA 10.8	DA 2
	SA 10.5	SA 13

Nervous system

Stimulus and response

Living things need to be able to respond to the environmental changes that go on around them for many reasons. A change in the **environment** that affects living things is called a **stimulus** and the reaction by the animal or plant is called a **response**. In humans most responses to a stimulus are brought about by electrical impulses which pass along nerves. There are two important types of responses:

- voluntary responses. These require thought and thus involve the **brain**. Speech is an example of a voluntary response.

- automatic responses called **reflex actions**. These do not involve the brain directly. The knee jerk reflex is an example.

The nervous system

Nerves are made up of vast numbers of nerve cells called **neurones**. The structures we call nerves within the body are made up of bundles of these neurones.

There are several types of neurones. **Sensory neurones** receive information from **receptors** that are found in all parts of the body. Receptors are cells which can detect stimuli. They may be scattered all over the body, such as touch receptors, or they may be grouped into sense organs.

Each receptor is sensitive to one type of stimulus only, so we have receptor cells that are sensitive to:

- changes in position; these help with balance, and are found in the ear
- light, found in the eye
- chemicals, found in the nose; these enable us to smell
- chemicals, found in the tongue; these enable us to taste
- sound found in the ear
- touch, pressure and temperature, all of which are found in the skin.

Effectors carry out responses. They are usually muscles or **glands. Motor neurones** carry the information which tells the effectors how to respond.

Figure 3.1
*The human
nervous system*

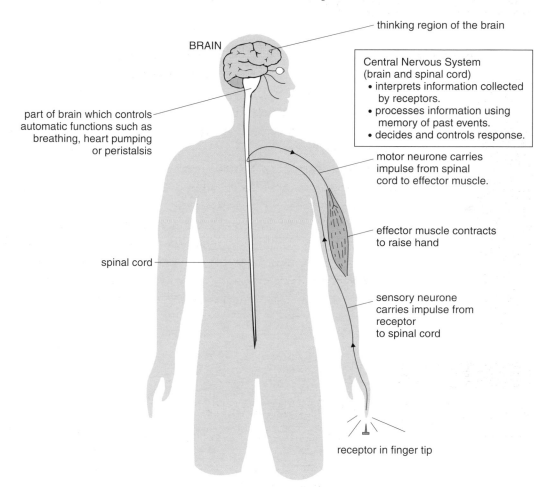

BRAIN

thinking region of the brain

Central Nervous System
(brain and spinal cord)
- interprets information collected by receptors.
- processes information using memory of past events.
- decides and controls response.

part of brain which controls automatic functions such as breathing, heart pumping or peristalsis

motor neurone carries impulse from spinal cord to effector muscle.

effector muscle contracts to raise hand

spinal cord

sensory neurone carries impulse from receptor to spinal cord

receptor in finger tip

Reflex actions

Examples of reflex actions include:

1 the knee jerk reflex – the purpose of which is to maintain the posture of the body

2 removing the hand or fingers from a hot object – the purpose of which is to protect the skin from burning

3 the dilation and constriction of the **pupil** in the eye to protect the **retina** from damage by bright light.

Reflex actions usually

- protect the body from harm.

- follow a fixed pathway – **nerve impulses** are sent by receptors through the nervous system to effectors. This pathway is called a **reflex arc**.

The reflex arc

In a simple reflex action electrical impulses are carried from receptors, such as the pressure receptors in your hand, along sensory neurones to the spinal cord. The spinal cord contains **relay neurones** (sometimes called **connector neurones**). These link the sensory neurones with the **motor neurones**. These neurones carry the electrical impulses to the **effectors**. It is the effectors that carry out the responses. Effectors can be muscles or glands.

Where the ends of the neurones come close together in the spinal cord are small gaps called synapses. When an electrical impulse reaches the end of the sensory neurone a chemical change takes place in the synapse. The chemicals released into the gap between the neurones cause the connector neurone to conduct the electrical impulse. A similar action takes place between the ends of the connector neurone and the effector neurone so allowing a continuous flow of impulses to occur between the receptor and the effector through the reflex arc. If the effector is a muscle then it contracts, if the effector is a gland then it secretes a chemical.

A reflex arc is shown in Figure 3.2.

Figure 3.2
Section through the spinal cord showing a reflex arc from the hand

Key

1 Pressure receptor
2 Sensory neurone
3 Sensory neurone cell body
4 Synapse between sensory neurone and relay neurone
5 Relay neurone
6 Motor neurone
7 Spinal cord

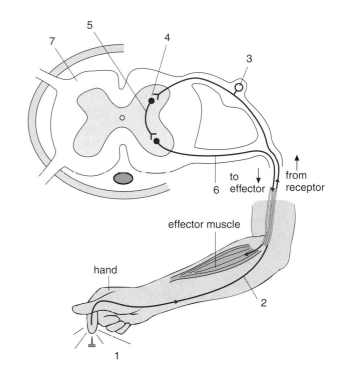

In this reflex action,

- the pin pressing on the skin is the stimulus,
- the pressure receptor detects the pressure of the pin,
- the spinal cord is the co-ordinator,
- the muscle is the effector,
- pulling the arm away is the response.

The eye and light

Receptors in the retina at the back of the eye change light into electrical impulses. These impulses are carried along the **optic nerve** to the brain. The brain then changes the impulses into pictures.

Figure 3.3
A vertical section through a human eye

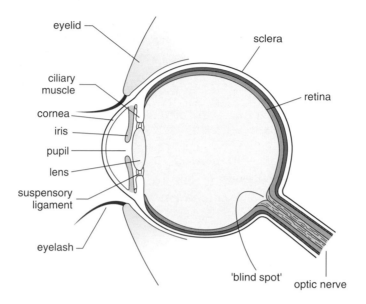

Some important features of parts of an eye

Part	Some important features
sclera	The tough outer layer of the eye to which eye muscles are attached.
cornea	The transparent part of the sclera that allows light into the eye. It is curved and so helps the light to bend inwards on its pathway to the lens.
iris	Coloured muscular layer that controls the amount of light reaching the retina.
pupil	The black opening in the middle of the iris through which light passes on its way to the lens.
lens	Made of a tough transparent jelly, held in place by ciliary muscles and suspensory ligaments. The lens causes the light to bend even more so that it is focused on the retina.
ciliary muscles	• As these muscles contract they squash the **suspensory ligaments** making them go slack and forcing the lens to become fatter so helping to focus light from a nearby object. • As these muscles relax the suspensory ligaments become stretched. This makes the lens become thinner, so helping to focus light from a distant object.
retina	It is on this layer that the light is focused. It contains receptor cells that are sensitive to light.
optic nerve	Links the retina with the brain. The receptor cells in the retina send electrical impulses to the brain along the sensory neurones in the optic nerve.

What happens when light enters the eye

Parallel rays of light enter the eye through the transparent window at the front called the **cornea**. The surface of the cornea is curved and this bends or refracts the light rays helping to **focus** them. The light rays then pass through the **lens**. The surface of the lens is also curved and the light rays are bent even more. The **ciliary muscles** can alter the curve of the lens. This helps the eye to focus on either near or distant objects (Figure 3.4).

Figure 3.4
The shape of the lens changes in order to focus either close up or far away

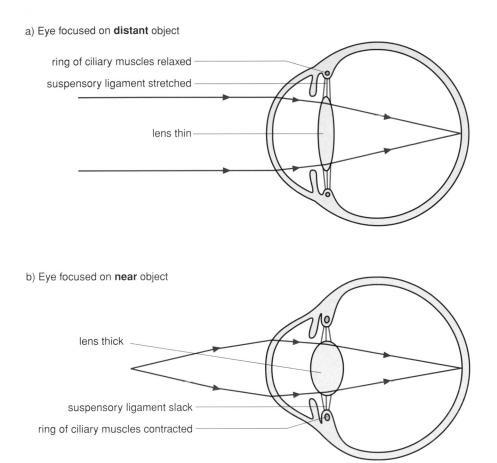

a) Eye focused on **distant** object

ring of ciliary muscles relaxed

suspensory ligament stretched

lens thin

b) Eye focused on **near** object

lens thick

suspensory ligament slack

ring of ciliary muscles contracted

The light rays focused by the cornea and the lens fall on to the **retina** where they form a sharp **image**. The retina is very sensitive to light and the cells change the light into nerve impulses that are carried to the brain along the optic nerve.

The **iris** controls the amount of light that reaches the retina. The iris is a ring of muscle in front of the lens. When the circular muscles in the iris are contracted, the gap or pupil through which the light can pass is very small and only a little light reaches the retina. When the circular muscles are relaxed, the pupil is large. The light is controlled in this way to protect the very delicate light receptor cells of the retina from damage by bright light.

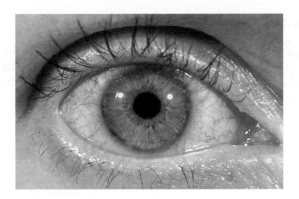
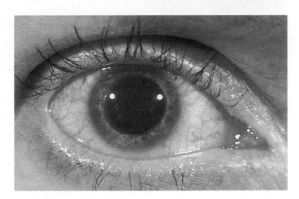

The pupil is small in bright light . . .   *. . . but larger in dim light*

Summary

◆ Cells called **receptors** detect **stimuli**.

◆ These stimuli are any changes that occur in the environment.

◆ The human body has a variety of receptors, each of which can detect a different stimulus.

◆ **Electrical impulses** from receptors pass along nerve cells called **neurones** to the brain which co-ordinates any **response** needed.

◆ **Reflex actions** are automatic, rapid responses that do not involve the brain coordinating them.

◆ Simple reflex actions involve an electrical impulse passing from the receptor along a **sensory neurone** to the spinal cord, then along a **motor neurone** to a muscle or a gland which brings about a response.

◆ The muscle or gland producing the response is called an **effector**.

◆ The spinal cord contains relay (connector) neurones which link the sensory neurones with the motor neurones during a reflex action.

◆ In a reflex action the spinal cord acts as the co-ordinator.

◆ A reflex action can be summarised as: stimulus → receptor → co-ordinator → effector → response

◆ The small gaps between the neurones in the spinal cord are called synapses.

◆ Chemicals are released at each synapse which allow the electrical impulses to pass from one neurone to the next.

◆ The important parts of the eye are the **sclera, iris, pupil, lens, suspensory ligament, retina** and **optic nerve**.

◆ The **cornea** and the **lens** together focus light onto the retina.

◆ The shape of the lens is controlled by the **ciliary muscles**.

Topic questions

1 Where in our bodies are receptors which are sensitive to:
   a) light
   b) sound
   c) changes in body position
   d) chemicals
   e) touch
   f) pressure
   g) temperature changes?

2 What travels along neurones?

3 What is the name of the muscles or glands that carry out responses?

4 a) What is a reflex action?

   b) Put the following in the order electrical impulses flow in a reflex action.

   effector   motor neurone   receptor
   sensory neurone   spinal cord

5 Copy out then complete the gaps in the following table.

Part	Some important features
	This is the tough outer layer of the eye to which eye muscles are attached.
	This is transparent and allows light into the eye.
	This is curved and so helps the light to bend inwards on its pathway to the lens.
iris	
	This is the black opening in the middle of the iris through which light passes on its way to the lens.
	This is made of a tough transparent and causes the light to bend so that it is focused on the retina.
retina	
	This links the retina with the brain.

6 Describe how the action of the ciliary muscles helps the lens to focus on a distant object.

7 a) What is a synapse?
  b) How are electrical impulses passed across a synapse?

8 Put the following in the order the electrical impulses flows in a reflex action.

**relay neurone   effector   motor neurone
receptor   response   sensory neurone   stimulus**

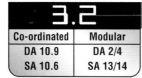

Hormones and the menstrual cycle

Co-ordinated	Modular
DA 10.9	DA 2/4
SA 10.6	SA 13/14

Not all the information within the human body is transmitted through nerves. It is also transmitted by **hormones**. Hormones are chemical substances secreted directly into the blood plasma by special glands called **endocrine glands**. They are carried in the blood plasma to a **target organ** in another part of the body. Hormones often control our long-term responses to changes in the body or the environment.

Sex hormones

The monthly release of an egg from a woman's ovaries and the changes in the thickness of the lining of her womb are controlled by hormones secreted by her **pituitary gland** and ovaries. The stages in this menstrual cycle are shown in Figure 3.5

The hormones involved include oestrogen, secreted by the ovaries, follicle-stimulating hormone (FSH) and luteinising hormone (LH) which are both secreted by the pituitary gland.

Hormones and fertility

Artificial sex hormones are now manufactured and given to some females who do not produce enough of their own or who are infertile. They help the ovaries to make more eggs than usual, increasing the chances of one of these eggs being fertilised by a sperm from a male. This increases the fertility of the female and are called fertility drugs.

Figure 3.5
Hormones and the menstrual cycle

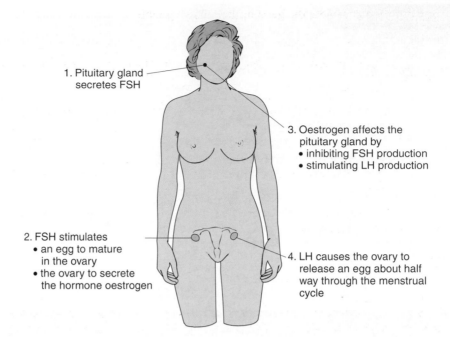

1. Pituitary gland secretes FSH

3. Oestrogen affects the pituitary gland by
   • inhibiting FSH production
   • stimulating LH production

2. FSH stimulates
   • an egg to mature in the ovary
   • the ovary to secrete the hormone oestrogen

4. LH causes the ovary to release an egg about half way through the menstrual cycle

Some women do not produce enough FSH to stimulate eggs in the ovary to mature. Doctors may give these women FSH. FSH acts as a fertility drug by stimulating eggs to mature. If the FSH dose is too high, several eggs may be released at the same time. If these eggs are fertilised, several babies will develop in the woman's womb. This is called a multiple pregnancy. Often few of these babies survive.

Other hormones can be given that prevent the release of eggs from the ovaries. Such hormones are **oral contraceptives** or birth control pills.

The birth control pill contains hormones, including oestrogen that inhibit FSH production. Without FSH no eggs mature, therefore the woman cannot become pregnant.

The use of oral contraceptives rather than barrier methods (condoms) may result in the spread of sexually transmitted diseases.

Summary

◆ **Hormones** are chemicals that control many processes within the body. They are secreted by **glands** and are transported by the bloodstream to **target organs**.

◆ In women, hormones secreted by the **pituitary gland** and the ovaries control the monthly release of an egg and the changes in thickness of the lining of the womb.

◆ **FSH** (follicle stimulating hormone), secreted by the pituitary gland, causes an egg in the ovaries to mature and this stimulates the ovaries to produce hormones including **oestrogen**.

◆ High levels of oestrogen stop the further production of FSH, but stimulate the production of LH.

◆ **LH** (luteinising hormone) secreted by the pituitary gland, stimulates the release of an egg about half way through the menstrual cycle.

◆ **Fertility drugs** are hormones that stimulate the release of eggs from the ovary.

◆ **Oral contraceptives** are hormones that prevent the release of eggs from the ovaries.

◆ There are benefits and problems regarding the use of hormones to control fertility.

Topic questions

1 Complete the gaps in the following sentence. Hormones are secreted by _____ and are transported to their _____ organs by the _____.

2 Where are the hormones that control the monthly release of an egg from a women's ovary and the changes in thickness of the lining of her womb produced?

3 a) What is the action of a fertility drug?
   b) What is the action of an oral contraceptive?

4 A couple are finding it difficult to conceive. After tests have been carried out, the man is found to have fewer live sperm than normal. The woman is given an artificial hormone to increase the number of eggs she produces. Explain why this may help her to conceive even though the man has a low sperm count.

5 a) What do the initials FSH mean?
   b) Which gland secretes FSH?
   c) Describe two functions carried out by FSH.

6 a) What do the initials LH mean?
   b) Which gland secretes LH?
   c) What function is carried out by LH?

7 What effect does oestrogen have on:
   a) the action of FSH?
   b) the action of LH?

3.3	Co-ordinated	Modular
	DA 10.9	DA 2
	SA n/a	SA n/a

Hormones and diabetes

Controlling blood sugar

The glucose concentration of the blood is both monitored and controlled by the pancreas. Blood sugar concentration is controlled by two hormones which are secreted by the **pancreas – insulin** and **glucagon**.

A rise in blood glucose concentration stimulates the cells in the pancreas to secrete insulin into the blood. One effect of insulin is to cause the liver to convert excess glucose into insoluble **glycogen** and store it.

A fall in blood glucose concentration stimulates other cells in the pancreas to secrete the hormone glucagon into the blood. This causes the liver to convert glycogen into glucose and release it into the blood.

Unfortunately there are some people who cannot produce insulin either at all or in sufficient quantities to control blood glucose. These people suffer from **diabetes**. Mild cases of diabetes are treated by carefully controlling fat and carbohydrate intake or by injecting insulin into the blood.

Figure 3.6
Feedback system for the control of blood glucose concentration

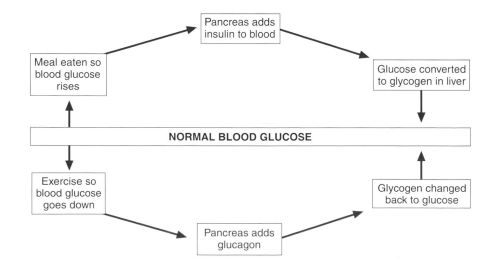

Summary

- Blood glucose concentration is controlled by **hormones** called **insulin** and **glucagon**. These are secreted by the **pancreas**.

- The pancreas monitors and controls the blood glucose concentration.

- If the blood glucose concentration is too high the pancreas secretes insulin. This causes the liver to convert glucose into insoluble glycogen and store it.

- If the blood glucose concentration is too low the pancreas secretes glucagon. This causes the liver to convert glycogen into glucose which is released into the blood.

- **Diabetes** is caused when a person's pancreas does not secrete enough insulin.

- Diabetes can be treated by a carefully controlled diet or by insulin injections.

Topic questions

1 a) Which two hormones control the concentration of glucose in the blood?
   b) Where are these two hormones produced?

2 What happens to the concentration of blood glucose if not enough insulin is produced?

3 a) What happens in the pancreas and the liver if the blood concentration becomes too high?
   b) What happens in the pancreas and the liver if the blood glucose concentration becomes too low?

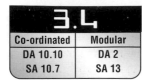

3.4

Co-ordinated	Modular
DA 10.10	DA 2
SA 10.7	SA 13

Homeostasis

In order to remain healthy, the conditions inside our bodies (the internal environment) must be kept the same. If the conditions are not kept the same, our body systems will not work properly. The processes through which the body monitors and controls its internal environment are called **homeostasis**.

The body produces several waste products including:

- carbon dioxide from respiration, which leaves the body via the lungs when we breathe out,

- urea which is produced in the liver by the breakdown of excess amino acids. Urea is removed by the kidneys in urine.

It is important that the body is able to monitor and control the amounts of these waste products. If they build up they will poison the body.

Four conditions that need to be kept the same are:

- the amount of water in the blood
- the temperature of the body
- the amount of glucose in the blood (see section 3.3).
- the ion content of the blood.

Controlling the amount of water in the blood

Each day different volumes of liquids are drunk, different amounts of exercise are taken, different volumes of sweat are produced and different volumes of water are breathed out and yet the water in the blood remains at a constant concentration.

Figure 3.7
Daily water inputs and outputs

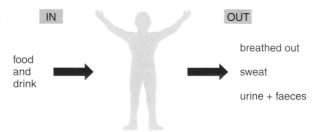

The **kidneys** play an important part in keeping the water concentration in the blood constant. The kidneys are part of the excretory system.

Figure 3.8
The excretory system in humans

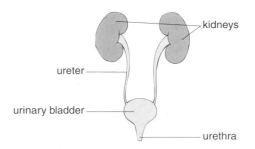

The excretory system

Arterial blood arrives at the kidney under high pressure. Blood leaves the kidney through the vein, joining the vena cava on the journey back to the heart. The venous blood will not only have lost pressure but will also have lost any excess water, ions and urea in the kidney.

Figure 3.9
The blood supply to the kidneys

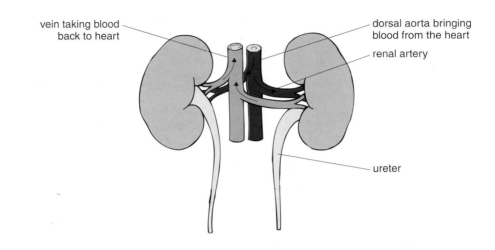

vein taking blood back to heart

dorsal aorta bringing blood from the heart

renal artery

ureter

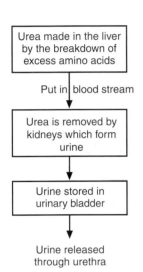

Urea made in the liver by the breakdown of excess amino acids

Put in blood stream

Urea is removed by kidneys which form urine

Urine stored in urinary bladder

Urine released through urethra

Figure 3.10
The process of removal of urea

The kidney makes **urine** with the excess water, excess ions and **urea**. Adults make about 1.7 litres of urine a day, depending on fluid intake, sweat etc. The urine is stored in the **bladder** before being excreted.

What happens in the kidneys?

As blood is circulated round the body it passes through the kidneys where it is filtered, the urea is removed, turned into urine and the concentration of urine is controlled.

Each kidney has many millions of kidney tubules. Each tubule is served by a network of capillaries. As the blood flows through the capillaries, it is filtered by the tubules. As the filtrate flows through each of the tubules, the urine is gradually formed. Blood in the arteries is under high pressure. Because the artery divides suddenly into the many narrow capillaries which surround each tubule, the tubule is also under high pressure. It is this pressure which results in filtration (Figure 3.11).

The high pressure causes a large volume of liquid and the molecules that are small enough to pass through the cell membranes, to enter the tubule. The problem is that too much water and too many useful molecules may have left the blood by filtration, so there is now a system to reclaim what the body needs. The tiny blood vessels that are wrapped around the tubule selectively reabsorb the water required to maintain the blood concentration, and the glucose and ions needed by the body. The blood then continues on to the vein leaving the kidney and the liquid in the tubules continues its journey to join with the contents of all the other tubules and finally to the bladder as urine.

What if the kidneys go wrong?

If the kidneys fail to remove the excess water and ions, the body retains the fluid and it will begin to build up in the tissues. This puts a considerable strain on the heart (due to impaired circulation) and the person can become very ill.

Figure 3.11
Filtration by a tubule

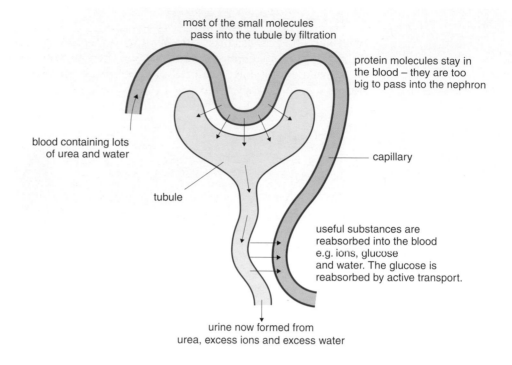

most of the small molecules
pass into the tubule by filtration

protein molecules stay in
the blood – they are too
big to pass into the nephron

blood containing lots
of urea and water

capillary

tubule

useful substances are
reabsorbed into the blood
e.g. ions, glucose
and water. The glucose is
reabsorbed by active transport.

urine now formed from
urea, excess ions and excess water

Did you know?

Kidneys and dialysis

One way of treating the failure to remove water and salts is to take the blood from a vein and pass it through a system that will do the job of the kidneys, by removing the water and salts. The blood is then put back into the vein and continues its journey around the body. One problem with this system is that it can only work while the person is linked to the 'dialysis' machine; the rest of the time the blood and body fluids continually move away from the right concentration. To minimise these effects, patients with failing kidneys are very careful with their diets and try to ensure that they do not put too great a strain on their body systems. However dialysis still has to take place regularly and is very time consuming.

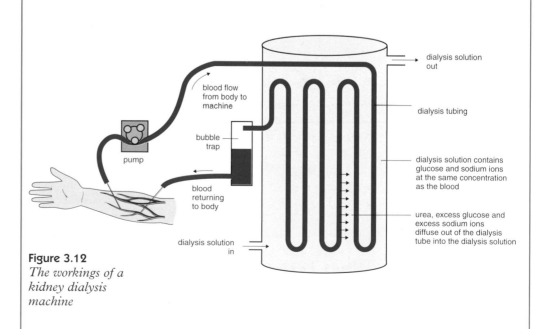

dialysis solution
out

blood flow
from body to
machine

dialysis tubing

bubble
trap

dialysis solution contains
glucose and sodium ions
at the same concentration
as the blood

pump

blood
returning
to body

urea, excess glucose and
excess sodium ions
diffuse out of the dialysis
tube into the dialysis solution

dialysis solution
in

Figure 3.12
The workings of a kidney dialysis machine

The blood and dialysis fluids are separated in the dialysis machine by fine layers of dialysis membrane (a bit like 'Clingfilm'). The blood is under quite high pressure flowing through the machine. The dialysis fluid is at the same concentration as the blood plasma (including sodium ions and glucose). The fluid does not have urea or excess water – both of which are present in the blood – this means that the urea, excess water and sodium ions diffuse out of the blood into the dialysis fluid and are thereby removed from the body. The 'cleaned' blood can then be returned to the patient.

Kidney transplants

If, as a result of kidney failure, a person needs dialysis regularly, one alternative is to have a kidney transplant. The donated kidney can come from a living person – generally a relative – or from an anonymous donor who has recently died. The kidney has to be a good blood and tissue match or the body's immune system will reject it. Even so, drugs are given to help the body accept the transplanted kidney.

Once the kidney is in place and hasn't been rejected by the body, the person can once again live a normal life with no need for dialysis.

How the water content of the blood is controlled

In order to keep the internal environment of the body stable, a number of automatic control systems are at work. The self-regulation requires the system to be continuously monitored. For example, if the body gains too much water then the system tries to restore the original balanced conditions by losing water. This automatic response is called negative feedback. Negative feedback is the system by which the desired factors are kept constant – it means that when one of the constants is changed, a response resulting in the opposite effect occurs.

Water concentration in the blood is controlled by the reabsorption of water by the kidneys. As blood flows through the blood vessels at the base of the pituitary gland in the brain, specialised cells measure the water content of the blood. If the blood is getting short of water because the person has not had a lot to drink or has lost a lot of water by exercise, then:

- the pituitary gland releases a hormone into the blood. The hormone is known as ADH (anti-diuretic hormone)

- when the blood containing the ADH reaches the kidneys, the cells of the kidney tubules let more water be reabsorbed into the capillaries.

- the urine then contains less water.

If, however, the person has had a lot to drink or has not lost a lot of fluid via exercise, then:

- the specialised cells will measure a high water concentration in the blood as it flows through the brain

- the pituitary gland stops releasing ADH into the blood

- when the blood reaches the kidneys the absence of ADH causes the tubules to reduce the amount of water being reabsorbed into the blood

- a large volume of urine is then produced, containing a lot of water.

Figure 3.13
The processes that occur when the blood contains a) too little and b) too much water

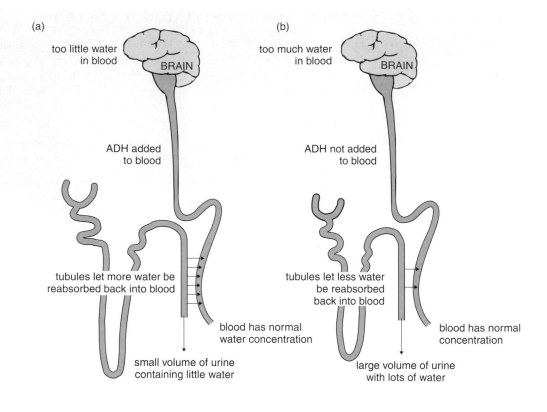

Figure 3.14
Feedback system used to control the production of ADH

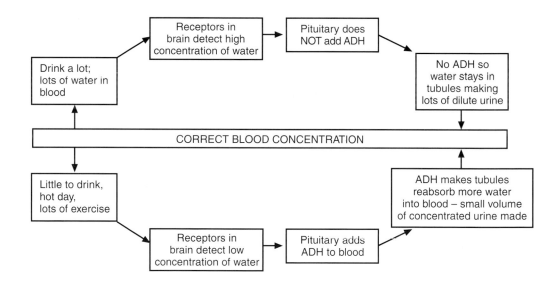

Keeping cool and keeping warm – the skin and its blood vessels

The **skin** has three important functions:

- it protects the body from invasion by bacteria, viruses and fungi.
- it is waterproof as a result of the outer layer.
- it helps to maintain a constant body temperature of about 37°C.

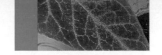

Keeping our temperature constant

It is vital to humans and all warm-blooded animals that a constant body temperature is maintained to enable all the enzymes to work efficiently.

Body temperature is monitored and controlled by the thermoregulation centre in the brain. Temperature sensors in the body monitor the temperature of the blood.

If the blood is too hot or too cold the body has a system for keeping cool and another for keeping warm. Blood flows from the arteries towards the surface of the skin in tiny vessels that further divide into smaller vessels, eventually recombining to become veins.

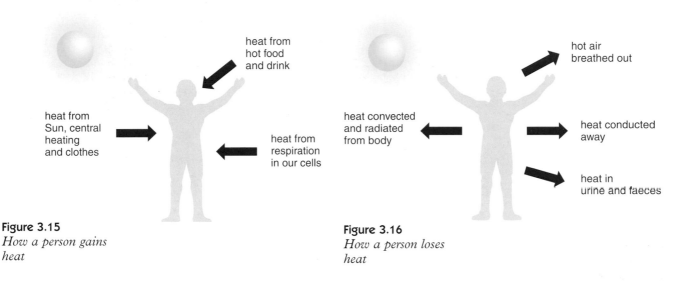

Figure 3.15
How a person gains heat

Figure 3.16
How a person loses heat

What happens at the skin surface to maintain a constant internal temperature?

When our body starts to get too hot:

- blood vessels supplying the capillaries near the surface of the skin get wider (dilate). This allows more blood to flow through the capillaries near the surface of the skin causing it to redden. Heat from this blood is transferred to the air by radiation.

- more sweat is produced by the sweat glands. The water in sweat evaporates. In order to evaporate, the water needs heat energy. The heat energy needed to cause the evaporation comes from the body, so the body cools down.

When our body starts to get too cool:

- blood vessels supplying the tiny capillaries get narrower (constrict), reducing the blood flow to the capillaries near the surface of the skin. This causes the skin to go white and limits the heat loss by radiation.

- almost no sweat is produced, reducing heat loss further. The reduction of blood flow towards the surface of the skin ensures that the blood for the vital organs is at the correct temperature.

At no time either in keeping cool or staying warm do the blood vessels actually move!

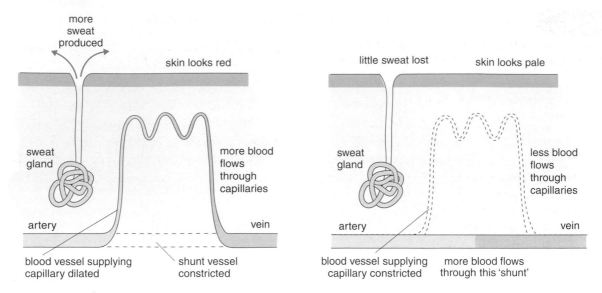

Figure 3.17 *What happens in the skin when the body gets too hot*

Figure 3.18 *What happens in the skin when the body gets too cool*

Shivering

When the core body temperature is too low, the body may shiver. Shivering is involuntary contraction of several of the muscles. Muscle contraction uses energy from respiration. Some energy from respiration in muscles is released as heat so blood flowing through these muscles is warmed.

Therefore thermoregulation of the skin keeps the body temperature constant.

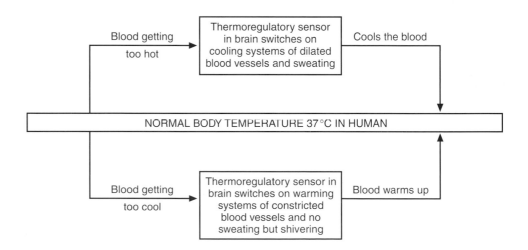

Figure 3.19
Feedback system of body temperature control

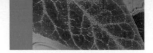

Summary

- Carbon dioxide produced by respiration is removed via the lungs during breathing out.

- **Urea** produced in the liver by the breakdown of excess amino acids is removed via the kidneys in the **urine**.

- **Homeostasis** is the process by which the internal conditions are self-regulated.

- The water content of the body needs to be kept constant. Water is lost via the skin as sweat, via the lungs and via the kidneys in urine. To balance these losses more water needs to be taken in as food and drink.

- The ion content needs to be kept constant. Ions are lost via the skin as sweat, and via the kidneys in urine.

- The kidneys maintain the internal environment by:

  - **filtering** the blood

  - **reabsorbing** all the glucose

  - reabsorbing the amount of dissolved ions needed by the body

  - reabsorbing as much water as the body needs

  - releasing urea, excess ions and excess water as urine.

- The internal temperature needs to be kept constant. This maintains the temperature at which enzymes work best.

- The thermoregulatory centre in the brain monitors the temperature of the blood. If the blood temperature is too high:

  - the blood vessels supplying the skin capillaries dilate so more blood flows through the capillaries so losing heat

  - the sweat glands release more sweat which cools the skin as it evaporates.

  If the blood temperature is too low:

  - the blood vessels supplying the skin capillaries constrict so less blood flows through the capillaries so losing less heat

  - the muscles 'shiver' because the muscles repeatedly contract and relax. Each contraction needs the energy transferred during respiration. Some of this energy is transferred as heat.

- Receptors in the brain monitor the concentration of water in the blood. If there is too little water in the blood:

  - the pituitary gland secretes a hormone called ADH (anti-diuretic hormone) into the blood.

  - the kidneys reabsorb more water, resulting in urine being more concentrated.

  If there is too much water in the blood:

  - less ADH is secreted and the concentration of water in the blood increases.

Topic questions

1. Why do we need to get waste products removed from our body?

2. Complete the gaps in the table.

Waste product	How produced	How removed
Carbon dioxide		
	breakdown of excess amino acids	

3. Why does our body temperature have to remain constant at about 37°C?

4. a) Where is urea made?
   b) What is urea made from?
   c) Where is urine made?
   d) What is urine made from?

5. Explain, using your knowledge of ADH, why on a hot day when James had been running in a cross-country race he did not produce much urine and was very thirsty.

6. What does the body do to keep the internal organs warm on a cold day? Include as much detail as possible.

Co-ordinated	Modular
DA 10.11	DA 1
SA 10.8	SA 13

3.5

Fighting disease

Bacteria and viruses

Most diseases are caused by two types of microorganisms, **bacteria** and **viruses**.

Although there are many different types of bacteria, the cells of each contains:

- cytoplasm
- a cell membrane
- a cell wall
- genetic material

Bacterial cells do not have a recognisable nucleus.

Diseases caused by bacteria include tetanus, diarrhoea, cholera and food poisoning.

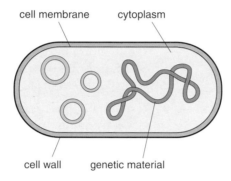

Figure 3.20
A bacterium

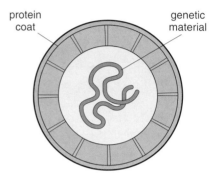

Figure 3.21
A virus

Viruses are very much smaller than bacteria.

The cell of each virus consists only of a protein coat which surrounds a few genes.

Viruses are unaffected by antibiotics. Because they survive only in living cells they are difficult to attack without destroying the living cells as well.

Different viruses attack different cells.

- The virus that causes the common cold and 'flu' attacks the cells lining the respiratory system.

- The AIDS virus attacks the blood cells which help our bodies fight disease.

- The rubella (German measles) virus can seriously damage the nervous system of a fetus during the first few months of its development.

The body's defences against disease

The body has several methods of defending itself against the entry of microorganisms.

- The skin acts as a barrier to prevent the entry of disease-causing microorganisms. Any cuts should be cleaned and covered with a plaster or bandage. This covering stops the microorganisms from getting into the bloodstream and speeds up the healing process.

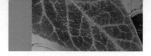

- The trachea and the air passages in the lungs are lined with special cells that have hair like structures called cilia. In between these cells are other types of cells that produce a sticky liquid called **mucus**.

Figure 3.22
Cross-section of the mucus-producing cells that line the air passages of the respiratory system

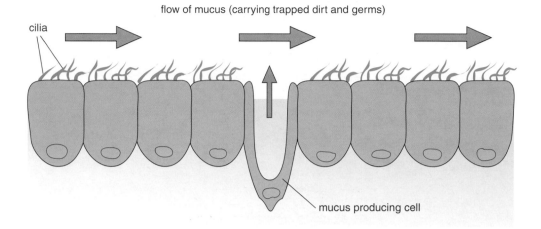

flow of mucus (carrying trapped dirt and germs)

cilia

mucus producing cell

This mucus traps many of the microorganisms that enter the body every time air is breathed in. The cilia are constantly beating in a direction away from the lungs. Their beating motion moves the mucus and any microorganisms that have been trapped towards the mouth where it gets swallowed.

- The blood produces clots which seal cuts.

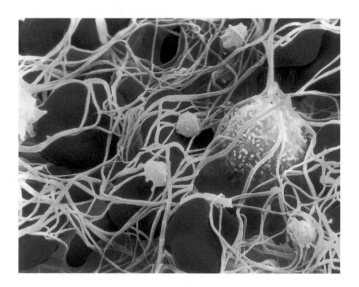

Figure 3.23
Fibrin threads trap the blood cells to form a clot

Platelets are cell fragments which play a very important role in the formation of blood clots. At a wound, platelets and cut capillaries release a chemical which starts a chain of enzyme reactions. The final reaction causes a protein in the plasma to form fine threads that then trap the platelets and cells forming a clot that eventually dries to make a scab. The clot serves two functions – it stops further bleeding and prevents entry of harmful microorganisms.

The work of white blood cells

Some white blood cells fight against infection by ingesting and destroying bacteria.

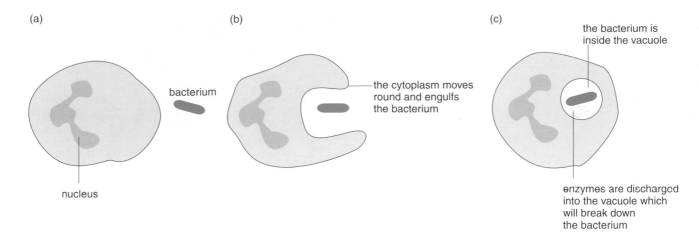

(a)

nucleus

(b)

bacterium

the cytoplasm moves
round and engulfs
the bacterium

(c)

the bacterium is
inside the vacuole

enzymes are discharged
into the vacuole which
will break down
the bacterium

Figure 3.24
*A white cell
engulfing and
destroying a
bacterium*

Another type of white cell engages in 'chemical warfare' in our bodies. Invading bacteria have 'marker proteins' on their cell walls and our bodies recognise these as 'foreign'. White cells produce specific **antibodies** which cause the invading bacteria to stick together so that they can be attacked and destroyed by bacteria-engulfing white cells.

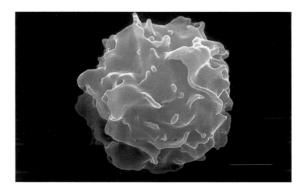

Figure 3.25
*White blood cells
have a very large
round nucleus.
these particular
white cells produce
antibodies*

Certain invading bacteria produce chemical **toxins**. Another group of white cells produce chemicals called **antitoxins** to neutralise these toxins.

How the blood fights infection		
By attack	white blood cells	take in and destroy bacteria
By chemical action	white blood cells	• destroy bacteria by protein chemicals called antibodies. • neutralise poisons produced by bacteria by making antitoxins.
By clotting	platelets and cells are trapped by fibrin threads	this forms a clot which dries to a scab preventing • further loss of blood • entry of bacteria.

Protection by immunisation

Figure 3.26
This girl is being immunised against German measles (Rubella)

It is now possible to provide **immunity** against many diseases by introducing into a person's bloodstream dead or mild forms of the particular strain of disease-producing organism. These dead or mild forms are called **vaccines**. When these dead or mild forms enter the blood they cause some of the white blood cells to produce **antibodies** that attack and destroy the particular strain of bacterium or virus. The antibody-making cells are then ready and available to produce more antibodies very quickly should the same strain of bacterium or virus get into the bloodstream at a later date.

Did you know?

The making of a mild form of a particular strain of disease-producing organism may take many years. It usually involves the selective sub-culturing of the organism over many generations.

Flu can be caused by over 100 different strains of virus. This is why it is possible to catch flu more than once.

Did you know?

The work of Edward Jenner (1749–1823)

Edward Jenner was a doctor who worked in a country practice in Berkeley in Gloucestershire.

During the 18th century smallpox was a cause of many deaths. Even those that recovered were scarred for life by the blisters that formed on the skin.

Cowpox is a disease of cows, which shows up as small blisters containing pus on their udders and teats. Humans, such as milkmaids, who touched these parts often caught cowpox but survived. Jenner noticed that those people who had once had cowpox were not affected by smallpox. He believed that in some way having cowpox prevented people from having smallpox. In 1796 he inoculated an eight-year-old boy with fluid from a person who suffering from cowpox. Six weeks later Jenner inoculated the boy with the pus from the blisters of someone who had smallpox. The boy survived. Although there were many people who thought Jenner had taken too many risks to prove his beliefs the procedure was soon adopted and deaths due to smallpox decreased.

Both cowpox and smallpox are now known to be caused by a virus. The procedure of inoculating people with the cowpox virus in order to protect them against smallpox was the first example of **immunisation**. Smallpox has now been eradicated worldwide – no immunisation is used anywhere in the world.

Did you know?

Ring a ring of roses
A pocket full of posies
A tishoo! A tishoo
We all fall down

This nursery rhyme is supposed to have been linked with the Great Plague. Between 1664 and 1666 the plague killed over 10% of the population of London. The plague was carried by rats that infested the houses and streets and was passed to humans by fleas. The 'ring of roses' refers to a rash which was an early symptom of the plague. People carried posies (bunches) of flowers to hide the smells of the rat-infested streets. They thought that the smell of the flowers would stop them catching the disease. Sneezing was also a sign of having the plague. The posies did not prevent the plague and so many of them 'fell down dead'.

What they did not realise was that in order to stop the spread of such diseases there was a need to improve the way sewage was got rid off, clean up the drinking water and improve the living conditions.

At the time of the Great Plague, poor living conditions caused the rapid spread of disease. Figure 3.27 shows the conditions which still existed in slums in London in 1900. Even today people in many parts of the World live in overcrowded conditions with no adequate sewage or clean drinking water.

Figure 3.27

Slums in London during Victorian times

Some health problems of today

- Even though food-hygiene regulations have been brought into action in many countries of the world, people are still careless in the way food is grown, prepared and cooked so there are always outbreaks of food poisoning.

- Smoking cigarettes, excessive drinking of alcohol, overeating, lack of exercise and unprotected sex are recognised as lifestyles which cause disease.

- BSE (mad cow disease) was first diagnosed in cattle in Britain in 1986. In an attempt to stop the spread of BSE and variant CJD in Britain the feeding to cattle of recycled animal tissues was banned in 1988. It was not until 1996 that the eating of beef products from cattle infected with BSE was officially recognised as causing a brain disease in humans – new variant CJD (Creutzfeldt-Jakob disease).

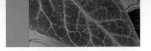

Figure 3.28
Some milestones that have reduced the spread of disease

Date	Milestone
1790s	Edward Jenner developed the technique of immunisation
1860s	Louis Pasteur showed that food could be prevented from going bad if microorganisms from the air were kept away from the food
1865	Joseph Lister carried out the first operation in which carbolic acid was used to kill the microorganisms on the instruments and bandages. Carbolic acid was the first antiseptic to be used.
1859–1875	The first attempt to stop London's sewage going directly into the river Thames.
1928	Alexander Fleming developed the first antibiotic – penicillin.
1945	DDT was used world-wide to fight insect-borne diseases such as malaria and typhus. Its use is now been banned in many countries.
1948	National Health Service was introduced into Britain to provide free health care for all
1956	Clean Air Act was introduced which banned the use of coal for burning in fires in London. This dramatically reduced the number of deaths from respiratory diseases.
1960s	Large scale of immunisation of children with measles, mumps and rubella vaccine reduced dramatically the number of children affected with these diseases.

Summary

◆ A **bacterial cell** consists of cytoplasm, a membrane, a cell wall and genes. There is no distinct nucleus.

◆ A **virus** is smaller than a bacterium and consists of a protein coat which surrounds a few genes.

◆ Viruses can only reproduce in living cells.

◆ Some methods by which the body defends itself from disease-producing micro-organisms include:
  – the skin acting as a barrier
  – the cells lining the respiratory tract produce mucus that traps microorganisms
  – the blood producing clots that seal cuts.

◆ Some white cells:
  – ingest microorganisms
  – produce **antibodies** that destroy particular bacteria and viruses
  – produce **antitoxins** which counteract the toxins (poisons) produced by bacteria and viruses.

◆ During immunisation, mild or dead forms of the infecting organism are introduced into the bloodstream. Some white blood cells respond by producing antibodies. These antibodies can be rapidly produced if the same strain of infecting organism invades the body on another occasion. This is immunity.

◆ Life styles are related to the spread of disease.

Topic questions

1 In what ways is a bacterium cell different from the cell of a virus?

2 Describe three ways in which the body prevents infection by bacteria or viruses.

3 Describe three ways the white blood cells defend against bacteria or viruses.

4 How does immunisation help you to be protected from further infection by the same disease?

61

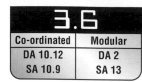

3.6	
Co-ordinated	**Modular**
DA 10.12	DA 2
SA 10.9	SA 13

Drugs

Our senses and co-ordination can be affected if we take certain substances into the body either accidentally or deliberately. Many of these substances are **drugs**. They alter the way the body works.

Many drugs are, of course, helpful. These include pain killers such as aspirin, and antibiotics like penicillin which slow the growth of bacteria so that the body can destroy them quicker. This helps sick people to recover more rapidly.

All these drugs are useful if used correctly but can be harmful if not. Pain killers only take the pain away, they do not cure the cause of the pain. Antibiotics are becoming less effective because some bacteria are becoming resistant to them.

Some drugs are harmful because they can be misused. Some examples of these are given below.

- Alcohol is a drug because it alters the way the body works. It is a small soluble molecule which is very quickly absorbed in the mouth and stomach. This means that its effects are also very quick – it slows down the reactions of the body and leads to a loss of self control. Excess alcohol can have long-term effects including permanent damage to brain and liver cells.

- Certain **solvents** such as those used in various glues, can be very dangerous. They cause hallucinations so that a person can become violent. Like alcohol, they can cause damage to the liver and brain. The problem is that both alcohol and solvents are **addictive** which means that with prolonged or heavy use, the body feels as though it cannot do without them.

- **Nicotine** is another drug. It is taken in when the tobacco in cigarettes is smoked. It has a very powerful effect on the body and, like alcohol and solvents, it is addictive, which makes giving up smoking very difficult. Babies born to mothers who smoked during pregnancy are sometimes born with nicotine in their blood.

> **Did you know?**
>
> The effect of an absence of nicotine after smokers have 'quit' can be reduced by wearing 'patches' containing small amounts of the drug. The body absorbs the drug through the skin. Gradually the amount in the patches is reduced so that the person has less nicotine in the blood and this lessens the addiction to it.

The effects of smoking

The passages in the nose cavity are lined with tiny hairs called cilia and the cells produce mucus (see section 3.5). Dust and microbes get caught in the sticky mucus and the cilia. The particles are expelled from the body during sneezing.

The tobacco smoke from cigarettes can stop the cilia beating and this allows the mucus to build up. The only way the body can get rid of the mucus, the dust it has trapped and bacteria is to cough it up. The smoke also contains a large number of chemicals, including tar, and some of these are known carcinogens (**cancer**-causing chemicals). Frequent coughing damages the lungs and can lead to severe lung diseases, such as bronchitis and emphysema. Emphysema results from the breakdown of the thin membranes of the alveoli and results in a reduction of gaseous exchange surface. This damage eventually prevents the lungs from working properly and the person may become disabled by breathing difficulties.

Another problem caused by smoking is the carbon monoxide produced by the burning cigarette. Carbon monoxide joins irreversibly to haemoglobin and prevents it carrying oxygen.

A reduction in the amount of oxygen carried by the blood is particularly dangerous to the fetus carried by a smoking mother as the baby is often born underweight. The cause of 'low birth weight' babies to smoking mothers could be that the fetus receives less oxygen for respiration.

Figure 3.29
The facts about smoking

Here are the facts for you to make up your own choice about smoking.

Nicotine:
Dose – per cigarette – 3 mg of which 1 mg is absorbed and reaches the brain in 30 seconds

Effect – increases heart rate and blood pressure putting extra strain on heart and capillaries in lungs, kidney etc.; causes stickiness of platelets leading to blood clotting which can occur anywhere in the body

Psychological effects – smoker feels more relaxed and capable of facing stressful situations; makes boring times more tolerable

Nicotine is addictive both psychologically and physiologically

Carbon monoxide:
Dose – 5% of cigarette smoke is CO

Effect – lowers oxygen-carrying capacity of blood haemoglobin by 3 – 7% permanently. This is serious for people with heart problems or for pregnant mothers as their fetus receives less O_2 and grows slowly

Tars:
Dose – enters lung as an aerosol in tobacco smoke. 70% is deposited in bronchioles and alveoli

Effect – tar causes chronic bronchitis and carcinogens greatly increase chances of developing lung cancer, cancer of the mouth and cancer of the throat

Irritants: including tar
Effect – cause extra secretion of mucus, coughing, breakdown of alveolar walls

The links between smoking and lung cancer

In 1585 Sir Francis Drake brought tobacco to England and Sir Walter Raleigh introduced the practice of pipe smoking among the Elizabethan courtiers. For more than 300 years tobacco was smoked only in pipes or as cigars. It was not until early in the 20th century did the smoking of cigarettes become common. By 1935 cigarette smoking was the most popular way of smoking tobacco and the smoking of cigarettes was socially acceptable. Up until the late 1950s the majority of the population were smokers.

Because medical records in the 19th century showed that deaths from lung cancer were very rare, no connection was made linking lung cancer to the smoking of tobacco. However as deaths from lung cancer began to increase some scientists began to look for links with the increase of cigarette smoking. Today, lung cancer kills about as many people each year as died from *all* forms of cancer in 1900.

In the 1950s scientists began to carry out detailed research to prove links between the incidence of lung cancer and smoking. The results of one very large survey are shown in the table.

	Number in sample	Number of deaths from lung cancer	Lung cancer rate per 100 000 people
Non smokers	32 460	2	6.0
Smokers	107 897	152	145

These results show that smokers increase their chances of dying from lung cancer by 24 times. Further research has shown that smokers who smoke at least 40 cigarettes a day increase their chances of dying from lung cancer by 90 times.

Because much of the research carried out *does* link an increase in deaths from lung cancer to smoking, TV tobacco commercials have been banned, tobacco products have health warnings on them, and smoking is banned in many public places. However governments in many countries collect a very large amount of money from taxes on tobacco. So rather than prohibit smoking most governments prefer to provide their people with enough information to make their mind up about whether to smoke or not.

Summary

◆ Harmful **drugs** change the chemical processes in a person's body so that he or she may become dependent or addicted to them and suffer withdrawal without them.

◆ **Solvents** affect behaviour and may damage the lungs, liver and brain.

◆ Alcohol affects the nervous system by slowing down reactions which can lead to a lack of self-control, unconsciousness or even coma.

◆ Alcohol can damage the liver and brain.

◆ Nicotine is the addictive substance in tobacco.

◆ Tobacco contains substances which can cause lung cancer, bronchitis, emphysema, diseases of the heart and blood vessels.

◆ Tobacco smoke contains carbon monoxide which reduces the oxygen-carrying capacity of the blood. In pregnant women this can lead to babies being underweight.

◆ The link between smoking and lung cancer only gradually became accepted.

◆ Carbon monoxide combines irreversibly with the haemoglobin in the red blood cells.

Topic questions

1 Name two organs of the body damaged by drinking too much alcohol.

2 Name three organs of the body damaged by solvent abuse.

3 What is the name of the addictive substance in tobacco?

4 Drugs such as alcohol, solvents and nicotine can be addictive. What does this mean?

5 Why should pregnant women not smoke?

6 What effect does smoking have on the cilia that line the air passages of the respiratory system?

7 What gas in tobacco smoke lowers how much oxygen the red blood cells can take around the body?

8 Explain why alcohol acts so quickly after it has been taken into the body whereas most other drugs take about 4 hours before they act.

9 Explain why it is dangerous to drive a motor car after even a small amount of alcohol has been taken in.

10 Why is carbon monoxide such a dangerous gas to breathe in?

Examination questions

1 The chart shows the effect of smoking on the annual death rate in men.

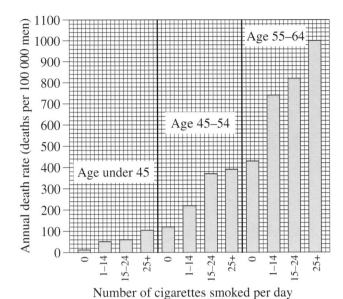

Number of cigarettes smoked per day

a) The death rate of men aged 45–54 who smoke more than 25 cigarettes a day is higher than the death rate for non-smokers. How much higher is it? Give your answer as a number per 100 000. *(1 mark)*

b) Explain, as fully as you can, why the death rate for smokers is higher than the death rate for non-smokers in each age group. *(3 marks)*

2 The diagram shows a reflex pathway in a human.

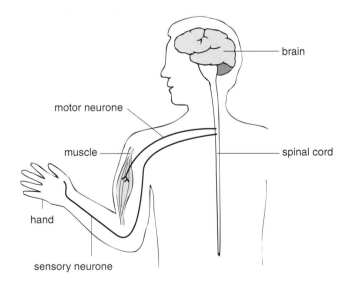

a) Label the *receptor* on the diagram. *(1 mark)*
b) Label the *effector* on the diagram. *(1 mark)*

c) i) Suggest a stimulus to the hand that could start a reflex response. *(1 mark)*
   ii) Describe the response that this stimulus would cause. *(1 mark)*
d) Put arrows on the diagram to show the direction of the path taken by the nerve impulses. *(1 mark)*

3 The diagram shows the human skin.

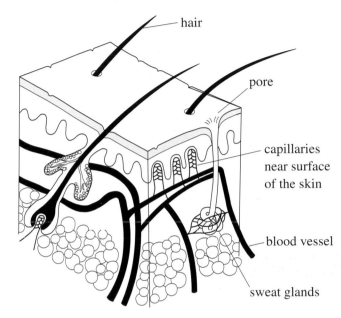

a) i) In hot weather the diameters of the blood vessels supplying the capillaries in the skin change. Explain how this change keeps us cool. *(3 marks)*
   ii) Give **one** other way that the skin can keep us cool. *(1 mark)*
b) Give **one** other function of the skin. *(1 mark)*

4

> ## Coordination of the body can be affected by chemicals called hormones

a) i) Where are hormones produced? *(1 mark)*
   ii) How do hormones move around the body? *(1 mark)*

b) Insulin and glucagon are hormones.

   i) Where are insulin and glucagon produced? *(1 mark)*

   ii) Explain the roles of insulin and glucagon in controlling blood sugar levels. *(6 marks)*

c) A hormone can be used to treat infertility in women. Explain how. *(2 marks)*

5 The diagram shows some of the processes which control the composition of blood.

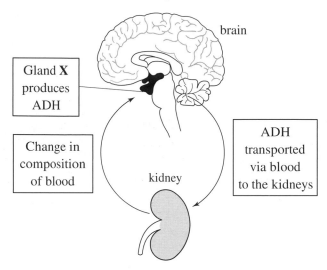

a) i)   Name gland **X** *(1 mark)*

  ii)   What is the stimlus which causes gland **X** to produce ADH? *(1 mark)*
 iii)   What type of substance is ADH? *(1 mark)*

b) Describe the effect of an increase in ADH production on the kidney and on the composition of the urine. *(3 marks)*

Chapter 4
Green plants as organisms

4.1 — Plant nutrition

Co-ordinated	Modular
DA 10.13	DA 2
SA n/a	SA n/a

One of the most important differences between green plants and animals is the way they feed. Plants can make their own food from simple raw materials. Animals have to obtain their food by eating plants and other animals. The process by which plants make their own food is called **photosynthesis**.

Photosynthesis is a process in green plants which produces **biomass** from the simple raw materials of carbon dioxide and water. The biomass is initially in the form of carbohydrate such as glucose, but from this, plants can make every substance they need. Glucose is stored in the plant as starch. The chemical reactions which convert the simple raw materials into more complex ones need a source of energy. This energy is supplied by light from the Sun which is absorbed by **chlorophyll**, a green pigment found in the chloroplasts of all plant cells which carry out photosynthesis.

Did you know?

Only 1% of the light from the Sun absorbed by the plant is used in photosynthesis.

Photosynthesis can be summarised as the word equation:

$$\text{carbon dioxide} + \text{water [+ light energy]} \longrightarrow \text{glucose} + \text{oxygen}$$

The need for carbon dioxide, water, light and chlorophyll in photosynthesis can be investigated by depriving plants of each of those materials in turn and seeing whether they can still make starch. Other experiments can show where the carbon dioxide goes when plants make starch and where the oxygen comes from. These experiments usually require the use of radioactive materials.

In most plants, photosynthesis takes place in the leaves. Leaves are adapted to absorb as much light as possible and to enable the plant to convert the maximum possible amount of light into chemical energy.

Carbon dioxide enters the leaves by **diffusion**. It moves from a higher concentration, outside the leaf, to a lower concentration, inside the leaf (see section 1.4).

Adaptation of the leaf for photosynthesis

- The leaves are flat and thin to provide a large surface area for the absorption of light.

- **Palisade cells**, which contain the most chloroplasts, are near the upper surface of the leaf. This enables the chlorophyll to absorb as much of the light as possible.

- Carbon dioxide enters and oxygen exits the leaf through **stomata**. **Guard cells** control the opening and closing of the stomata and therefore control the movements of gases into and out of the leaf. In most plants, stomata open when it is light, and close when it is dark.

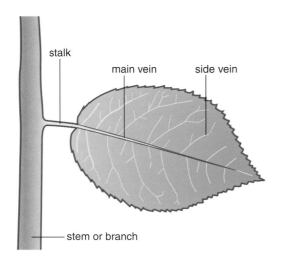

Figure 4.1
The structure of a leaf

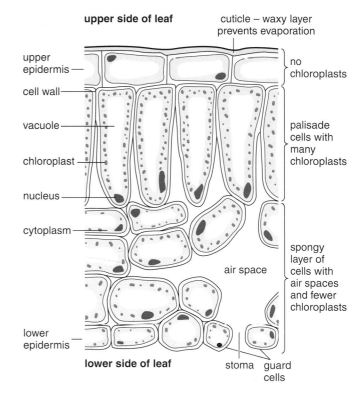

Figure 4.2
Cells in the cross section of a leaf

Limiting factors

Since photosynthesis needs the raw materials carbon dioxide, water and light energy, it is clearly going to be affected by their availability. They become **limiting factors**, i.e. the amount of photosynthesis that takes place is determined by whichever raw material is in shortest supply.

It would be expected that plants make more glucose on a warm sunny day than on a cold dull one. This can be investigated by measuring how much photosynthesis is going on. There are several ways of doing this:

- measure how much glucose or starch is being produced, although the plant has to be destroyed to make the measurements.

- measure how much oxygen is being made, this can be done easily by counting the oxygen bubbles coming off from a plant in water.

- measure how much carbon dioxide is being used up. However, a plant does not always take the carbon dioxide out of the air. Some of the carbon dioxide is produced by the plant itself by respiration.

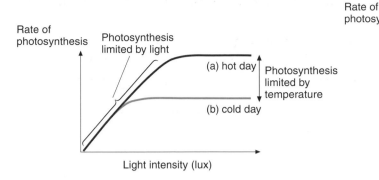

Figure 4.3
Graph to show the effect of increasing light intensity (i.e. increasing brightness) on the rate of photosynthesis on a) a hot day and b) a cold day

Figure 4.4
Graph to show the effect of increasing light intensity on the rate of photosynthesis at a) high CO₂ and b) low CO₂ concentrations in the air

Figure 4.3 shows the effect of the increase in light intensity on the rate of photosynthesis as a day progresses on a) a warm sunny day and b) a day that is just as sunny but much colder.

The rate of photosynthesis increases with the intensity of the light. The light must have been a limiting factor. Beyond a certain point there is no further increase however bright the light becomes. This indicates that light is no longer a limiting factor but that something else is – it could be the amount of carbon dioxide. Figure 4.4 shows what happens when the effect of light is investigated at two different levels of carbon dioxide but at the same temperature. The level of carbon dioxide was the limiting factor because when the level of carbon dioxide is increased, the rate of photosynthesis again increases with light intensity.

It is possible to apply the same principle to any of the factors which affect the rate at which plants make glucose. Experiments show that water can also be a limiting factor. In dry summers and in hot climates the availability of water is likely to be the most important limiting factor of all.

Did you know?

Plants which can make their own food by photosynthesis are green because they contain chlorophyll. This is a vital substance in converting light energy into chemical energy. Chlorophyll appears green because it reflects green light. This means that it cannot absorb very much green light and must absorb mostly red and blue.

Figure 4.5
Absorption of different wavelengths of light by chlorophyll

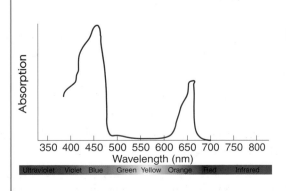

Figure 4.5 shows how much of each colour of light is absorbed by green plants and how much photosynthesis goes on in each colour. Maximum absorption is in the blue/violet and orange/red ranges.

There is a close link or correlation between the colour that is absorbed and the colour that is used by the plant. This suggests that plants produce most food when light in the red and blue parts of the spectrum is shone at them and very little in the green part.

Uses of the products of photosynthesis

The equation for photosynthesis shows that both glucose and oxygen are made by photosynthesis. This is only part of the story, however, because plants can make everything they need from photosynthetic products. The glucose can be used to make every other substance the plant needs.

The energy released from glucose during respiration is used to build smaller molecules into larger molecules. For example:

1 Starch – Some of the glucose is made into starch and stored in the roots, stems and leaves. Starch is insoluble and can easily be stored without upsetting the water balance of cells, so the amount that can be stored by plants is almost unlimited.

2 Cellulose – Another important material made is **cellulose**. This is important for the manufacture of cell walls which all plant cells have. It is a very tough material and helps the plant cell walls to support the weight of the plant.

3 Lipids – Plants can convert glucose into lipids for storing in seeds. Like starch, lipids are insoluble but by weight they contain about twice as much energy as starch. A lot of energy can therefore be stored in a smaller space. The seeds need the stored energy so that they can remain dormant in the winter until conditions favourable to growth return in the spring. The lipids are then used to supply the young seedling with the energy it needs until it can make its own food.

4 Amino acids – All the substances described so far contain only the atoms carbon, hydrogen and oxygen. Plant cells can also make amino acids from glucose. Amino acids contain the element nitrogen in addition to carbon, hydrogen and oxygen, and to make them, plants must absorb nitrate from the soil. Amino acids are then made into protein, a vital building block of cells. This is why the healthy growth of plants is often dependent on the application of nitrate fertiliser to the soil.

Did you know?

Sucrose – Some of the glucose is converted to sucrose for storage in fruits. Sucrose is soluble so the amount that can get stored is limited but it makes the fruits taste sweet and therefore attractive to animals so that the seeds the fruits contain can be dispersed.

Did you know?

A strand of cellulose is stronger than a strand of steel of the same width.

Cotton is nearly pure cellulose – try breaking a cotton thread.

Because of its tensile strength, cellulose is used in fabrics such as cotton and rayon, for tyre cords, paper packaging and the manufacture of cellophane.

Mineral ions

Besides needing water from the soil to photosynthesise, plants need mineral ions to make other compounds. Plants absorb these mineral ions from the soil. In farming these mineral ions are removed from the soil when the crop is harvested. To maintain crop yields, the mineral ions must be replaced by fertilisers. The most common artificial fertiliser is known as NPK because it contains nitrogen, phosphate and potassium. What does a plant need these three ions for?

Nitrate

Plants need **nitrates** to make proteins. Without proteins plants cannot grow. The symptoms of nitrate deficiency are stunted growth and yellowing of the older leaves.

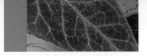

Phosphate

Plants need **phosphates** for the energy transfers that take place in photosynthesis and respiration. Without phosphate the plant cannot obtain enough energy for normal growth. The symptoms of phosphate deficiency are poor root growth and purple younger leaves.

Potassium

Potassium is needed to produce some of the enzymes involved in photosynthesis and respiration. The symptoms of potassium deficiency are yellow leaves with dead spots.

Summary

◆ **Photosynthesis** can be summarised by the equation:

carbon dioxide + water + [light energy] ⟶ glucose + oxygen

◆ During photosynthesis:

- light energy is absorbed by a green substance called chlorophyll. **Chlorophyll** is found in the chloroplasts of green plants

- the light energy is used by converting carbon dioxide and water into a sugar (glucose)

- oxygen is released as a by-product.

◆ The rate of photosynthesis may be limited by such factors as:

- low temperatures

- a shortage of carbon dioxide

- a shortage of light.

These are called **limiting factors**.

◆ Much of the glucose produced is often converted into and stored as insoluble starch.

◆ Some of the glucose is used by the plant cells for respiration.

◆ The energy transferred during respiration is used by plants to build smaller molecules into larger molecules, such as:

- glucose into starch

- glucose into **cellulose** for cell walls

- glucose, nitrates and other nutrients into amino acids which are then built up into proteins

- glucose into lipids (fats or oils) for storage in seeds.

◆ Plant roots absorb mineral salts including nitrates all of which are needed for healthy growth.

◆ For healthy growth plants need mineral ions such as:

- **nitrates**, for the synthesis of proteins

- **phosphates**, which play an important role in the energy transfers involved in photosynthesis and respiration

- **potassium**, which helps to produce some of the enzymes involved in photosynthesis and respiration.

◆ Where these mineral ions are missing from the growing conditions the following deficiencies will be observed:

- a lack of nitrate ions results in stunted growth and a yellowing of the older leaves

- a lack of phosphate ions results in poor root growth and young leaves going a purple colour

- a lack of potassium ions produces yellow leaves with dead spots.

Topic questions

1 Why are most leaves wide and flat?

2 a) Write down the word equation for photosynthesis.
   b) What are the two reactants in this equation?
   c) What are the two products in this reaction?

3 Photosynthesis takes place in cells containing chlorophyll.
   a) What colour is chlorophyll?
   b) Where in a cell is chlorophyll found?
   c) What is the function of chlorophyll?

4 The graph shows how the rate of photosynthesis changes as the light intensity changes.

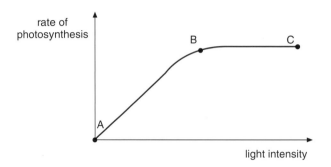

   a) What trend is shown between A and B?
   b) What trend is shown between B and C?
   c) Suggest why the rate of photosynthesis is not increasing beyond B?

5 What is some of the glucose made during photosynthesis stored as?

6 Much of the energy transferred during respiration is used by plants to build large molecules from small molecules. Name four examples of large molecules made by plants. For each example state what use it is to the plant.

7 Copy and complete the table

Mineral ion	Why needed by a plant	Effect if missing from the growing conditions
nitrate		
phosphate		
potassium		

4.2 Plant hormones

Co-ordinated	Modular
DA 10.14	DA 2
SA n/a	SA n/a

Plants are sensitive to light, moisture and gravity. Plant shoots grow *towards* the light and *away* from the force of gravity. The advantage of this is that the plant's leaves are more likely to be in the light where they can absorb energy for photosynthesis.

Figure 4.6
Cress plants grown with light from above (left) and with light from only one direction (right)

Plant roots grow *towards* the force of gravity and *towards* water. The advantage of this is that the roots are more likely to anchor the plant firmly in the soil and be able to absorb water.

Animals have chemicals produced by glands circulating in their bloodstream which can bring changes in cells, tissues and organs a long way from the gland which produced them. These chemicals are called **hormones**. Plants also produce hormones. One type of plant hormone is produced at the shoot tip. The hormones pass down the stem from the shoot tip and cause the cells just behind the shoot tip to elongate, making the shoot grow.

Response of plant shoots to light

In growing shoots some hormones respond to the effects of light. These are considered to encourage cell growth. If a plant is lit equally from all sides the hormones produced by the growing tip diffuse down the shoot evenly. If the plant is lit only from one side then more of the hormones produced by the growing tip diffuses down and collects on the unlit side of the growing shoot. The hormone encourages faster growth on the unlit side so as the shoot grows, it bends towards the light.

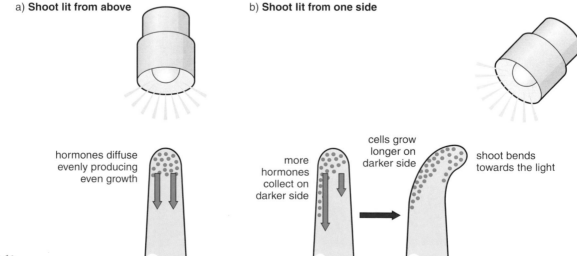

Figure 4.7
Plant responses to light

73

Response of plant roots to gravity

In roots some of the hormones produced by the growing tip respond to the effects of gravity. These hormones slow down the growth rate of the cells in the growing tip. So if a root tip is growing downwards then the hormones are evenly distributed and the root tip continues to grow straight down. If the tip is growing sideways then more hormones collect on the underside of the growing tip, causing the cells on the underside to slow down their rate of growth. This causes the root tip to grow downwards.

Figure 4.8
Plant responses to gravity

a)
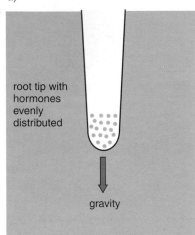

root tip with hormones evenly distributed

gravity

b)
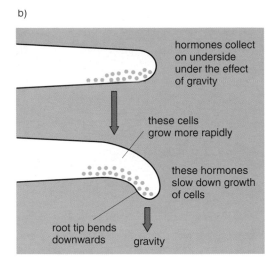

hormones collect on underside under the effect of gravity

these cells grow more rapidly

these hormones slow down growth of cells

root tip bends downwards

gravity

Commercial uses of plant hormones

It is now possible to make artificial plant hormones.

* If plants are given too much hormone, they can grow too quickly and die. Broad-leaved plants, such as some weeds (dandelions), are more affected by this than narrow-leaved plants, such as grasses. If the hormone is sprayed onto a lawn, the weeds are killed and the grass is left unaffected. The hormone is being used as a selective weed killer.

Figure 4.9
A dock plant before (left) and after treatment with a weed killer

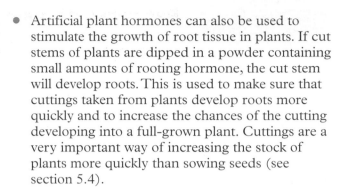

- Artificial plant hormones can also be used to stimulate the growth of root tissue in plants. If cut stems of plants are dipped in a powder containing small amounts of rooting hormone, the cut stem will develop roots. This is used to make sure that cuttings taken from plants develop roots more quickly and to increase the chances of the cutting developing into a full-grown plant. Cuttings are a very important way of increasing the stock of plants more quickly than sowing seeds (see section 5.4).

Figure 4.10
Cuttings dipped in rooting hormone quickly develop roots

- Hormones in plants are important in the development of fruits after fertilisation. Artificial hormones can be sprayed on the plant at flowering time so that the fruits will grow without pollination and fertilisation taking place. This leads to the production of fruits that do not contain seeds (so-called seedless fruits such as seedless grapes).

Summary

- The shoots of a plant grow *towards* the light and *against* the force of gravity.

- The roots grow *towards* water and *towards* the force of gravity.

- Growth in plants is co-ordinated by **hormones.**

- The responses of roots and shoots to light, water and gravity are due to the unequal distribution of hormones. This unequal distribution causes unequal growth rates.

- Hormones controlling the growth and reproduction in plants can be used by humans to:

  – produce large numbers of plants quickly by stimulating root growth in cuttings
  – regulate the ripening of fruit
  – kill weeds.

Topic questions

1 To what three things are plants sensitive?

2 a) Towards what do plant shoots grow?
  b) How does this help the plant survive?

3 a) Towards what do plant roots grow?
  b) Why is this an advantage for the plant?

4 What do plants produce that causes them to control how they grow?

5 If a plant has light coming from one side only it bends towards the light. Explain why.

6 If a root is growing sideways it will bend downwards under the effect of gravity. Explain why.

7 Give three commercial uses of plant hormones.

4.3 Transport and water relations in plants

Co-ordinated	Modular
DA 10.15	DA 2
SA n/a	SA n/a

Water is vital to plants for the following:

● Photosynthesis – The process of photosynthesis cannot proceed without water which is a raw material for the reaction.

● Transport – Plants need to transport materials, such as glucose or mineral ions, in solution from one part of the plant to another.

● Mineral ion uptake – Materials such as nitrates, phosphates and other mineral ions from the soil can only be absorbed from the soil if they are in solution.

● Support – If plants suffer from a lack of water, they may **wilt** and eventually die because the cells are no longer swollen due to a lack of **turgor** caused by a lack of water.

Transpiration

Water vapour evaporates from the surfaces of the leaves. This process is called **transpiration**. Transpiration will occur most rapidly in hot, dry and windy conditions. In order to reduce the amount of water loss many plants have a waxy surface. Plants, such as cacti, that grow in very hot, dry areas often have leaves with a very thick waxy surface and a very small surface area. The water evaporates through the **stomata**, the opening through which gases enter and exit the leaf (see section 4.1). The size of the opening is controlled by the pairs of **guard cells** that surround each stoma.

Figure 4.11
Stomata and hairs on the surface of a leaf

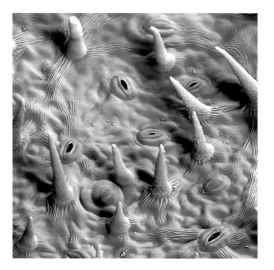

If plants lose water faster than the roots can take it from the soil the stomata will close to stop the plant from wilting.

How does water move from the roots to the leaves?

The evaporation of water from the leaves causes a transpiration pull. This is powerful enough to suck the water from the roots all the way up the plant. Water travels up the stem and into the leaves through **xylem** vessels.

> **Did you know?**
>
> Transpiration can raise water from the roots to the leaves of the tallest plants which can be over 100 m tall?

Did you know?

How stomata open and close

The wall of each guard cell is thickest and therefore less likely to change its shape where it lines the gap. When the vacuoles of a pair of guard cells fill with water, by osmosis, the guard cells swell and bend. They bend because the guard cells are joined at each end. This causes the gap between them to widen. When guard cells lose water they shrink and straighten. This causes the gap between them to get narrower.

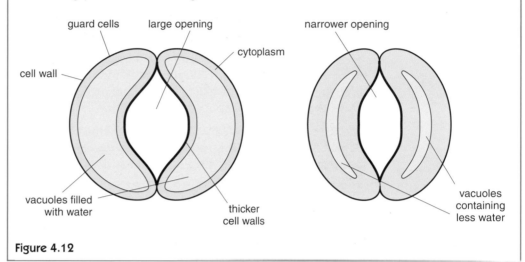

Figure 4.12

Uptake of water and mineral salts by plants

Plants take in nearly all the water and mineral salts they need through the roots. The water is absorbed by **osmosis** (see section 1.4).

The mineral salts are absorbed in solution in the form of ions e.g. NO_3^-, PO_4^{2-}, K^+ (see section 1.4).

The roots of the plants are covered with cells called **root hair cells**.

Figure 4.13

Diagram to show the role of the root hair cell in the uptake of water and the movement of water through the root to the xylem

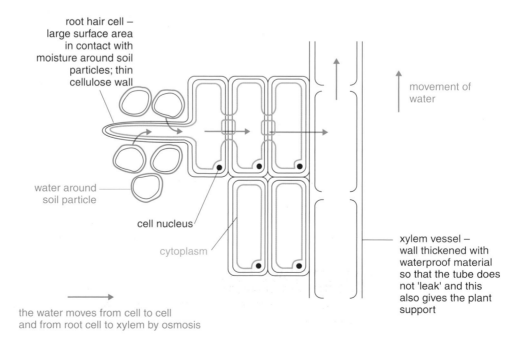

These root hair cells have very thin walls and greatly increase the surface area for absorbing mineral salts and water from the soil.

The salts get into the root hairs by **diffusion** or by active transport (see Chapter 1). To be absorbed by diffusion, there must be a higher concentration of mineral salts in the solution outside the cell than inside. This enables a concentration gradient to exist and then mineral salts diffuse into the root hair cells down the concentration gradient.

Active transport is needed if there is a higher concentration of salts inside the cells than outside. In such cases, the cell can only take in the salts if it uses cellular energy.

Once inside the root hair cells, the water and mineral salts are passed to the **xylem** tissue in the centre of the root. Xylem contains differentiated cells that form vessels. The xylem vessels transport the mineral ions up the root and stem to all cells.

The carbohydrate made by photosynthesis must also be moved to other parts of the plant. The soluble food is transported in the **phloem** tubes to the roots for storage or to the tips of the shoots where it is used as a source of energy for growth.

Figure 4.14
The passage of water through a plant

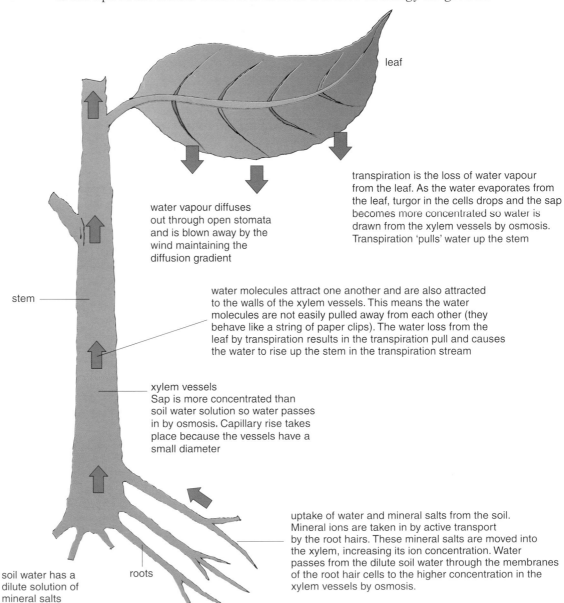

leaf

transpiration is the loss of water vapour from the leaf. As the water evaporates from the leaf, turgor in the cells drops and the sap becomes more concentrated so water is drawn from the xylem vessels by osmosis. Transpiration 'pulls' water up the stem

water vapour diffuses out through open stomata and is blown away by the wind maintaining the diffusion gradient

stem

water molecules attract one another and are also attracted to the walls of the xylem vessels. This means the water molecules are not easily pulled away from each other (they behave like a string of paper clips). The water loss from the leaf by transpiration results in the transpiration pull and causes the water to rise up the stem in the transpiration stream

xylem vessels
Sap is more concentrated than soil water solution so water passes in by osmosis. Capillary rise takes place because the vessels have a small diameter

uptake of water and mineral salts from the soil. Mineral ions are taken in by active transport by the root hairs. These mineral salts are moved into the xylem, increasing its ion concentration. Water passes from the dilute soil water through the membranes of the root hair cells to the higher concentration in the xylem vessels by osmosis.

soil water has a dilute solution of mineral salts

roots

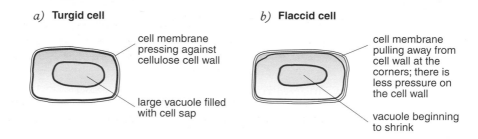

Water and the cell

Figure 4.15
a) A turgid plant cell,
b) a flaccid plant cell

a) **Turgid cell**

cell membrane pressing against cellulose cell wall

large vacuole filled with cell sap

b) **Flaccid cell**

cell membrane pulling away from cell wall at the corners; there is less pressure on the cell wall

vacuole beginning to shrink

Cells normally contain as much water as possible. In this state the cells are said to be turgid. The vacuole in the cell which contains the sap exerts a pressure on the walls of the cell, helping to support it. When all cells are turgid this helps to support the whole plant. This is particularly important in non-woody plants.

Sometimes water evaporates from the surface of the leaf faster than it can be replaced by osmosis. This means that cells lose water faster than they can gain it. The vacuole of the plant cell begins to shrink and the pressure on the wall of the cell decreases. The cell becomes flaccid. If this happens in many cells, the leaf may begin to wilt. If the plant is given water, the cell becomes turgid again and the leaf recovers.

Summary

- **Transpiration** is the loss of water vapour from the surface of a leaf.

- Transpiration is most rapid in hot, dry and windy conditions.

- Most leaves have a waxy coating (cuticle) that prevents the loss of too much water.

- Transpiration takes place through the **stomata**.

- The size of the stomata is controlled by **guard cells**.

- Water in a plant provides support. Plants **wilt** if they lose too much water.

- **Xylem** tissue transports water and minerals from the roots to the stem and leaves.

- **Phloem** tissue transports nutrients, such as glucose, from the leaves to the rest of the plant

- Water moving into a cell by osmosis increases the pressure inside the cell.

- Cell walls are strong enough to withstand this pressure.

- This pressure keeps the cell rigid (maintains its turgor). This turgor provides the support.

Topic questions

1 a) What is transpiration?
   b) In which conditions would transpiration be slowest?

2 Through which part of a leaf does the water evaporate?

3 What controls how much water evaporates?

4 Why do plants wilt on a hot day?

5 What draws water all the way up from the roots to the leaves?

6 What is transported along the xylem tissues?

7 What is transported along the phloem tissues?

8 Describe how cells become rigid but do not burst.

Examination questions

1 a) Copy and complete the following sentences.

   Green plants produce their own food by a process called photosynthesis. In this process the raw materials are _____ and carbon dioxide. Glucose and _____ are produced. _____ energy is absorbed by the green substance called _____ .   *(4 marks)*

   b) Name **two** things that can happen in the plant to the glucose produced in photosynthesis.
        (1 mark)

   c) Plants need mineral salts.
      i) Through which part do mineral salts get into the plant?   *(1 mark)*
      ii) Explain why water is important in this process.   *(2 marks)*

   d) Some students set up water cultures to find out how plants use nitrates.
      They had two sets of nutrient solutions.
      A full solution provided the plant with all the required nutrients.
      The results table shows the average mass of the seedlings after 28 days growth.
      i) Give a conclusion you could make from these results.   *(1 mark)*
      ii) Calculate the difference in average mass caused by the addition of nitrates to the culture solution.   *(1 mark)*
      iii) What are nitrates used for in the seedling?
        (1 mark)
      iv) Some factors need to be controlled to keep this test fair. Name **two** of them.   *(2 marks)*
      v) Suggest **one** way you could improve the experiment.   *(1 mark)*

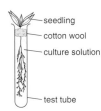

Culture solution	Average mass of seedling in g
distilled water	0.14
full solution with no nitrates	0.29
full solution	0.43

2 The diagram shows a plant leaf during photosynthesis.

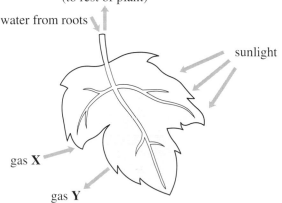

   a) Name:
      i) gas X;
      ii) gas Y.   *(2 marks)*
   b) Name the tissue which transports:
      i) water into the leaves;
      ii) sugars out of the leaves.   *(2 marks)*
   c) Why is sunlight necessary for photosynthesis?
        (1 mark)

d) Some of the sugars produced by photosynthesis are stored as starch in the roots. Explain, as fully as you can, why it is an advantage to the plant to store carbohydrate as starch rather than as sugar. *(3 marks)*

3 A potted plant was left in a hot, brightly lit room for ten hours. The plant was not watered during this period. The drawings show how the mean width of stomata changed over the ten hour period.

a) Why do plants need stomata? *(1 mark)*
b) Name the cells labelled **X** on the drawing. *(1 mark)*
c) The width of the stomata changed over the ten hour period. Explain the advantage to the plant of this change. *(2 marks)*

d) Explain, in terms of osmosis, how cells in a young plant are kept rigid. *(2 marks)*

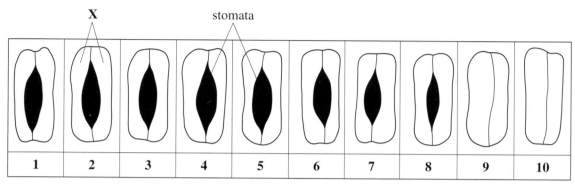

Time (hours)

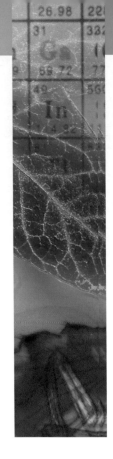

Chapter 5
Variation, inheritance and evolution

Key terms adaptation · allele · artificial selection · asexual reproduction · cancer · carrier · cell division · characteristic · chromosomes · clone · DNA · dominant · extinct · fertile · fossils · fusion · gamete · gene · genetic · genetic engineering · genotype · growth · heterozygous · homozygous · meiosis · mitosis · mutation · natural selection · recessive · reproduction · selective breeding · sexual reproduction · species · variation

5.1	
Co-ordinated	Modular
DA 10.16	DA 4
SA 10.10	SA 14

Variation

People are all different because each person has a unique set of genetic information. The only exceptions are identical twins which are formed from a single fertilised egg cell.

The children in the photograph can easily be divided into groups of males and females. But it's not so easy to group them according to variations in height, eye colour or whether they have ear lobes.

What causes variation?

Differences in the **characteristics** of individuals of the same kind (species) are caused by differences in:

- **environmental factors**. For example, plants grown in plenty of light, fertile soil and which are well watered, will grow sturdy and strong. But if plants are grown in poor light and are not watered properly they will be pale and spindly. For both plants and animals, nutrients will determine whether or not the organism reaches its genetic potential size. However, environment cannot change some inherited characteristics. Living in the dark or only eating chips will not change a person's blood group.

- **genes** they have inherited (**genetic factors**). Cell division to produce **gametes** (**meiosis**) is a source of variation as each individual receives a unique genetic combination.

- **Mutation** – A mutation is a change in the genetic information. Mutations can be of two types:

  1 a chromosomal mutation is a major change such as a piece of **chromosome** breaking off and either getting lost or attached to another chromosome, or the loss of a whole chromosome, changing the chromosome number. These changes produce serious results in animals but are less catastrophic in plants.

  2 a gene mutation in which the information within a gene is changed.

Figure 5.1
Variation as seen in a group of children

A mutation that happens when the gamete is formed will be passed on to the offspring. Queen Victoria received a mutation that she passed onto some of her sons resulting in them having haemophilia. In this condition, the blood is very slow to clot.

What causes mutations?

Mutations can happen spontaneously, but exposure to gamma radiation, strong X-rays, ultraviolet rays and certain chemicals can trigger mutations. Food additives and household chemicals are always tested to try to make sure that they do not cause mutations. Excess exposure to sunlight can cause mutations in skin cell nuclei and cause skin **cancer**. A cancer develops when cells multiply in an uncontrolled way.

Are mutations harmful?

Most mutations are harmful if they occur in reproductive cells. This is because the young may develop abnormally or even die at an early stage of their development. Mutations are also harmful if they occur in body cells because they are likely to develop into cancers.

Some mutations do not affect the individual. In some cases mutations may increase the chances of an individual surviving. If such a mutation is genetic and therefore heritable, then offspring produced by such an individual are likely to have their chances of survival increased.

Variation and reproduction

There are two forms of reproduction:

- **sexual reproduction**; which involves the **fusion** of male and female gametes
- **asexual reproduction**; where there is no fusion of cells and only one individual gives rise to offspring.

Asexual reproduction produces offspring with genetic information that is identical, not only to each offspring, but is also identical to that of the parent. **Genetically** identical individuals are called **clones**.

Sexual reproduction produces offspring that have a mixture of genetic information from both parents. These individuals show variation between themselves and each of the parents.

Differences caused by the environment

The potential to grow to a particular size is determined by the genetic material present at fertilisation or, in the case of an asexually produced organism, by the genetic material present in the parent. However whether an individual reaches their potential is due mainly to the effects of environmental factors such as diet, living conditions etc. Plants show the importance of the environment very clearly. Seeds sown in poor soil will certainly not grow and develop their full potential.

Variation in hydrangeas

> **Did you know?**
>
> There is a plant called a hydrangea that has flowers that can be blue, white or pink. The colour of the flower is not controlled by genes but by the type of soil in which the plant grows. If the soil is acidic then the flowers are blue, if the soil is alkaline then the flowers are white or pink.

Why does reproduction sometimes produce variation?

The cells of offspring produced by asexual reproduction are produced by mitosis from the parental cells (see section 1.5). Because mitosis produces identical genes the offspring contain the same genes as the parent.

Sexual reproduction gives rise to variation because:

- there is a random assortment of the chromosomes during meiosis (see section 1.5) so the gametes do not contain identical genetic material,
- there is random fertilisation – it is pure chance which gametes fuse with each other, therefore the offspring have a new assortment of genetic material.

The work of Gregor Johann Mendel (1822–1884)

Mendel was an Austrian monk who was very interested in plant breeding. He was particularly interested in breeding pea plants which showed clear variation in a number of characteristics. For example, some pea plants had short stems and others had long stems, some plants produced white flowers others produced red flowers, some plants had round seed pods others had wrinkled seed pods. He started his investigations in 1856 and grew the pea plants in the garden of the monastery.

Figure 5.2
Gregor Mendel

He cross-pollinated those plants that showed the characteristics in which he was interested, taking the greatest care to ensure that only he and not the insects or the wind pollinated the flowers. Each year he collected the seeds from his selected plants and grew them. When the new plants grew he recorded their characteristics and by 1863 he had bred over 25 000 pea plants. For seven years he had taken very careful notes of all the crosses he had made together with the results of those crosses.

In the 19th century the accepted view was that the characteristics of offspring are the result of a blending of the characteristics of the parents. Mendel gradually realised that the idea of characteristics being inherited through a blending process did not match his results. He proposed the following two ideas:

- That there were hereditary factors (his word for gene) that showed either dominant or recessive properties. When a pollen grain meets an egg, a factor pair is formed in which a dominant factor will mask a recessive factor.
- That hereditary factors do not blend but remain unchanged from one generation to the next.

These two ideas form the basis of what we now know as the laws of heredity. In 1866 Mendel published his results and conclusions in a local scientific journal. For the next 34 years very few scientists took any notice of his work because:

- the journal in which it appeared was not read by many scientists
- the complexity of his results meant that the work was difficult to understand.

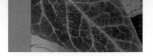

Mendel died with his contribution to genetics unrecognised. It was not until 1900, when a Dutch scientist came across the journal in which Mendel's work was published, that the great importance of Mendel's conclusions was realised. Several years later Mendel's ideas were used to help explain Darwin's theory of evolution.

Summary

- All young resemble their parents because of genetic information passed on through the sex cells (**gametes**).

- The information is carried by **genes**.

- Different genes control different characteristics.

- Mendel proposed the idea of genes.

- Differences in characteristics within a species may be due to differences in:

  - the genes inherited (genetic causes)
  - the conditions in which they have developed (environmental causes)
    or a combination of both.

- New forms of genes (**mutations**) result from changes in existing genes.

- Mutations can occur naturally, but can occur more frequently by exposure to:

  - ionising radiation and from the radiation from radioactive substances
  - certain chemicals.

- Most mutations are harmful if they occur in:

  - reproductive cells
  - body cells.

- Some mutations are neutral in their effects.

- In rare cases, a mutation may increase the chances of survival of an organism and any offspring if the mutation is genetic.

- There are two forms of reproduction:

  - **sexual reproduction**; which involves the **fusion** of male and female gametes
  - **asexual reproduction**; where there is no fusion of cells and where only one individual gives rise to offspring.

- Asexual reproduction produces offspring with genetic information that is identical to that of the parent. **Genetically** identical individuals are called **clones**.

- The cells of offspring produced by asexual reproduction are produced by mitosis from the parental cells. They contain the same genes as the parent.

- Sexual reproduction produces individuals that have a mixture of genetic information from both parents. These individuals show variation between themselves and each of the parents.

- Sexual reproduction gives rise to variation because:

  - the gametes are produced from parental cells by meiosis
  - when gametes fuse, one of each pair of alleles comes from each parent
  - the alleles in a pair may vary and therefore produce different characteristics.

- Mendel's discovery about separately inherited factors only gradually came to be accepted.

Topic questions

1. What part of a nucleus carries the genetic information for a particular characteristic?

2. a) Why are people different?
   b) Which people will have identical genes?
   c) What does a single gene control?

3. a) Which type of reproduction produces identical offspring?
   b) What is the name given to the offspring that are genetically identical to each other and to the parent?

4. a) What is a gene mutation?
   b) Give four different ways a mutation can occur.

5. Give two reasons why sexual reproduction produces offspring with different genetic material?

6. Why is it harmful if a mutation occurs in
   a) the reproductive cells
   b) body cells?

7. When could a mutation be useful?

5.2

Co-ordinated	Modular
DA 10.17	DA 4
SA 10.11	SA 14

Genes

Sex determination

Humans have 23 pairs of **chromosomes** in body cells which means that the egg and sperm cells each contain 23 single chromosomes.

Of the 23 pairs, one pair is made up of the sex chromosomes that determine the sex of the person. These are labelled X and Y on Figure 5.3. Males have an X and a Y chromosome (as in the diagram) while females have two X chromosomes, XX.

Figure 5.3

A set of male human chromosomes

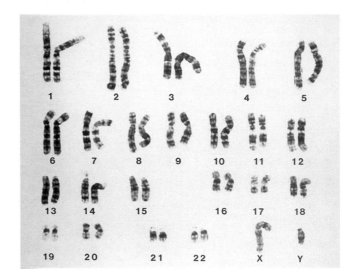

Will the baby be a boy or a girl?

- A sperm will contain either one X sex chromosome or one Y chromosome.

- Every egg will contain one X sex chromosome.

During fertilisation one sperm will fuse with one egg. After fertilisation the cells of the baby will contain two sex chromosomes. So the possible combinations for the baby's sex chromosomes will be:

	sperm		egg	
sex chromosomes	X or Y		X or X	

baby's sex chromosomes	X + X	or	Y + X	
	(from a sperm) (from an egg)		(from a sperm) (from an egg)	
	a baby girl		a baby boy	

So there is a 50% chance of a baby being a boy or a girl.

Figure 5.4
Sex determination in humans

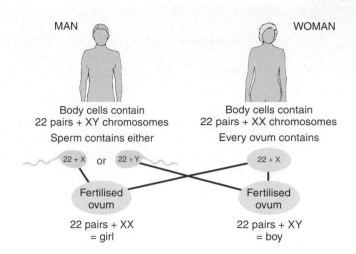

From now on only the X and Y chromosomes will be considered, although it must be remembered that there are 22 other chromosomes in each ovum and sperm.

The only possible combinations are XX and XY as shown below.

Figure 5.5
A Punnett square crossing XX and XY

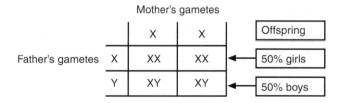

As a result it can be seen that the chances of having a baby boy or a baby girl are equal. Again there is an equal chance at each conception. However, not all families will have equal numbers of sons and daughters.

Genes, chromosomes and alleles

Chromosomes are made of long molecules of a substance called **DNA**. Each chromosome is divided into genes. A gene is a length of chromosomal DNA which codes for one characteristic. It is helpful to use the word gene to refer to the position of the information on the chromosome.

Alleles can be considered to be different forms of a gene found in the same position on the chromosome. They control the same characteristic but code for a different detail. There are very many alleles in plants and animals, each controlling different features of the organism. It is much easier to understand with an example.

Alleles for eye colour

A chromosome has a gene position for eye colour but the allele at that position could be either the allele coding for blue eyes or the allele coding for brown eyes.

Figure 5.6a

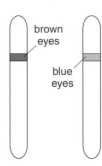

This individual has inherited the allele for brown eyes from one parent and that for blue eyes from the other parent.

- The alleles for eye colour are different so the individual is said to be heterozygous for eye colour.

- The allele for brown eyes is **dominant** over the allele for blue eyes so the instruction 'have brown eyes' hides, but does not destroy the instruction 'have blue eyes'. The individual will therefore have brown eyes.

Figure 5.6b

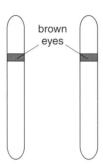

This individual has inherited one of the alleles for brown eyes from each parent.

- The alleles for eye colour are the same so the individual is said to be homozygous for eye colour.

- Because both alleles are for the same instruction the individual will have brown eyes. The individual would have blue eyes if both alleles had been for this instruction. An allele which controls the development of characteristics only if the dominant allele is not present, is called a **recessive** allele. The allele for blue eye colour is recessive.

Some dominant and recessive characteristics in humans

Characteristic	Dominant	Recessive
Eye colour	brown eyes	blue eyes
Freckled skin	freckles	no freckles
Tongue-rolling	can tongue roll	cannot tongue roll

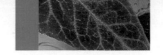

Using genetic diagrams to find out more about inheriting eye colour

- The word genotype describes the genetic composition of the organism – that is the alleles present for the characteristic being studied. When setting out a cross, the first letter of the phenotype is usually used as a capital letter to represent the dominant allele and as a lower case letter to represent the recessive allele.

Figure 5.7
A Punnett square showing the results of a homozygous cross between brown eyes (BB) and blue eyes (bb)

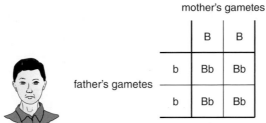

mother's gametes

	B	B
b	Bb	Bb
b	Bb	Bb

father's gametes

100% of offspring have brown eyes

100% of offspring are heterozygous Bb

But what if the brown-eyed parents is heterozygous. This cross – heterozygous with homozygous recessive – gives a 1:1 ratio in the offspring.

Figure 5.8
A Punnett square showing the results of a cross between brown eyes (Bb) and blue eyes (bb)

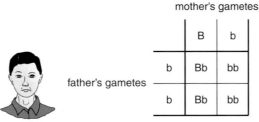

mother's gametes

	B	b
b	Bb	bb
b	Bb	bb

father's gametes

50% of the offspring are heterozygous Bb and will be brown-eyed

50% are homozygous bb and will be blue-eyed

If the person has blue eyes, their genotype *must* be bb, as the allele for blue eyes is homozygous recessive. Consider what the outcome might be if a brown-eyed man and his brown-eyed wife have children. They do not know if one or both of them are heterozygous for eye colour (Bb) or homozygous (BB).

Figure 5.9
A Punnett square showing the results of a homozygous cross between brown eyes (BB)

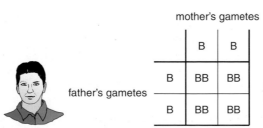

mother's gametes

	B	B
B	BB	BB
B	BB	BB

father's gametes

100% of the offspring have genotype BB and are brown-eyed

If one parent is homozygous brown eyed and the other is heterozygous, they will both look the same and the offspring will all look the same but look at the ratio of the offspring genotypes.

Figure 5.10
A Punnett square showing the results of a cross between brown eyes (BB) and brown eyes (Bb)

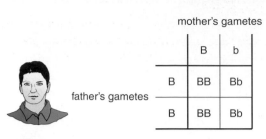

mother's gametes

father's gametes

50% are homozygous brown-eyed (BB) and 50% heterozygous brown-eyed (Bb)

The third possibility is that both parents are heterozygous, and this gives the important 3:1 ratio in the results.

Figure 5.11
A Punnett square showing the results of a heterozygous cross between brown eyes (Bb) and brown eyes (Bb)

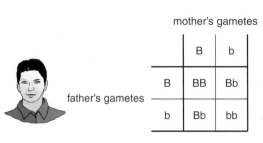

mother's gametes

father's gametes

75% of the offspring have brown eyes

25% of the offspring are homozygous bb and have blue eyes

This means that there is a 75% (3 in 4) chance of these two heterozygous brown-eyed parents having a brown-eyed child, and a 25% (1 in 4) chance of offspring with blue eyes. This gives the ratio 3:1 brown:blue.

Remember that the % chance applies at each conception, so it is possible for all of their children to have blue eyes.

Plant breeding

When considering plant breeding, it is quick and easy to see the results of crosses (when one plant fertilises another). When the parents are pure homozygous RR (red) and rr (white) the following result is obtained.

Figure 5.12
A Punnet square showing the results of a cross between a red plant, RR, and a white plant, rr

ovule

		R	R
pollen	r	Rr	Rr
	r	Rr	Rr

All the offspring from this cross are heterozygous, Rr, and will have red flowers. If the offspring are then crossed, the following result is obtained.

Figure 5.13
A Punnet square showing the results of a cross between Rr and Rr

ovule from offspring

		R	r
pollen from offspring	R	RR	Rr
	r	Rr	rr

In this generation, 50% of the offspring are heterozygous, Rr, 25% are homozygous RR and 25% are homozygous, rr. Therefore 75% are red flowers and 25% are white flowers, a 3:1 ratio.

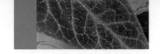

Genetic diseases

Cystic fibrosis

Cystic fibrosis is an inherited disorder of the lungs and digestive system. In cystic fibrosis the mucus made in the lungs is abnormally thick which results in more lung infections, coughing and wheezing. The problems with digestion occur because the duct that transports the digestive enzymes from the pancreas to the digestive system becomes blocked by sticky mucus. Because of this, a person with cystic fibrosis must take supplementary enzymes. The varied conditions that are recognised as cystic fibrosis are all the result of a recessive mutation in one allele.

Inheriting cystic fibrosis

The allele carrying the instruction 'have cystic fibrosis' can be called '**c**'. This allele is recessive. The allele for the instruction 'do not have cystic fibrosis'can be called '**C**'. This allele is dominant.

A person will only inherit one instruction from each parent. The possible combinations in an individual are **CC**, **Cc**, or **cc**.
An individual with the combination **CC** will **not** have cystic fibrosis.
An individual with the combination **Cc** will **not** have cystic fibrosis. However this individual is described as a **carrier**, because they still contain the instruction 'have cystic fibrosis' which he or she could pass onto the next generation.
An individual with the combination **cc** will have cystic fibrosis because the dominant allele is not present.

Figure 5.14
A Punnett square showing the results of a cross between two heterozygous individuals who are carriers of the c allele

Let C = normal allele
and c = allele for cystic fibrosis

Mother

		C	c
Father	C	CC	Cc
	c	Cc	cc

Result: 50% chance offspring will be heterozygous carriers of the c allele
25% chance offspring will be homozygous normal
25% chance offspring will be homozygous for the c allele and will have the disease

Therefore, if both parents are carriers, there is a 1 in 4 chance at every pregnancy that they will have a child with cystic fibrosis.

Sickle cell anaemia

Sickle cell anaemia is an inherited disorder of red blood cells. In a healthy person, haemoglobin in red blood cells binds with oxygen from the air we breathe into our lungs and carries it to all parts of the body. These cells are able to squeeze through tiny blood vessels because they are soft and round.

People with sickle cell anaemia make an abnormal form of haemoglobin which causes the red blood cells to be sickle shaped. These sickle-shaped red blood cells become stiff and distorted so they can block small blood vessels. When this occurs, less oxygen-carrying blood can reach that part of the body, causing considerable pain and tissue damage.

The allele that causes sickle cell anaemia is the result of a mutation. The allele is described as being abnormal. For a child to develop the disease it must have inherited the abnormal allele from both parents – making the child homozygous for the abnormal gene.

Children who are heterozygous have what is called the sickle cell trait. Their blood will contain a mixture of the two types of haemoglobin. In countries where malaria is common, this mixture provides some protection because the malarial parasite cannot survive in red cells containing abnormal haemoglobin. About 1% of the heterozygous individual's red cells are sickled; this compares with 5% of red blood cells in a person who is homozygous for the abnormal gene.

Figure 5.15
The red cells at the top of the photo have a sickle shape

Did you know?

Normal haemoglobin red blood cells live about 120 days before being replaced.

Inheriting sickle cell anaemia

The allele carrying the instruction 'do not have sickle cell anaemia' can be called 'Hb^A'. This allele is dominant. The allele for the instruction 'have sickle cell anaemia' can be called 'Hb^S'. This allele is recessive.

A person will only inherit one instruction from each parent. The possible combinations in an individual are Hb^A and Hb^A, Hb^A and Hb^S, or Hb^S and Hb^S.
An individual with the combination Hb^A and Hb^A will **not** have sickle cell anaemia.
An individual with the combination Hb^A and Hb^S will **not** have sickle cell anaemia, because the instruction 'do not have sickle cell anaemia' is dominant. However, this individual will be a carrier and will have the sickle cell trait.
An individual with the combination Hb^S and Hb^S has two copies of the 'sickle cell anaemia' gene and will develop the disease.

For the Punnett square, let Hb^A represent the allele for normal haemoglobin, and let Hb^S represent the allele for abnormal haemoglobin.

Figure 5.16
A Punnett square showing the results of a cross between two people with the sickle cell trait, $Hb^A Hb^S$

Mother's gametes

	Hb^A	Hb^S
Father's gametes Hb^A	$Hb^A Hb^A$	$Hb^A Hb^S$
Hb^S	$Hb^A Hb^S$	$Hb^S Hb^S$

Result: 25% (1 in 4) $Hb^A Hb^A$ normal haemoglobin
50% (2 in 4) $Hb^A Hb^S$ sickle cell trait – these will have mostly normal haemoglobin
25% (1 in 4) $Hb^S Hb^S$ these will have full sickle cell anaemia

Huntington's disease

Huntington's disease is a disorder of the nervous system. It is caused by a dominant allele of a gene and can therefore be passed on by only one parent who has the disorder.

Inheriting Huntington's disease

The allele carrying the instruction 'have Huntington's disease' – call this allele '**H**', is dominant. The allele for the instruction 'do not have Huntington's disease' – call this '**h**', is recessive.
A person will only inherit one instruction from each parent. The possible combinations in an individual are **HH**, **Hh**, or **hh**.
An individual with the combination **HH** will have Huntington's disease.
An individual with the combination **Hh** will have Huntington's disease, because the instruction 'have Huntington's disease' is dominant.
An individual with the combination **hh** will **not** have Huntington's disease.

Let **H** = allele for Huntington's disease and **h** = normal allele.

Figure 5.17
A Punnett square showing the results of a cross between one individual who is heterozygous for the Huntington's disease gene and one individual who does not carry the gene at all

Mother's gametes

	H	h
h	Hh	hh
h	Hh	hh

Father's gametes

Therefore if only one parent is a carrier, there is a 50% chance of a child being **Hh**.

Summary

◆ In human body cells one of the 23 pairs of chromosomes carries the genes that determine sex.

◆ In females the sex chromosomes are the same (XX).

◆ In males the sex chromosomes are not the same (XY).

◆ Most characteristics are controlled by one gene.

◆ Some genes have different forms called **alleles**.

◆ An allele which controls the development of a characteristic when it is present on only one of the chromosomes is the **dominant** allele.

◆ An allele which controls the development of a characteristic only if the dominant allele is **not** present is a **recessive** allele.

◆ If both chromosomes in a pair contain the same allele of a gene, the individual is **homozygous** for that allele.

◆ If both chromosomes in a pair contain the different alleles of a gene, the individual is **heterozygous** for that allele.

◆ Huntington's disease (a disorder of the nervous system) is caused by a dominant allele.

◆ A recessive gene causes cystic fibrosis (a disorder of cell membranes). Parents can be carriers without actually having the disorder.

◆ Sickle-cell anaemia is a disorder of the red blood cells which reduces their oxygen-carrying capacity. Carriers of this disorder are less likely to be affected by malaria.

◆ Genetic diagrams can be constructed and used to predict and explain the outcomes of crosses for each possible combination of dominant and recessive alleles of the same gene.

Topic questions

1 a) Which pair of sex chromosomes are found in human males?
   b) Which pair of sex chromosomes are found in human females?
   c) Which sex chromosomes could be in a sperm?
   d) Which sex chromosome is in all eggs?

2 The chances of a baby being a boy or a girl are 50%. Explain why.

3 The allele for 'have freckles' is dominant over the allele for 'do not have freckles'. What does this mean?

4 a) Name a disease caused by a dominant gene.
   b) Name two diseases caused by a recessive gene.

5 A person could be a carrier of a disease. What does this mean?

6 How is being a carrier of sickle cell anaemia beneficial?

7 In a pea plant the allele for being a tall plant (T) is dominant over the allele for being a dwarf plant (t).

   a) Explain using a genetic diagram what plants would be produced if:
      i) a homozygous tall plant is crossed with a homozygous dwarf plant.
      ii) two plants from the crossing in (i) were crossed.

   b) Why is it impossible to have a heterozygous dwarf plant?

93

5.3 DNA

Co-ordinated	Modular
DA 10.17	DA 4
SA n/a	SA n/a

Chromosomes carry a sequence of genes (Figure 5.18). A **gene** is a length or section of the chromosomal **DNA**. It codes for one **characteristic**.

The chromosomes are made up of two chains of DNA wrapped around a protein molecule and look like a twisted ladder (Figure 5.19). DNA can be considered as a polymer of base monomers (see section 10.4).

Each group of three bases is a code for a particular amino acid. So the sequence of bases in DNA controls the sequence of amino acids in a protein and therefore the type of protein produced.

The 'code' in each gene is made by the order of the chemical molecules called bases. These are represented by the coloured bars and their initial letters A, C, T, G.

Each 'gene code' has instructions for making proteins such as enzymes, hormones, pigment or protein structures. When this code is translated, we see the results as a physical characteristic such as eye colour or the production of a hormone such as insulin.

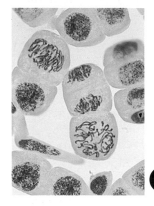

Figure 5.20
These chromosomes are visible in a slide of the tip of a root using a school microscope

Did you know?

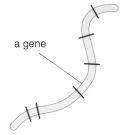

a gene

Figure 5.18
A chromosome containing seven genes

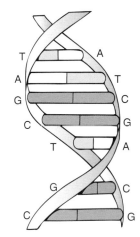

Figure 5.19
Section of DNA showing the sequence of four bases

Summary

◆ Chromosomes have long molecules of a substance called DNA.

◆ A gene is a section of DNA.

◆ DNA contains coded information that determines inherited characteristics.

◆ DNA is made of long strands made up from four different compounds called bases.

◆ A sequence of three bases is the code for a particular amino acid.

◆ The order of the bases controls the order in which amino acids are assembled to make a particular protein.

S.4	
Co-ordinated	Modular
DA 10.18	DA 4
SA 10.12	SA 14

Controlling inheritance

Reproduction is the process of making new individuals. Asexual reproduction is the term used when a single individual produces offspring that are genetically identical to each other and the parent.

Cloning plants from cuttings

Florists or garden centres often display rows of almost identical plants. These were probably produced by asexual reproduction from one parent and therefore the parent plant and the new plants produced from it are genetically identical. If cuttings are taken from a favourite plant, such as a Fuchsia or African Violet, and rooted, the cuttings should grow and have exactly the same leaf shape and flower colour as the parent plant.

Such plants are called **clones** and are genetically the same as the parent plant.

Cloning plants – tissue culture

Figure 5.21
A dish of plant cells can be the equivalent of a field of thousands of plants

The cloning of plants through tissue culture involves the taking of a very small piece of plant and placing it in a jelly containing various nutrients.

↓

Within a short time a layer of thick new tissue forms.

↓

Cells are removed from this new tissue and placed in a different jelly, this time it contains not only nutrients but also hormones that speed up the growth of roots and shoots.

↓

Once roots and shoots have developed the tiny plant is moved to a greenhouse where further growth continues.

Selective breeding (or artificial selection) in plants

The aim of plant breeding is to produce plants with particular desirable characteristics selected from existing plants of the same species. For example, a specific aim could be to produce short (but strong) stalked wheat plants with large seed heads. In order to try and produce these plants the plant breeder has to select the plants which have the required characteristics, plant the seed, cross pollinate them and then collect their seeds.

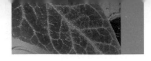

Figure 5.22

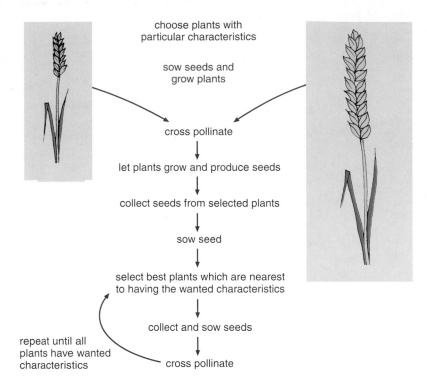

choose plants with
particular characteristics

sow seeds and
grow plants

cross pollinate

let plants grow and produce seeds

collect seeds from selected plants

sow seed

select best plants which are nearest
to having the wanted characteristics

collect and sow seeds

repeat until all
plants have wanted
characteristics

cross pollinate

If the seeds from a short stalk are sown, and the seeds from a tall, large-headed stalk are sown and cross pollinated, when the plants grow to maturity, the best of the crop can be selected and grown. These are again cross pollinated and allowed to make seed.

Figure 5.23
*Wheat in fields
used to consist of
tall stalks with
some short ones The
number of seeds in
the heads of the
wheat also varied*

Again the shortest and largest headed stalks of wheat are cross pollinated. Their seeds are collected and sown.

Figure 5.24
*Variety in wheat
plants*

This procedure is repeated and repeated until all the seeds produce short stalks, carrying big heads of corn.

Figure 5.25
Over a number of generations, the wheat plants become more similar

In a modern field of wheat, the plants are almost identical. All the stalks are uniformly short. All the heads of wheat are big with lots of seeds.

Figure 5.26
Eventually all the wheat plants are almost identical

This means that:

- the wheat is less likely to be wind damaged
- the combine harvester can work easily in the field
- there is a higher yield per field
- less plant energy has gone into making a tall stalk and more into making seeds.

Figure 5.27
A field of modern wheat, showing that all the stems are of similar short height with heads of large seeds

Selective breeding in animals

One problem with this process is that most animals take longer to grow to maturity and produce offspring than plants. The second problem is that a single head of wheat will produce perhaps 20 seeds, but a cow, for example, produces only one or two calves each year.

The principle of selective breeding in animals is the same as that for plants – animals with the desired characteristics are cross bred and the offspring observed and perhaps crossed again until the final 'cross' is achieved. Obviously this process takes a long time.

Did you know?

Artificial insemination has meant that sperm can be taken from bulls and put into the best cows much more easily than when the cow or bull had to be taken to a farm!

Did you know?

Australia originally had no wild or domestic cattle. The cattle found there now were brought in by settlers.

The selective breeding programme to produce a drought-resistant breed of cattle in Australia took eight years before enough offspring of several crosses were available. Farmers then chose the variety which best suited the conditions. It was as a result of this that the farmers selected the cross now known as 'Droughtmaster'.

'Shorthorn' bred in Australia from cattle the British settlers took with them, were used to establish the cattle industry in the northern areas of Australia. Their good qualities are that they are fertile, docile, easy to milk and cross breed well. However, they were very susceptible to cattle tick (insect pests).

The Brahman breed of cattle came originally from the 'Bos indicus' in India. These cattle have short thick glossy hair which reflects sunlight, deeply pigmented skin which keeps out the Sun's rays, loose skin which increases the surface area for cooling and sweat glands which allow the body to cool quicker and seem to deter insects!

Cross breeding programmes resulted in the cattle and their descendants being carefully monitored for eight years before being made available to ranchers in 1941. Further studied selection and planned mating fixed the characteristics required to survive in the Northern Territory. In 1956 the breed was named Droughtmaster. This contains approximately 50% Shorthorn and 50% Brahman. The Droughtmaster breed is basically red. They have a high resistance to ticks, they can tolerate heat and drought, they have good fertility, calve easily, have a quiet temperament and are a good beef animal.

It has been found that the Brahman and Shorthorn cattle can both survive well at temperatures down to −13°C and up to 21°C. However temperatures from 21°C up to 24°C cause the European Shorthorn cattle to suffer. The Indian Brahman can tolerate temperatures up to 41°C.

Figure 5.28
Selective breeding in cattle

Variation and selective breeding

Humans have been selectively breeding crops and animals ever since they have been used as sources of food. For example, grasses with desirable characteristics have been crossed repeatedly for thousands of years to produce wheat with ever increasing yields or resistance to disease. Care has to be taken that selective breeding does not end up producing, a limited number of varieties of wheat. If this were to happen then the varieties of wheat are likely to be genetically very similar. Should environmental conditions change in some unforeseen way or some new disease appear, then because of the genetic similarity of the wheat most could be affected as only a few might have the allele that makes them resistant to the changes.

Cloning genes – genetic engineering

Genetic engineering involves altering the genetic make-up of an individual. The principles in all examples and the technique involved are the same – a section of DNA is moved from one species to another to manufacture useful biological products. Using this technique, proteins and drugs can be produced, as well as hormones such as insulin.

The genetic engineering of insulin is described below.

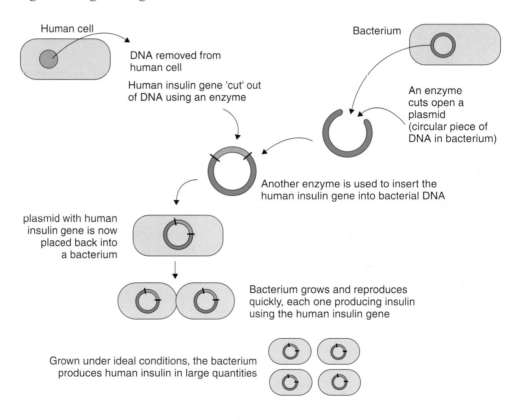

Figure 5.29

Cloning animals – embryo transplants

Embryo transplantation is one of the latest cloning techniques being attempted in the breeding of mammals including mice, sheep, cattle and horses. Cells are taken from embryos that are only a few days old. At this stage the embryo is made up of a mass of cells that have not yet become specialised into any particular body cell. These cells can be separated, introduced into the wombs of adult females where they can develop normally.

Figure 5.30
Dolly the sheep

Cloning quickly and cheaply produces a large number of genetically identical organisms. Cloning reduces the genetic variation in a species. So if at some time in the future all the sheep in the world were clones from the same parent and conditions changed, all could die as none might have the allele to make them resistant to the change. As with selective breeding, any reduction in the variety of organisms in a particular population decreases the number of different alleles available. This reduces the chances of new varieties being produced, either naturally or artificially, when conditions change and threaten the existence of the population.

Genetic engineering – gene transplants

Genetic engineering means that scientists can now isolate specific genes that carry out a specific function. Such functions can include the production of a particular insecticide or resistance to a weedkiller. These genes can be transferred into plant cells at an early stage of development so that the plant develops with the desired characteristic.

The genetic modification of crops is often carried out in order to produce greatly increased yields as cheaply as possible. One such modification is to make the plants produce an insecticide that kills any insects that try to eat the plants, so reducing the cost of spraying with insecticides. There is concern that this could destroy a wide range of insects, many of which are beneficial in pollination or form part of a food chain. Another concern is when the modification is to make the crop resistant to the action of herbicides. This means that the fields can be sprayed with a weedkiller that will get rid of the weeds but not harm the growing crop. The concern here is that the herbicide-resistant allele could spread into other plants or weeds, perhaps through cross-pollination.

Summary

◆ By taking cuttings new plants can be produced quickly and cheaply.

◆ Cuttings will be genetically identical (**clones**) to the parent.

◆ **Artificial selection (selective breeding)** is used to produce organisms with a particular desirable characteristic.

◆ Artificial selection and cloning reduce the number of alleles available in a population.

◆ A reduction in the number of alleles in a population could reduce the chances of producing new varieties if conditions change to threaten a particular population.

◆ Modern cloning techniques include:

- tissue culture
- embryo transplants.

◆ Genes can also be transferred to the cells of animals so that they develop with desired characteristics. This is called **genetic engineering**.

◆ The culturing of genetically engineered bacteria on a large scale produces large quantities of useful products e.g. insulin.

◆ There are economic, social and ethical issues concerning cloning and genetic engineering.

Topic questions

1 Many types of plants can be grown by taking cuttings.

a) Why do the cuttings from a particular plant all produce the same colour of flower as the parent plant?

b) Why do the cuttings from a particular plant all produce the same colour of flower as each other?

2 What is selective breeding?

3 What are the advantages of producing plants using a tissue culture technique?

4 Why is important to make sure that a variety of alleles are always available in any particular population?

5 Describe what happens during the production of genetically engineered insulin.

6 Embryo cloning uses embryos that are only a few days old. Why?

5.5		Evolution
Co-ordinated	**Modular**	
DA 10.19	DA 4	
SA 10.13	SA 14	

Evolution

A **species** is a group of organisms which look similar and which can breed together producing **fertile** offspring. Examine the birds below, all of which can be seen in gardens. The species differ not only in size and colour but also in behaviour, call sounds and nest building. These factors keep the species separate.

Figure 5.31
Variety in bird species

What are fossils and how were they formed?

Fossils are the remains or imprints of plants and animals that lived millions of years ago.

Fossilisation

When an organism dies, it is normally decomposed by bacteria and fungi very quickly, leaving nothing to see except possibly a few bones. Fossils are formed where the organism fell into a marsh or bog in which the conditions – no oxygen and water with a low pH – prevented decay. The remains were buried under layers of vegetation and became compressed with age.

Alternatively, fossils can also form where organisms have been covered with layers of sand, volcanic ash or silt. The remains get buried under layers of vegetation and become compressed with age and impregnated with mineral salts from water, turning them into stone. There are other interesting remains such as insects preserved in fossilised tree resin called amber, dinosaurs footprints, petrified trees, remains preserved in ice and coprolites (fossilised droppings).

Fossils can show the time sequence of the existence of, and changes to, organisms over a very long period of time. In sedimentary rocks the material is laid down in layers so the deeper down you go, the older the material and so from top to bottom the fossils are arranged in order of increasing age. These layers can be investigated at cliffs, gravel pits and in spectacular scenery such as the Grand Canyon in Arizona.

The depth of the face is 1700 m and this represents a time span of 500 million years of material.

How fossil evidence supports the theory of evolution

Geologists can trace changes in climate and rock movements and at the same time palaeontologists (fossil experts) can follow changes in the shape and structure of some animals and plants which could result from environmental changes.

Figure 5.33
The Grand Canyon

Figure 5.32
*An insect trapped
in amber*

The fossil record presents us with the following information.

- There were only a small variety of the simplest marine invertebrates in the bottom layers (oldest rocks) and these represent several million years without much change.

- The presence of fossils in one layer but not in any of the layers above it shows that many forms have become **extinct**.

- There are gaps in the fossil record. There was great excitement when the Archaeopteryx was found, as this is thought to be the 'link' between birds and reptiles. Fossil experts hope that more links will be found.

- There is a definite order in which the fossils appear.

    1 Marine invertebrates – many of these would have been soft bodied and so only the impressions remain. There were many trilobites which are an extinct form of arthropod, some as large as 40 cm. There were also crustaceans, molluscs and primitive starfish. The only plant life was seaweed.
    2 Land plants – the first of these were dependent on water for support and reproduced by means of spores.
    3 Land invertebrates that were air breathing, such as wingless insects, millipedes, spiders and crustaceans that looked like woodlice, have been found in deposits 350 million years old.
    4 The first fish had bodies covered with bony plates; later fossils show scales. Some of these fish have paired limb-like fins that seem to be a transition towards amphibians. Sharks were also around.
    5 Winged invertebrates developed.
    6 As there was more dry land, reptiles evolved. There were both herbivores and carnivores with distinctively-shaped teeth like mammals today. Other reptile groups included turtles and primitive crocodiles. These were followed by the dinosaurs.
    7 Reptiles evolved into birds which had feathers but also reptile-like teeth, solid bones and jointed tails.
    8 Small mammals appeared.
    9 Flowering plants developed.
    10 Development of larger mammals.

Remember, just as today, the majority of plants and animals that died would have been eaten or been decomposed. Only a very few will have died in conditions that would result in fossilisation.

103

Extinction

An animal or plant is extinct if there are no living specimens. The evidence of these specimens having lived in the past is in the fossil record. The fossil record shows several episodes of mass extinction when many species of plants and animals disappeared, and these have been related to catastrophic changes in climate. Fossil evidence indicates that the dinosaurs became extinct 65 million years ago. A popular explanation for this is that an asteroid or comet struck the Earth causing acid rain and clouds of dust, which blocked out the sunlight leading to global cooling and vegetation changes. Other scientists have suggested massive volcanic activity as the cause. Both ideas leave us with the problem of explaining how the ancestors of dinosaur groups, such as crocodiles, frogs, snakes and insects, survived.

A less dramatic explanation could be that the climate was warmer in the dinosaur age and these large creatures warmed themselves in the Sun whilst browsing on the bushy vegetation. The early herbivorous dinosaurs had large bodies, small brains and, skeletal evidence of their breathing systems suggests, a slow metabolism. The more recent dinosaurs were smaller and had a faster metabolism and these would have been selected against by environmental change, as the climate became cooler and drier. At this time, the active species of insects, reptiles and mammals were increasing and conditions would have been more favourable for these organisms.

Figure 5.34
A reconstruction of an Archaeopteryx

Figure 5.35
A woolly mammoth which became extinct at the time of the last Ice Age

Another much more recent extinction was the woolly mammoth that died out about 10 000 years ago. We know much more about mammoths because some were trapped in ice crevasses and have been preserved as frozen specimens. They had adaptations that allowed them to survive in the cold:

- short dense under-fur with a longer coarse top coat

- skin with an insulating layer of fat and a hump-like fat reserve on the back

- deep scratch marks on the lower side of the huge tusks indicated that they could be used for clearing snow from the grass for feeding

- short ears, tail and legs to reduce heat loss.

Although mammoths were hunted, it is thought that they also became extinct as a result of selection against them after a climate change because there was a reduction in grassland at the time of the Ice Age.

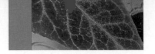

Extinction may be caused by:

- changes in the environment
- successful *new* predators
- successful *new* competitors
- *new* diseases.

Unless evolution occurs so that species become better adapted to these changes, they may become extinct.

The rise of antibiotic-resistant bacteria

Some microorganisms produce substances that prevent bacteria from reproducing. Such substances, which include penicillin, are called antibiotics. Antibiotics have no effect on the action of viruses. The ability of penicillin to prevent the reproduction of bacteria was first observed in 1928. It was not used on humans until 1940 when it was used to fight bacterial infection in wounded soldiers. For the next few years penicillin was considered to be a 'wonder-drug' as it was able to control almost every bacterial infection. In the 1950s strains of bacteria were appearing that were unaffected by penicillin. This was not considered a problem as more and more varieties of antibiotics were produced. Some of the main problems during this time were that:

- too many patients thought that antibiotics were the cure for any illness, including those caused by a virus

- too many doctors were willing to prescribe antibiotics, often inappropriately

- farmers and vets were using large quantities of antibiotics on food-producing animals.

It was this over-use of antibiotics that has resulted in a large majority of the disease-causing bacteria now being unaffected by the action of the majority of antibiotics.

How do bacteria become resistant?

Bacteria can, in an ideal environment reproduce asexually every 20 minutes. This rapid rate of reproduction can produce changes in the genetic material as the DNA tries to make copies of itself during mitosis (see section 1.5). In some instances, the mutation in the DNA results in a change to an enzyme structure in the bacterium that makes the bacterium resistant to the action of the antibiotic. This resistant bacterium will reproduce and very quickly form clones of antibiotic-resistant bacteria.

Charles Darwin and his theory of evolution

Charles Darwin (1809–1882) was a well-travelled naturalist and a breeder of fancy pigeons. He kept very careful records of the parents and offspring of his pigeon crosses. He made a 5 year voyage to South America and the Galapagos Islands from 1831 to 1836, during which he kept very careful observations. These observations were the basis of the paper he presented with another scientist called Alfred Wallace to the Linnean Society in London in 1859. This paper was called 'On the Origin of Species by Means of Natural Selection'. At the time that Darwin proposed the ideas of **natural selection,** there was no knowledge of genetics and his theory was based on breeding results and observations only.

He made the following propositions:

- All organisms produce more young than the environment can easily support – there is often a shortage of food.

- Variations exist within populations and few individuals are exactly alike in any measurable variable but some of the features seen in parents are also seen in their offspring.

- There is a struggle for existence – those organisms best suited to the environment survive to reproduce and have more offspring than those less well adapted and therefore the beneficial characteristics become more common. (The reverse argument can be used to explain the extinct forms found in the fossil record only.)

- The result of the 'struggle for existence' and limited food and shelter sites in the environment means that populations remain approximately constant in a balanced ecosystem until they are affected by man or natural disasters.

Reactions to Darwin's theory of evolution

When Darwin published his theory of evolution in 1859 there was a great deal of opposition to the ideas contained in it. Most opposition came from those people who considered that Darwin's ideas were a serious challenge to the religious teachings about the story of creation. Even many scientists criticised the ideas because:

- there was no laboratory proof so it must remain a theory

- natural selection did not explain the big changes shown by fossil records

- there was no explanation as to how variations within a species arise, nor how they were passed on to future generations.

Gradually during the rest of his lifetime Darwin's ideas became more acceptable. But it was not until early in the 20th century, when Mendel's work on genetics (see section 5.1) was used to help explain the process of natural selection, that Darwin's ideas were almost universally accepted.

Did you know?

The popular idea that human beings were descended from apes was not part of Darwin's theory. What he did say was that human beings, apes and monkeys all descended from the same primitive ancestors.

How Mendel's work helps to explain natural selection

Knowledge of genetics enables some of the gaps in Darwin's theory of evolution to be explained.

- Variations in the population can be explained by spontaneous mutation that may not be caused by environmental factors. Unless these mutations are lethal, they will be passed on to the offspring.

- As a result of the formation of gametes and fertilisation in sexual reproduction, offspring will receive a selection of alleles from both parents and will have a different combination from their parents. (They may have Mum's nose and build but Dad's height and shape of fingers.)

- Some combinations of alleles are better suited to particular environments and have a selective advantage. Individuals with such combinations will survive and breed successfully.

- Over many generations, natural selection results in an increase in the spread of beneficial mutations and a reduction in that of harmful mutations.

Conflicting theories of evolution

The fossil record provides evidence of change over very long periods of time and the development of antibiotic resistance in bacteria gives evidence of changes still taking place in organisms. Darwin explained these changes through his theory of evolution, however a scientist, called Lamarck, had a different explanation.

Jean Baptiste Lamarck (1744–1829) was a French natural philosopher. Lamarck's theory was related to people and was meant to please the politicians. He suggested that a desire or need for change caused that change to happen in the organism and that this would be passed on to the offspring. Structures that are constantly in use are well developed and characteristics acquired during life would be passed on. For example, the blacksmith needed big muscles and would pass these on to his son. His 'evidence' was that when the environment causes the need for a structure, this induces the structure that will help the organism adjust to the environment.

His famous example was the giraffe. Lamarck suggested that the giraffe acquired its high shoulders and long neck by straining to reach the leaves in the tree, and that these characteristics were passed on to the offspring which by continual stretching extended them more. Gradually, generation by generation, giraffes got longer necks.

How Darwin's theory of natural selection accounts for the long necks

- Although most of the ancestors of giraffes had short necks there would be some with longer necks – there was variation and the characteristics were inherited.

- If the food supply became in short supply – a struggle for existence would occur.

- Those animals best able to find food would survive. In this case those with long necks could reach leaves higher up in the trees. These would survive and breed offspring with long necks – survival of the fittest.

- Nature has selected those most suited to the changing conditions – natural selection.

- If this continues over very many generations the present day giraffe with its very long neck would evolve – evolution.

Summary

- **Fossils** are the evidence of plants or animals from many millions of years ago.

- Fossils provide evidence of how organisms have changed.

- The theory of evolution states that all species of living things alive today – and many others that have become **extinct** – have evolved from simple life forms which first developed more than three billion years ago.

- Evolution occurs through **natural selection** in which:

  - individuals within a species may show variation because of genetic differences
  - predation, disease and competition cause large numbers of individuals to die
  - individuals with characteristics most suited to the environment are more likely to survive and breed successfully
  - the genes which have enabled these individuals to survive are then passed on to the next generation.

- It is possible to explain:

  - how fossils provide evidence for the theory of evolution
  - how the over-use of antibiotics can lead to the evolution of resistant bacteria.

- The conditions in which a species survives may change. Unless evolution occurs and the species become better adapted to survive the changing conditions they may become extinct.

- It is possible to suggest why Darwin's theory of evolution was only gradually accepted.

- There are differences between Darwin's theory of evolution and those of Lamarck.

1 a) What are fossils?
  b) Give four different ways fossils can be formed.
2 What does the fossil record show?
3 Why do some organisms become extinct?
4 What are the key points in Darwin's theory of evolution?
5 What was Lamarck's theory of evolution?

Examination questions

1 The drawing shows some of the stages of reproduction in horses.

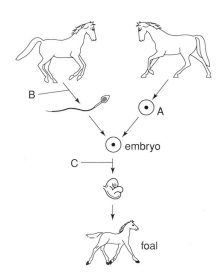

a) i) Name this type of reproduction. *(1 mark)*
   ii) Name the type of cell labelled **A**. *(1 mark)*
b) Name the type of cell division taking place at the stage labelled:
   i) **B**
   ii) **C**. *(2 marks)*
c) How does the number of chromosomes in each cell of the embryo compare with the number of chromosomes in cell **A**? *(1 mark)*
d) When the foal grows up it will look similar to its parents but it will **not** be identical to either parent.
   i) Explain why it will look similar to its parents. *(1 mark)*
   ii) Explain why it will **not** be identical to either of its parents. *(2 marks)*

2 This couple has just found out that the woman is pregnant. They wonder whether the child will be a boy or a girl.

Sex chromosomes	Sex chromosomes ....

a) Fill in the boxes to show the sex chromosomes of the woman and the man. *(2 marks)*
b) The couple already has one girl. What is the chance that the new baby will be another girl? Explain the reason for your answer. You may use a genetic diagram if you wish. *(3 marks)*

3 In the 1850s an Austrian monk, called Gregor Mendel, carried out a series of investigations on heredity.

a) i) What plants did he use for his investigations? *(1 mark)*

   ii) In his work he assumed that one gene controlled one characteristic. He started his investigations with pure breeding parents. Use a genetic diagram to show how he explained the following result. *(4 marks)*

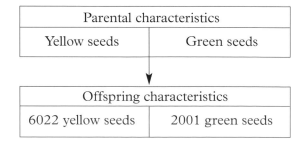

Parental characteristics	
Yellow seeds	Green seeds

Offspring characteristics	
6022 yellow seeds	2001 green seeds

4 Giraffes feed on the leaves of trees and other plants in areas of Africa. They are adapted, through evolution, to survive in their environment.

a) Use the information in the picture to give one way in which the giraffe is adapted to its environment. *(1 mark)*

b) Explain how Jean-Baptiste Lamarck (1744–1829) accounted for the evolution of the long neck in giraffes. *(3 marks)*

c) Another scientist, August Weismann (1834–1914) wanted to check Lamarck's explanation. To do this he cut off the tails of a number of generations of mice and looked at the offspring. His results did not support Lamarck's theory. Explain why. *(2 marks)*

d) Explain how Charles Darwin (1809–1882) accounted for the evolution of the long neck in giraffes. *(4 marks)*

5 Sickle cell anaemia is an example of a disease caused by a *mutation* affecting one of the genes involved in the production of haemoglobin.

- Hb is a gene that determines haemoglobin.
- Hb^A causes normal haemoglobin and is *dominant*.
- Hb^S causes defective haemoglobin and is *recessive*.
- In the *homozygous* recessive condition the person suffers acute anaemia and has a low life expectancy.
- In the *heterozygous* condition individuals suffer from the sickle cell trait but have increased resistance to malaria.

a) What is the role of haemoglobin in the body? *(1 mark)*

b) What is a *mutation*? *(1 mark)*

c) Use the information above to explain what is meant by the terms *homozygous* and *heterozygous*. *(2 marks)*

d) Use the information above to explain what is meant by the terms *dominant* and *recessive*. *(2 marks)*

e)

	Father	
	Hb^A	Hb^S
Mother Hb^A	**Child 1** $Hb^A Hb^A$	**Child 2** $Hb^A Hb^S$
Hb^S	**Child 3** $Hb^A Hb^S$	**Child 4** $Hb^S Hb^S$

i) Which child will have sickle cell anaemia?
ii) Which child will have sickle cell trait?
iii) Which child will not carry any sickle cell genes?
iv) Which child will be more resistant to malaria? *(4 marks)*

Chapter 6
Living things in their environment

6.1	
Co-ordinated	**Modular**
DA 10.20	DA 03
SA 10.14	SA 14

Adaptation and competition

All living things have to be able to cope with the changes going on in the environment around them. Living things reproduce within their **environment** and this creates a **population** of a particular organism which compete for the food, water and space within the area.

As the population of an organism grows, the **competition** for these **resources** increases until only those which are able to compete best can survive. A typical growth curve for a population is shown below.

Figure 6.1
A typical population curve

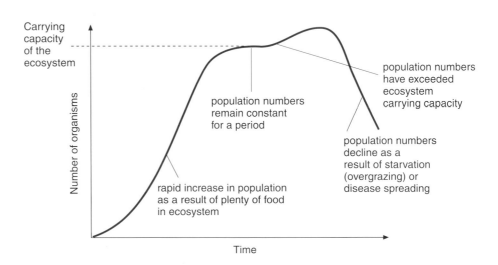

In the early stages the population grows rapidly because there is plenty of food, water and space available. As numbers increase, the food supply may begin to run out, some of the organisms may starve and the growth in the population begins to slow down.

Eventually a point is reached where as many organisms are dying as are being produced. The population becomes stable and remains so as long as the conditions of food, water and space also remain constant.

In some areas, animals may migrate to avoid overpopulating an area. As the population increases, the pressure on the food supply also increases. Animals may move around an area or move to a completely different area to ensure that the food supply is not completely exhausted or to give it time to recover. In places where there is a marked seasonal change – hot/cold or wet/dry – the **migration** is an annual occurrence.

Distribution and abundance of organisms

What happens in nature, however, is not always as simple as the situation described in Figure 6.1. There are often other forces that control populations, such as **predation** – one animal preying on another. The **predator** may reduce the population of its prey until there is not enough for it to feed on. The predator may then begin to starve. At this point, the **prey** population may rise again because there are not enough predators to control numbers. As the population of the prey rises, the predators again have enough to eat and their population grows. This pattern is often repeated over many generations.

The diagram below showing the relationship of the vole (prey) and the fox (predator) clearly demonstrates these changes.

Figure 6.2
Population curves to show the changes in vole and fox populations over time

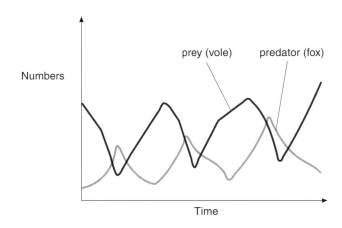

So the size of a population may be affected by:

- the total amount of food or nutrients available
- the total amount of water available
- competition for food or nutrients
- competition for water
- competition for light
- predation or grazing
- disease.

Adaptations for survival

Each type of environment poses problems to the plants and animals living in it. Those organisms that become adapted to the environment are the ones that will survive. Those that cannot adapt often become extinct. The arctic and desert environments represent two extremes in one sense but they are similar in another. One is very hot and the other very cold. In both, however, water is in short supply.

In the arctic it is very cold, much too cold for most organisms. Only those which are protected against the cold or can conserve heat can survive.

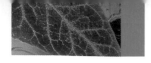

Since the Industrial Revolution there has been an increasing reliance on the use of fossil fuels as a source of energy. Fossil fuels were formed many millions of years ago from the remains of dead micro-organisms and plants. Normally the organisms would have decayed and they would have been returned to the environment for recycling as carbon dioxide, but the organisms did *not* decay and the carbon compounds became locked in the soil. The micro-organisms became oil and natural gas and the plants became coal.

When the fossil fuels are burned, the chemicals they contain are released into the atmosphere. One of the gases formed in the process of combustion is carbon dioxide. This is a gas which is naturally found in the atmosphere. It is vital for photosynthesis, however there is evidence that the amount of carbon dioxide in the atmosphere is increasing.

There are also compounds of sulphur and nitrogen in fossil fuels. When fossil fuels are burned they release sulphur dioxide and nitrogen oxides, both of which are poisonous gases. Sulphur dioxide and nitrogen oxides are strongly acidic. These gases are soluble and when they dissolve in rain they make the rain acidic. This 'acid rain' falls on soil where it causes mineral salts in the soil vital to plant growth to be removed. The acid rain also has a direct effect on the plants by damaging roots and leaves. The rain eventually finds its way into rivers, streams and lakes, making the water too acidic for organisms to live.

Acid rain can also attack the stonework of buildings causing it to dissolve and crumble.

Figure 6.6
Acid rain has severely damaged these trees

The increase in the levels of carbon dioxide in the atmosphere is causing concern in another way. Carbon dioxide absorbs infra-red light (heat rays) from the Sun. If the amount of carbon dioxide increases, more heat will be absorbed and become trapped in the upper atmosphere. The atmosphere will behave exactly like the inside of a greenhouse and it will heat up (the '**greenhouse effect**'). The effect of this will be to cause the temperature of the whole Earth to increase. Possible effects of this '**global warming**' might include:

- the melting of the ice caps at the poles, increasing sea levels and causing widespread flooding of low-lying coastal regions.

- climatic changes which might alter the wind and rainfall patterns and change the distribution of plants and animals on the Earth, perhaps increasing the numbers of pests and the incidence of serious diseases.

Deforestation

As the human population has grown, it has demanded more and more living area, space for growing crops and raw materials. These demands have often been met by cutting down forests. The space created by this is used to grow crops, build homes and roads (this is called urban development) and the trees have been used to make furniture and other items. As long as the cutting down of the forests is carried out at the same rate as growth, there is no harm done, but recently the rate of **deforestation** has greatly exceeded the rate of growth of the forests (reforestation).

The areas used for agriculture often yield only one or two crops before the thin soil is exhausted and has run out of nutrients. The areas used for urban development are lost forever.

Less trees also means less carbon dioxide removed from the atmosphere for photosynthesis and so deforestation can contribute to global warming.

Problems resulting from deforestation as seen in Brazil

The vast rainforest of Brazil had its own climate but the removal of huge areas of trees has changed this.

- The overall rainfall has reduced, making the climate drier but now the rain often falls in heavy short bursts.

- The canopy of trees once prevented the rain from hitting the soil but now the trees have gone, the heavy rain falls straight on the soil washing it away.

- The soil has been eroded and cannot support the growth of crops. The crops fail and the land is abandoned.

- The removal of the trees has had a terrible effect on the wildlife in the area. The number of different species that can be supported by the dwindling vegetation has been greatly reduced.

- Reduction in species diversity has a destructive effect on food chains and webs and has led to the extinction of many species of plants and animals.

Greenhouse effect and global warming

Carbon dioxide is not the only gas responsible for the **greenhouse effect**. Methane, though much less common in the atmosphere than carbon dioxide, has a greater 'greenhouse effect'. A certain volume of methane has about 20 times the 'warming potential' than the same volume of carbon dioxide. Like carbon dioxide, it too is increasing due to human activities.

Where does the methane come from?

Much of the methane released into the atmosphere comes from rice fields and from the digestive system of cattle.

Did you know?

Although the amount of methane in the atmosphere is low, about 20% of it is produced from rice paddies and cattle.

Methane and rice paddies

The flooding of a rice field encourages the anaerobic fermentation of organic soil matter. This fermentation produces large amounts of methane that is released through the roots and stems of the growing rice plants.

Methane and cattle

The breakdown of grass in the digestive system of cattle (and other animals that 'chew the cud') produces large amounts of methane which the animal releases into the atmosphere.

Figure 6.7
Flooded rice paddy field

Food production

There is no doubt that humans have been very successful at increasing the amount of food produced by agriculture. Much of this has been achieved by increased efficiency through:

1 the elimination of pests which destroy food crops or compete with farm animals for food supplies

2 the elimination of weeds which compete with plant crops for water, light and nutrients

3 the use of **fertilisers** to make crops grow faster and give bigger yields

4 the use of larger fields so that more crops can be grown in a given area.

Farmers have a duty of care, however. This means that while they try to increase yields and eliminate pests, they must also be aware of the effects on the whole agricultural environment caused by pesticides, fertilisers and increase in field sizes by removal of hedgerows.

Sadly in the past some farmers have neglected this duty of care.

Damaging ecosystems

Pesticide use

The pesticides used have, in many cases, turned out to be toxic to other animals which might be helpful to the farmer or they have disrupted food chains causing the starvation of animals further up the food chain. Some pesticides will also kill pollinating species and reduce crop yield.

Excessive use of fertilisers

The use of fertilisers is often not well controlled. Farmyard manure is a very useful fertiliser but its quality cannot be guaranteed. It may not contain enough of the right kind of minerals or too much of another. Although farmyard manure does improve the water retention of the soil, its action is slow because bacteria and fungi have to break it down into the simpler substances the crops needs. Many farmers have resorted to the use of chemical fertilisers because the composition is known and the action is much faster. The problem is that they had little idea of how much fertiliser to apply and as a result they apply too much. The fertiliser which is not taken up by the crop plant is often leached out of the soil by rain into rivers.

The fertiliser causes a rapid increase in the growth of plants in the rivers and streams. The plants often grow so quickly that they completely block out the light for plants below the surface and prevent them producing oxygen by photosynthesis.

The plants die and start to decay. The aerobic bacteria and fungi which decompose the plants use up the oxygen from the water. The amount of oxygen in the water falls until there is not enough for larger animals. These also die and decay, using up more oxygen. Eventually the river or stream becomes devoid of oxygen and the water becomes lifeless.

The addition of excess plant nutrients which results in excess plant growth, death, decomposition of the plants by aerobic bacteria and the removal of oxygen from rivers and streams is termed eutrophication. It can be a serious problem in rivers and streams that flow through agricultural land.

Figure 6.8
Section through a stream showing the effects of eutrophication, causing light to be blocked out from the submerged plants

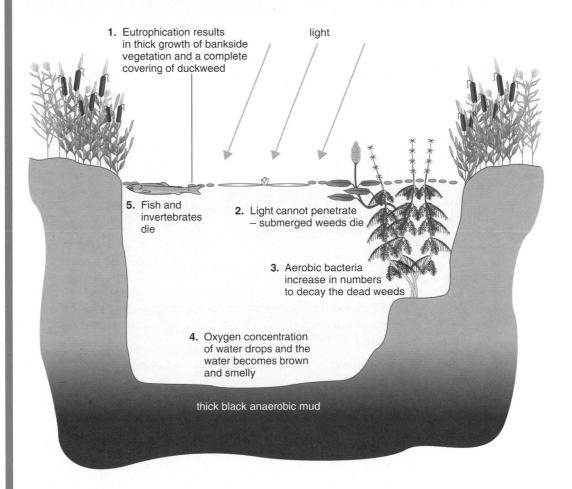

The dumping of raw sewage into a lake or river causes the same eutrophication effects.

Making decisions

The solutions to the problems caused by human conflict with the environment will not be easy to solve. There are many factors to be taken into consideration. The human population is rising rapidly and however much care is taken, the conflict with the environment is bound to increase. Humans are faced with a huge extra demand for food and without the use of chemicals, such as fertilisers, it is difficult to see how that demand can be met.

We are faced with a population requiring more leisure pursuits, and dwindling unspoiled areas of land and water. It is vital that these should be managed properly otherwise we could all be the losers as the number of species of animals and plants declines.

We are faced with the prospect of an atmosphere that is changing due to the massive amounts of pollutants being pumped into it. The increases in gases such as methane and carbon dioxide must be slowed down or reversed to prevent global warming becoming a more serious problem than it is now. All the decisions that have to be made will be complex. There will be winners and losers if the right ones are taken, but only losers if the decisions are wrong ones.

Summary

◆ Humans reduce the amount of land available for other animals and plants by building, quarrying, farming and dumping waste.

◆ Human activity may **pollute**:
 – water with sewage, **fertiliser** or **toxic** chemicals
 – air, with smoke and gases such as carbon dioxide
 – land, with toxic chemicals, such as **pesticides** and **herbicides**, which may get washed from the land into water.

◆ The burning of fossil fuels releases carbon dioxide into the air.

◆ The burning of fossil fuels may also release sulphur dioxide and oxides of nitrogen. These dissolve in rain to make it acidic.

◆ Rapid growth in the human population and an increase in the standard of living means that:
 – raw materials are being used up rapidly
 – large volumes of waste are produced
 – there is a rapid increase in pollution.

◆ Large scale **deforestation** for timber and for land for farming has:
 – increased the release of carbon dioxide into the air, because of burning
 – reduced the rate at which carbon dioxide is removed from the air and 'locked up' as wood.

◆ Increases in the number of cattle and rice fields have increased the amount of methane released into the atmosphere.

◆ Carbon dioxide and methane in the air absorb much of the energy radiated by the Earth. Some of this energy is re-radiated back to the Earth and keeps the Earth warmer than it otherwise might be.

◆ The concentrations of carbon dioxide and methane in the air are increasing. These increases enhance the '**greenhouse effect**'.

◆ An increase of only a few degree Celsius (**global warming**) may cause large changes in the Earth's climate and a rise in sea level.

◆ Farmers add fertilisers to replace nutrients which crops remove. Excess fertiliser may be washed into lakes and rivers.

◆ Pollution of water by fertilisers may cause eutrophication.

◆ Untreated sewage provides food for microorganisms. This in water has the same effect (eutrophication) as does dead vegetation.

Topic questions

1 When fossil fuels are burnt, carbon dioxide, sulphur dioxide and oxides of nitrogen are released into the air.
 a) Which of these gases is a 'greenhouse gas'?
 b) Which gases form acid rain?

2 In many parts of the world more trees are being cut down than are being planted. How will the fewer number of trees affect:
 a) the amount of carbon dioxide in the air? Give a reason for your answer.
 b) the amount of oxygen in the air? Give a reason for your answer.

3 Cattle and rice fields produce large amounts of another 'greenhouse gas'. What is its name?

4 Give two possible effects of global warming.

5 Explain how the use of too much fertiliser can cause eutrophication.

6 Explain how global warming is thought to occur.

Co-ordinated	Modular
DA 10.22	DA 03
SA n/a	SA n/a

Food chains and webs

All living things need energy. The law of conservation of energy states that energy cannot be created or destroyed. This means that the same amount of energy comes out of an **ecosystem** as goes into it.

An ecosystem is a unit made up of living components – plants and animals – and non-living components, for example a woodland with trees, animals and non-living components such as soil. There is a complex and often delicate balance between the organisms living in an area and the environment.

The primary source of energy for nearly all ecosystems is the Sun because it provides the energy needed for photosynthesis. The plants that can carry out photosynthesis are known as the **producers** because they produce all the food for themselves and are often eaten by other organisms. The organisms that feed on plants are called the primary **consumers**. The primary consumers are fed upon by the secondary consumers and these in turn may be fed upon by the tertiary consumers.

Food chains and pyramids of biomass

Food chains show the feeding relationships and direction of energy transfer between a number of organisms.

It is possible to produce pyramids by measuring the **biomass** in a food chain. These pyramids are usually based on the dry mass of the organisms in a food chain. A pyramid of biomass for the food chain would show a typical pyramid shape.

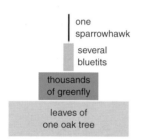

Figure 6.9
A pyramid of biomass for an oak tree

one sparrowhawk

several bluetits

thousands of greenfly

leaves of one oak tree

oak tree　→　insects　→　blue tits　→　sparrow hawk

The change in the size of the blocks from the base to the tip of a pyramid of biomass shows clearly the energy loss that occurs at each step.

A pyramid of biomass provides a quantitative display of biomass in a food chain. The simple link type of food chain does not show this.

Energy losses along a food chain

At every link in all food chains there is a loss of energy.

All the organisms in a food chain must respire to live. The respiration uses energy that cannot be passed to the next step in the food chain. The energy from respiration is used for movement, growth and heat production. These losses are especially large in mammals and birds. This is because their body temperatures are kept constant, at a level much higher than the temperature of the environment. This requires a lot of energy. None of this energy can be used by the organisms in the next link in the food chain. Eventually there is so little energy left there is not enough to go round to all the organisms in the population.

Notice in Figure 6.10 that the direction of energy flow is always away from the producer to the consumer.

The biggest loss of energy in any food chain is between the Sun and the producer. The producers trap only about 1% of the Sun's energy. The loss in the other steps in the food chain may be as much as 90% in each step.

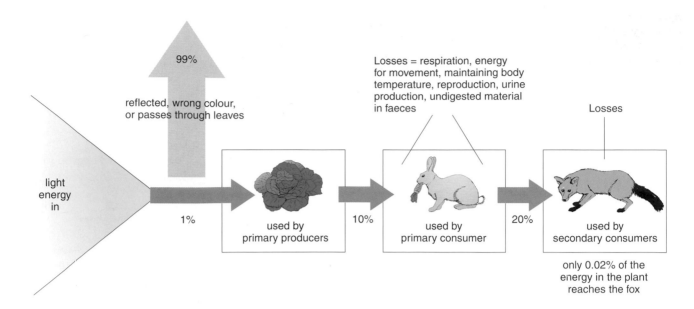

99%

reflected, wrong colour, or passes through leaves

light energy in

1%

used by primary producers

10%

Losses = respiration, energy for movement, maintaining body temperature, reproduction, urine production, undigested material in faeces

used by primary consumer

20%

Losses

used by secondary consumers

only 0.02% of the energy in the plant reaches the fox

Figure 6.10 ▲
A simple food chain showing where the energy is lost

Producing food for humans

There are many food chains which involve humans. In some of these humans are primary consumers.

In other food chains most humans are secondary consumers because most humans eat primary consumers such as cattle (Figure 6.11).

Because energy is lost at each link in a food chain, the fewer steps the food chain has the more energy reaches humans. It should be more efficient for humans to eat plants but unfortunately not all the parts of a plant can be digested by humans and so quite a lot of the energy is lost. It does show, however, that the same area of agricultural land can be used more effectively if used to grow crops than if it is used for the grazing of animals.

Problems involved with the large scale production of food

The cow in Figure 6.12 has transferred only 4% of the energy originally available in the grass into new tissue. If farmers wanted to increase this value then methods would need to be taken to decrease the energy losses due to the animal moving around, needing to keep warm and producing waste. Such methods could include:

- limiting how much the animal moves
- keeping the animal in warm surroundings, especially in cold weather
- controlling the diet to reduce the amount of waste produced.

Figure 6.11 ▶
In the food chain, wheat → pigs → humans, humans are the secondary consumers. The humans receive 20% of the energy from the chain – the remaining 80% is used by the pigs to keep warm and move, or is lost in their waste products

Figure 6.12
For every 100J of energy available in the grass, only 4J gets into the cow's tissues

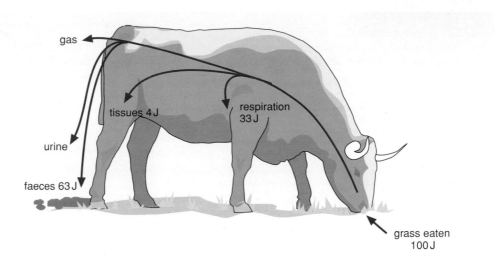

gas

tissues 4 J

respiration 33 J

urine

faeces 63 J

grass eaten 100 J

Much of our fruit comes from other parts of the world. It needs to be harvested, packaged and transported. The energy and resources used to package and transport foods over long distances makes food production inefficient. During each of these processes the quality can deteriorate especially through the action of decomposers. One technique used to stop fruit from decaying before it can be sold is to use special hormones that slow down the rate at which it ripens.

Many of the methods used to improve the efficiency of food production led to the practice of 'factory farming'. Many people consider the confining of animals, especially chickens, in a restricted space and feeding them a specially prepared diet to be cruel. Only recently has there been pressure on farmers to improve the conditions under which the animals are reared. This pressure has been triggered by increased outbreaks of food-poisoning from the eating of eggs and the concerns over the links between cattle food, BSE and CJD in humans.

Summary

- Radiation from the Sun is the source of energy for all organisms.

- Green plants use some of the solar energy that reaches them. This energy is stored in the substances from which the plant is made.

- The mass of living material (**biomass**) at each stage in a **food chain** is less than it was at the previous stage.

- The biomass at each stage can be drawn to scale and shown as a pyramid of biomass.

- Because at each stage in a food chain, less material and less energy are contained the biomass of the organisms, the efficiency of food production can be improved by reducing the number of stages in the food chain.

- The amounts of material and energy in the biomass of organisms is reduced at each successive stage in a food chain because:
  - some materials and energy are always lost in the organisms waste
  - respiration supplies the energy needs for living processes. Much of this energy is eventually lost as heat to the surroundings. These losses are large in warm-blooded animals.

- The efficiency of food production can be improved by:
  - restricting energy losses from food animals by limiting their movement and controlling the temperature of their surroundings
  - using hormones to regulate the ripening of fruits.

- There are positive and negative effects of managing food production and distribution.

Topic questions

1 In the food chain: grass → cattle → humans
  a) Which organisms are consumers?
  b) Which is the producer?
  c) Which is a primary consumer?
  d) Which is a secondary consumer?

2 What is the original source of energy for nearly all living organisms?

3 a) What is biomass?
  b) What is a pyramid of biomass?
  c) What happens to the biomass at each stage in a food chain? Give a reason for your answer.

4 It would be more efficient if humans obtained their energy from directly eating grass rather than from the eating of cattle. Why?

5 a) Why are heat losses greater from cattle and chickens than from salmon and other fish?
  b) Explain three ways in which the efficiency of beef production can be increased.
  c) How can fruit be prevented from over-ripening before it reaches the shops?

6.4 Nutrient cycles

Co-ordinated	Modular
DA 10.23	DA 3
SA n/a	SA n/a

Decomposers and their role in nutrient cycling

When living things grow, they must remove materials from the environment. During photosynthesis, plants take in carbon dioxide from the air and water from the soil and use these inorganic materials to make their food – organic compounds such as glucose and starch. They also take in mineral salts from the soil which help the plants to make other **organic** compounds called proteins.

Note: An inorganic substance generally does not contain carbon; organic substances are carbon-containing chemicals.

These organic compounds are transferred from plants to animals, and from one animal to another along food chains. As plants and animals live, some of the materials are returned to the environment as waste compounds such as carbon dioxide and urea. The process of **decay**, which begins as soon as plants and animals die, helps the return of all nutrients to the environment. The materials are firstly eaten by **detrivores** (detritus-eating species) such as worms, woodlice and maggots and then broken down by decomposers. The **decomposers**, which are bacteria and fungi, are most important in the decay process. They feed on the dead animals and plants or on their waste, turning it back into carbon dioxide, water, nitrogen-containing compounds and mineral salts.

The decay process is vital in the recycling of these materials and allows them to be taken up again by plants. If all the materials were not eventually returned to the environment, some would run out. This can happen when plants are removed, as in deforestation of rainforest areas.

The process of decay can be very slow but it can be speeded up by making sure that the decomposers have warm, moist conditions and a plentiful supply of oxygen. These are the conditions found in a compost heap or sewage farm.

Did you know?

About 300 million years ago, the organic materials did not decay but became coal and oil. We use the chemical energy locked away in coal and oil when we burn them.

Carbon cycle

One important element recycled by the process of decay is carbon. All plants and animals need a supply of carbon. Plants make glucose, during photosynthesis, from carbon dioxide in the air and use the glucose to make other complex carbon compounds such as cellulose. Animals cannot make complex carbon compounds from simple inorganic ones like carbon dioxide, but rely on plants to supply them with organic compounds that are taken in when the plants are eaten. The carbon compounds made by plants and taken in by animals are said to be 'fixed'. They are returned to the environment by animals and plants in the carbon cycle.

The key stages in this cycle are:

- atmospheric carbon dioxide is fixed in plants during photosynthesis. The carbon dioxide is removed from the atmosphere.

- animals eat the plants and the carbon compounds are fixed in the animals.

- respiration in plant and animal cells returns the carbon compounds to the atmosphere. Plant material can also be burnt during combustion and turned into atmospheric carbon dioxide. Fossil fuels are the remains of plants and animals that died many millions of years ago. These fuels are burnt returning carbon dioxide to the atmosphere.

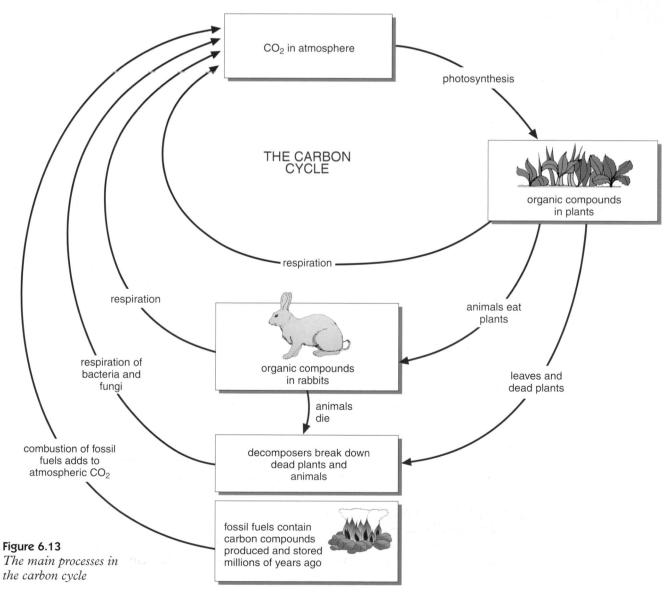

Figure 6.13

The main processes in the carbon cycle

Nitrogen cycle

Nitrogen is another element that must be recycled, as it is essential to the growth of animals and plants. Much of the nitrogen in plants and animals is in the form of amino acids and protein.

Plants absorb nitrogen, in the form of nitrate ions, from the soil. Inside the plant, nitrates are combined with the organic compounds made during photosynthesis to form nitrogen-containing compounds such as proteins. When plants are eaten by animals some of the nitrogen in the plants become part of the proteins in their bodies.

Animals produce waste in the form of a nitrogen-containing compound called urea and this, together with dead plants and animals, is broken down by bacteria during a process called putrefaction. These putrefying bacteria produce ammonia. Plants cannot use ammonia as it is poisonous but the nitrogen becomes available to them once the nitrifying bacteria have converted the ammonium compounds into nitrates.

The main stages in the nitrogen cycle are summarised in Figure 6.14.

Figure 6.14
The main stages of the nitrogen cycle

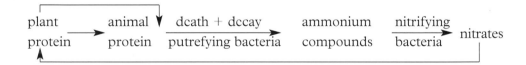

Summary

◆ Living things remove materials from the environment for growth and other life processes.

◆ Living things return the materials to the environment in their waste or when they die and **decay.**

◆ Materials decay by the action of micro-organisms.

◆ Decay is more rapid in moist, warm conditions where there is a plentiful supply of oxygen.

◆ Micro-organisms are used:
  – at sewage works to break down human waste
  – in compost heaps to break down plant material.

◆ The decay process releases substances which plants need to grow.

◆ In a stable community, the processes which use up materials are balanced by the processes which return materials. The materials are constantly being cycled.

◆ The constant cycling of carbon is called the carbon cycle.

◆ The constant cycling of nitrogen is called the nitrogen cycle.

Topic questions

1. a) What are detrivores? Name two.
    b) What are decomposers? Name two.

2. Organic matter decays because of the action of some types of microorganisms.
    a) Under what conditions do these microorganisms cause decay to be most rapid?
    b) Where are microorganisms used to break down human waste?
    c) Where are microorganisms used to break down plant waste?
    d) Why is it important that organic matter decays?

3. What must be happening if an aquarium containing plants and animals is to remain clean and clear?

4. a) Which organisms remove carbon dioxide from the air?
    b) What do green plants do with carbon dioxide?
    c) Animals need carbon. From where do they get it?
    d) How does carbon get released into the air?

5. a) In what form do plants absorb nitrogen from the soil?
    b) What use do animals and plants make of nitrogen?
    c) What do putrefying bacteria do?
    d) What do nitrifying bacteria do?

Examination questions

1. The diagram shows some of the stages by which materials are cycled in living organisms.
    a) In which of the stages, **A**, **B**, **C** or **D**:

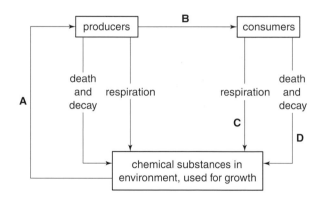

    i)   are substances broken down by micro-organisms.
    ii) is carbon dioxide made into glucose.
    iii) are plants eaten by animals? *(3 marks)*
    b) In an experiment, samples of soil were put into four beakers. A dead leaf was put onto the soil in each beaker. The soil was kept in the conditions shown.

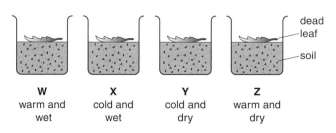

W	X	Y	Z
warm and wet	cold and wet	cold and dry	warm and dry

    In which beaker, **W**, **X**, **Y** or **Z**, would the dead leaf decay quickest? *(1 mark)*

2. A population of rabbits lived on a small island. The graph shows their population over the last 50 years.

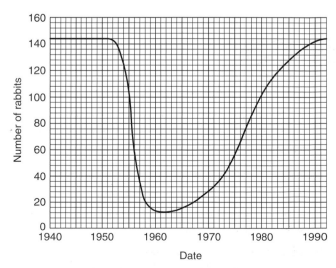

    a) i) How many rabbits were there on the island in 1950? *(1 mark)*
       ii) Give **one** year when there were 88 rabbits on the island. *(1 mark)*

125

b) i) Calculate the decrease in rabbit population between 1950 and 1960. *(1 mark)*
   ii) Suggest a reason why the rabbit population fell in these years. *(1 mark)*
c) The most rabbits on the island is always about 140. Suggest a reason for this. *(1 mark)*

3 Read the passage.

Glutton up a gum treet

Along the banks of the Cygnet River on Kangaroo Island, the branches of the dying gum trees stretch out like accusing fingers. They have no leaves. Birds search in vain for nectar-bearing flowers.

The scene, repeated mile upon mile, is an ecological nightmare. But, for once, the culprit is not human. Instead, it is one of the most appealing mammals on the planet – the koala. If the trees are to survive and provide a food source for the wildlife such as koalas that depend on them, more than 2000 koalas must die. If they are not removed the island's entire koala population will vanish.

Illegal killing has already started. Worried about soil erosion on the island, some farmers have gone for their guns. Why not catch 2000 koalas and take them to the mainland? "Almost impossible," says farmer Andrew Kelly. "Four rangers tried to catch some and in two days they got just six, and these fought, bit and scratched like fury."

a) Use the information from the passage and your own knowledge and understanding to give the arguments for and against killing koalas to reduce the koala population on Kangaroo Island. *(4 marks)*
b) The diagram shows the flow of energy through a koala.
   The numbers show units of energy.

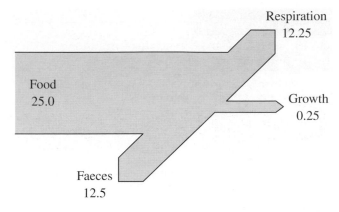

i) Calculate the percentage of the food intake which is converted into new tissues for growth. Show your working. *(2 marks)*
ii) Give **three** different ways in which the koala use the energy released in respiration. *(3 marks)*

4 The information in the table compares two farms. Both are the same size, on similar land, close to one another and both are equally well managed.
a) Use this information to work out the average daily human energy requirement in kilojoules (kJ) per day. *(2 marks)*
b) The figures show that farms like Greenbank Farm can be nine times more efficient at meeting human food energy requirements than farms such as Oaktree Farm.
   i) The food chain for Greenbank Farm is:

   vegetation → humans

   What is the food chain for Oaktree Farm? *(1 mark)*
   ii) Explain why Greenbank Farm is much more efficient at meeting human food energy requirements. *(3 marks)*
c) The human population has been increasing rapidly throughout this century. It is now about 6 billion and is still growing. What does the information in this question suggest about likely changes in the human diet which may need to occur during the coming century? Explain your answer. *(4 marks)*

Name of farm	Activity	Energy value of food for humans produced in one one year	Number of people people whose energy requirements can be met by this food
Greenbank Farm	Grows food for humans	3285 million kJ	720
Oaktree Farm	Grows food for animals on the farm which become food for humans	365 million kJ	80

Chapter 7
Getting together

Key terms

anions • atoms • atomic number • cations • charge • compound • covalent bond • electrolysis • electron • element • formula mass • free electrons • ion • ionic bond • isotope • Law of Conservation of Mass • mass number • molar volume • mole • molecule • negative • neutron • noble gas • nucleon • nucleus • Periodic Table • positive • proton • relative atomic mass • relative formula mass • relative molecular mass

7.1		Atoms
Co-ordinated	Modular	
DA 11.1/12.23	DA 8/12	
SA 11.1/12.15	SA 16/18	

Although the information provided in this chapter is based largely on work carried out on the theory of atomic structure developed in the last 80 years it is important to be aware of the historical background that led up to the modern ideas of atomic structure. More historical background on the development of atomic theory is found in section 18.2.

The story of the atom – from the Greeks to John Dalton

The idea that all substances could be made of atoms originated in Greece some 2000 years ago. At that time many Greek thinkers (philosophers) were interested in trying to determine what all substances were made from.

Figure 7.1
Democritus

Democritus (460–370 BC) was a philosopher who believed that if a lump of metal was cut into smaller and smaller pieces you would eventually end up with a very small piece that was too small to cut up further. He called these very small pieces 'atomos' (which means indivisible). He believed that all atomos were made of the same matter but were different in shape, size and speed.

However, another Greek philosopher called Aristotle (384–322 BC) believed that it was possible to keep dividing up something into smaller and smaller pieces. He did not believe in the idea of atoms. He believed that all substances were made up of four 'elements' – earth, air, fire and water and that each substance was made of a mixture of the four elements.

127

Democritus's idea was not accepted because it required people to believe in the concept of invisible particles whose behaviour could be explained and was therefore not influenced by the gods. Aristotle's ideas gained favour because they were easier to understand. They were accepted for nearly 2000 years.

Did you know?

For many centuries the ideas of Aristotle helped:

● Egyptian and Chinese philosophers and metalworkers to investigate ways to turn common metals into gold. The ability to turn common metals into gold was known as transmutation – later called alchemy. The Egyptians wanted the gold because of its monetary value. The Chinese wanted it because it was thought to give long, if not everlasting life for those who swallowed it.

● Arabic alchemists attempt to discover the 'philosopher's stone' which would not only help transmutation but would provide both wealth and health.

No-one succeeded in turning common metals into gold or in finding the philosopher's stone but a number of important chemical processes were discovered, a knowledge of practical chemistry was gained, as well as a better understanding of chemical reactions.

All these advancements in knowledge and experience began to undermine the theory of Aristotle.

In the 18th and 19th centuries chemists began to realise that many solids, liquids and gases could be broken down into simpler substances called elements. They also realised that elements could be joined to make more complicated substances called compounds.

Figure 7.2
John Dalton

Early in the 19th century John Dalton (1766–1844), a British chemist and physicist, developed his atomic theory which attempted to explain the structure of elements and compounds. This theory was based on the concept of atoms proposed by Democritus almost 2000 years before.

Dalton's atomic theory was made public in a series of lectures and in a book called *New System of Chemical Philosophy* which was published between 1808 and 1810.

The key parts of his theory proposed that:

● all elements are made up of very small solid spheres called atoms
● the atoms of a given element are identical and have the same weight
● the atoms of different elements have different weights
● chemical compounds are formed when atoms combine
● a given compound is always made up of the same number and type of atoms.

There were some important scientists, such as Humphrey Davy and Michael Faraday, who were still reluctant to accept the idea that all matter was made up of atoms. It was not until the late 1800s that the existence of atoms was finally accepted. By this time Dalton's ideas had been successfully used to explain the existence and structure of molecules and in the determination of the relative atomic masses of the elements.

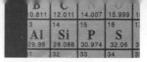

Through the work of John Dalton it is now known that the basic building blocks of all materials, whether they are solids, liquids or gases, are tiny particles known as **atoms**. Atoms rarely occur on their own – usually they join up with other atoms. There are about 100 different sorts of atoms.

If all the atoms in a material are identical, the material is an **element**, but if atoms of two or more different types are chemically joined together, the material is a **compound**.

Atoms are the smallest particles of an element that can exist on their own. Dalton believed that they could not be broken into anything simpler. Research carried out during the last 100 years has shown, however, that atoms consist of even smaller particles. The most important of these are called **protons**, **neutrons** and **electrons**.

The structure of the atom

The atom is now known to consist of a very small **nucleus** containing protons which have a positive **charge** and neutrons which are uncharged. Negatively-charged electrons orbit the nucleus in shells or energy levels. The negatively-charged electrons are attracted to the positively-charged nucleus.

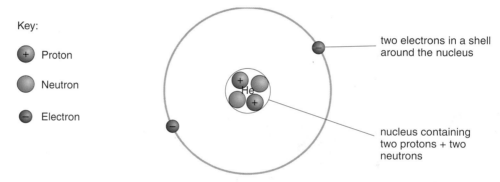

Figure 7.3
An atom of helium

Atomic number (Z)

Each element in the **periodic table** has a fixed number of protons in its atoms. For example, all hydrogen atoms have just one proton in the nucleus, all helium atoms have two, all lithium atoms have three, etc. This number of protons is known as the **atomic number** (proton number), Z. Since there are always equal numbers of protons and electrons in any atom, the atomic number also tells us the number of electrons present in the shells around the nucleus.

This atomic number is usually shown as a subscript to the left of the symbol used for the element e.g. $_1$H or $_2$He or $_3$Li etc.

More about the particles

Atoms have no electrical charge because the **positive** charges of the protons are cancelled out by the **negative** charges of the electrons. In all atoms of a particular element there are equal numbers of protons and electrons.

Most of the mass of the atoms is due to the protons and neutrons in the nucleus; electrons have very little mass as Figure 7.4 shows.

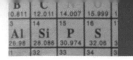
Figure 7.4

The relative masses and charges of the particles in atoms				
Name of particle	Symbol	Relative mass	Relative charge	Where found
proton	p	1	$+1$	nucleus
neutron	n	1	0	nucleus
electron	e	negligible	-1	in shells

Did you know?

If an atom was the size of Wembley Stadium, the nucleus would be smaller than the spot at the middle of the centre circle where the ball is placed for the kick off!

The radius of an atom is only about 10^{-8} cm (0.000 000 01 cm) and the radius of a nucleus is about 10^{-12} cm (0.000 000 000 001 cm).

Mass number (A)

To get some idea of the relative mass of different elements it is necessary to take into account the fact that most of the mass of the atom is due to protons and neutrons. The **mass number** (nucleon number), A, is obtained by adding together the number of protons and neutrons (**nucleons**) present in the nucleus. So in a helium atom, where there are two protons and two neutrons, the mass number is $(2 + 2) = 4$.

This mass number is usually shown as a superscript to the left of the symbol used for the element e.g. ^{4}He.

The atomic number and mass number of an element can both be displayed together, so an atom of sodium (Na) that has 11 protons and 12 neutrons in its nucleus would be shown as $^{23}_{11}$Na ($Z = 11$; $A = 11 + 12 = 23$).

Isotopes

Although the number of protons in the nucleus of an atom is fixed for that element, the number of neutrons present in the nuclei of atoms of the same element can vary from atom to atom. Atoms that have the same number of protons but different numbers of neutrons in their nuclei are called **isotopes** and most elements exist as mixtures of isotopes. An example of this is lithium which has two isotopes: $^6_3$Li and $^7_3$Li.

^{6}Li has three protons and three neutrons but ^{7}Li has three protons and four neutrons, so ^{7}Li is slightly heavier than ^{6}Li. Naturally-occurring lithium is a mixture of about 90% ^{7}Li and 10% ^{6}Li giving an average relative mass for a lithium atom as 6.9. This average relative mass which takes into account the different masses of the isotopes and their relative amounts in a naturally-occurring sample of the element is called the **relative atomic mass**. Usually the relative atomic mass is rounded to the nearest whole number.

Figure 7.5
Isotopes of lithium – atomic number 3

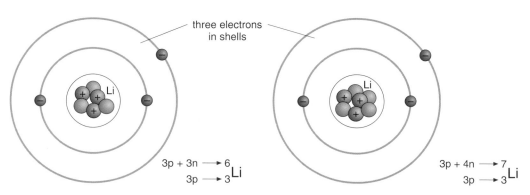

Remember that during chemical reactions the number of protons and neutrons in the nucleus *never changes*.

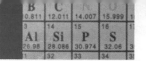
Arrangement of electrons

Each electron in an atom is at a particular energy level (shell). The electrons in any atom occupy the lowest available energy levels (the innermost available shells).

The simplest atom is an atom of hydrogen which has one proton in the nucleus and one electron at the lowest energy level (in the first shell) outside the nucleus (see Figure 7.6).

The second element is helium. It has an atomic number of 2 so it will have two electrons outside the nucleus. Both of these electrons go into the first shell (Figure 7.7).

Figure 7.6

The atomic structure of hydrogen (no neutrons present)

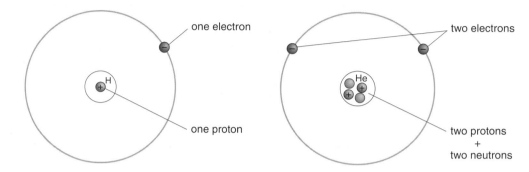

Figure 7.7

The atomic structure of helium

With two electrons in it, the first shell is full. The evidence for this is that helium is a **noble gas**. All noble gases exist as individual atoms and do not take part in chemical reactions because their electronic structures are very stable. If an atom of an element has more than two electrons, the extra electrons go into the second shell and this is what happens with element number 3 – lithium.

Figure 7.8

The atomic structure of lithium (neutrons not shown)

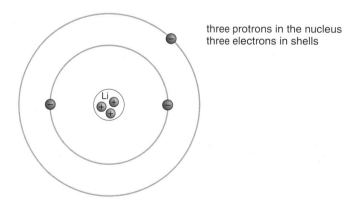

three protrons in the nucleus
three electrons in shells

As the atomic number continues to increase and the number of electrons also increases, the second shell begins to fill up (see Figure 7.9).

In element neon, atomic number 10, the second shell is full – neon is a noble gas. The second shell is stable and full when it contains eight electrons. Any further electrons go into the third shell until it too becomes 'full'. This happens with element number 18 – argon. Argon is also a noble gas so its outer shell must be full when it contains eight electrons. In elements 19 (potassium) and 20 (calcium) the fourth shell is used.

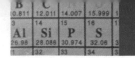

Figure 7.9
The electronic structures of elements 3 to 10. No details of the contents of the nucleus are shown

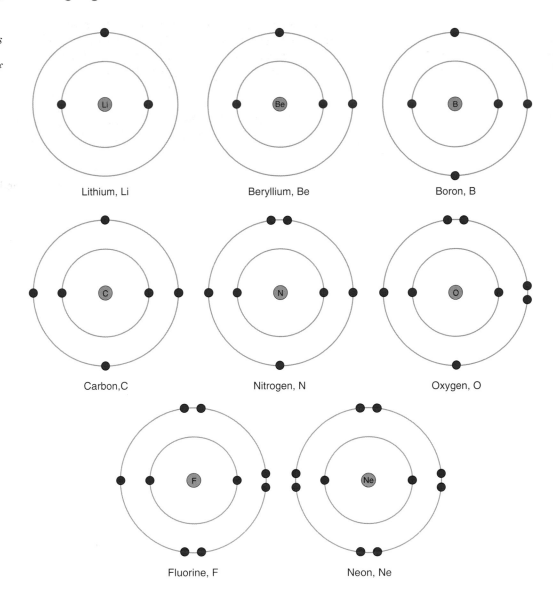

Lithium, Li

Beryllium, Be

Boron, B

Carbon, C

Nitrogen, N

Oxygen, O

Fluorine, F

Neon, Ne

The electronic structure of an element can also be shown as a 'dot and cross' diagram. Figure 7.10 shows the electronic structure of fluorine using the 'dot and cross' method.

Figure 7.10
The electronic structure of fluorine using the 'dot and cross' method

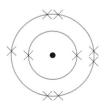

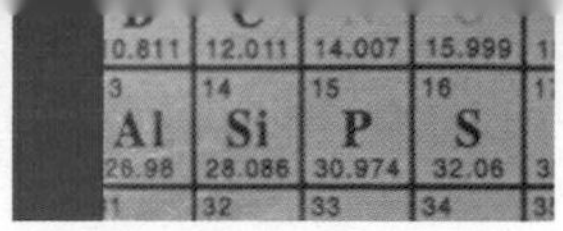

The electronic structures of these elements can be written in a shorter way to show the number of electrons present in each shell. For example, element number 17, chlorine, can be written as $_{17}Cl = 2,8,7$ showing that the 17 electrons in a chlorine atom are arranged as two in the first shell, eight in the second and seven in the third.

Figure 7.11 shows how the electrons are arranged in the first 20 elements.

Figure 7.11

Element	Symbol	Atomic no.	First shell	Second shell	Third shell	Fourth shell
hydrogen	H	1	1			
helium	He	2	2	First shell is now full		
lithium	Li	3	2	1		
beryllium	Be	4	2	2		
boron	B	5	2	3		
carbon	C	6	2	4		
nitrogen	N	7	2	5		
oxygen	O	8	2	6		
fluorine	F	9	2	7		
neon	Ne	10	2	8	Second shell full	
sodium	Na	11	2	8	1	
magnesium	Mg	12	2	8	2	
aluminium	Al	13	2	8	3	
silicon	Si	14	2	8	4	
phosphorus	P	15	2	8	5	
sulphur	S	16	2	8	6	
chlorine	Cl	17	2	8	7	
argon	Ar	18	2	8	8	Third shell full
potassium	K	19	2	8	8	1
calcium	Ca	20	2	8	8	2

Summary

◆ John Dalton reintroduced after 2000 years the Greek idea that matter was made of solid indivisible particles called atoms.

◆ The modern model of the atom has a small central **nucleus** made up of **protons** and **neutrons** around which are the **electrons**.

◆ Protons have a positive charge.

◆ Neutrons have no charge.

◆ Electrons have a negative charge and are found in energy layers or shells around the atomic nucleus.

◆ Atoms have no overall charge because the number of protons equals the number of electrons.

◆ The **atomic number** (proton number), Z, is the number of protons in the nucleus. It is the same as the number of electrons in shells around the nucleus.

◆ The **mass number** (nucleon number), A, is the sum of the number of protons and neutrons (nucleons) in the nucleus.

◆ **Isotopes** are atoms of the same element that have the same number of protons but different numbers of neutrons in their nuclei.

◆ The electronic structure of an atom of a particular element can be represented using either the 2, 8, 8, 1 or the 'dot and cross' convention.

◆ **Compounds** are substances in which two or more elements are chemically combined.

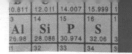

Topic questions

1 a) In what ways did the beliefs of Democritus and Aristotle differ about what matter was made of?
    b) Why were Democritus's ideas ignored?
    c) John Dalton believed that:
        ● all elements were made up of very small solid spheres called atoms
        ● the atoms in a given element are identical and have the same weight
        ● the atoms of different elements have different weights
        ● chemical compounds are formed when atoms combine
        ● a given compound is always made up of the same numbers and types of atoms.

For each of these points explain whether or not they hold true today.

2 Name the three particles present in atoms.

3 Use words from the following list to complete the sentences:

nucleus   protons   shells   electrons
positive   neutral   negative   neutrons

An atom consists of a small central part called the _____ surrounded by several _____. The central part contains _____ which have a _____ charge and _____ which have no charge. The _____ which are found in the shells around the nucleus have a _____ charge.

4 The symbol for an isotope of carbon is $^{14}_{6}C$. How many protons, neutrons and electrons does this form of carbon contain?

5 How are the electrons arranged in the following elements:
    a) sulphur S
    b) magnesium Mg
    c) neon Ne?
    d) Which one of these elements is a noble gas?

6 Draw a diagram to show how all the particles are arranged in an atom of $^{31}_{15}P$.

7 What is the maximum number of electrons that can be held in
    a) shell 1?
    b) shell 2?
    c) shell 3?

7.2 Bonding

Co-ordinated	Modular
DA 11.2	DA 8
SA n/a	SA n/a

Metals and non-metals

When elements are sorted into metals and non-metals it can be seen that their electronic structures have some noticeable similarities and differences.

Figure 7.12

| The electronic structures of selected elements ||||
Metals	Electron structure	Non-metals	Electron structure
lithium	2,1	carbon	2,4
beryllium	2,2	nitrogen	2,5
sodium	2,8,1	oxygen	2,6
magnesium	2,8,2	fluorine	2,7
aluminium	2,8,3	phosphorus	2,8,5
potassium	2,8,8,1	sulphur	2,8,6
calcium	2,8,8,2	chlorine	2,8,7

The first thing to notice is that the metals have either one, two or three electrons in their outer shell, but non-metals have four, five, six or seven electrons in theirs. All elements react so that their electronic structures become identical to that of the noble gas nearest to them in the periodic table (i.e. they try to get a full outer shell of electrons). In this way the elements become more stable (see section 11.3).

0.811	12.011	14.007	15.999	1
3	14	15	16	
Al	**Si**	**P**	**S**	
26.98	28.086	30.974	32.06	3
	32	33	34	3

Since metals and non-metals have quite different structures, they behave differently as they try to achieve the electronic structure of the nearest noble gas. Metals usually react by losing their outer electrons but non-metals try to gain more outer electrons to complete their outer shell. There are three main ways that elements can do this. Each method results in a chemical bond being formed. The methods produce **covalent bonds**, **ionic bonds** or **metallic bonds**.

Molecules

The only elements that exist in nature as individual atoms are the noble gases. This is because their outermost electron shells are already full and stable. The atoms of all other elements join together in some way to form molecules and the method used depends on the type of element they are and the type of element with which they join.

Remember: Atoms join together to make their electronic structures more stable. Each element tries to achieve the same electronic structure as the noble gas nearest to it in the periodic table.

Covalent bonds in elements

Covalent bonds are formed when atoms *share* electrons so that they all end up with a stable, noble gas structure in which all the electron shells are full. If the elements are non-metal elements, their outer shells are short of electrons and they achieve full shells by covalent bonding.

Hydrogen molecules, H$_2$

Hydrogen atoms have only one electron in their outer shell. The nearest noble gas to hydrogen is helium which has two outer electrons. Two hydrogen atoms join by sharing their electrons so that each has control over two (see Figure 7.13).

Figure 7.13
How two hydrogen atoms combine to make a hydrogen molecule

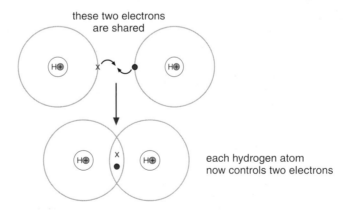

This produces a **molecule** of hydrogen containing two hydrogen atoms chemically joined together. It is a diatomic (two atom) molecule and the two atoms are joined by a pair of shared electrons. This is a covalent bond.

Sometimes the sharing of one pair of electrons is not enough to give all the atoms present completely full shells but elements always adapt so that stable structures are produced. Figure 7.14 shows how two oxygen atoms join together to make an oxygen molecule (O_2) by forming a double covalent bond. Each oxygen atom has gained control over eight electrons in its outer shell.

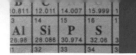

Figure 7.14
How two oxygen atoms combine to form an oxygen molecule with a double covalent bond

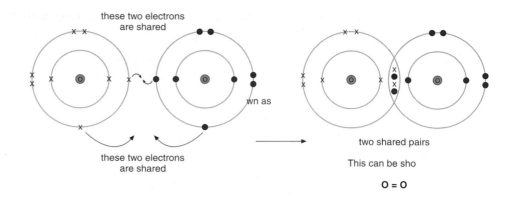

Covalent bonds in compounds

Covalent bonds can also be formed between atoms of different elements. In each case the atoms are linked by pairs of shared electrons – each atom usually donates one electron to each shared pair.

Figures 7.15 to 7.17 show how covalent bonds are formed in molecules of methane (CH_4), ammonia (NH_3) and hydrogen chloride (HCl), respectively.

Figure 7.15 ▼
The bonding in methane

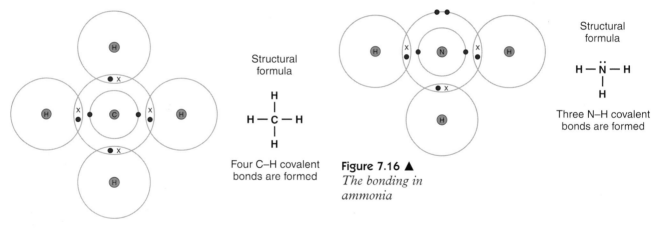

Structural formula

$$H - \overset{\displaystyle H}{\underset{\displaystyle H}{C}} - H$$

Four C–H covalent bonds are formed

Structural formula

$$H - \overset{\displaystyle \cdot\cdot}{N} - H$$
$$|$$
$$H$$

Three N–H covalent bonds are formed

Figure 7.16 ▲
The bonding in ammonia

Figure 7.17
The bonding in hydrogen chloride

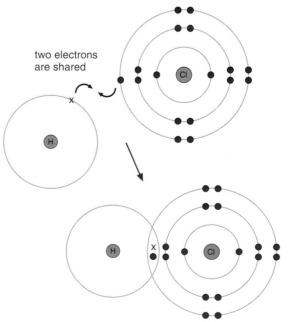

two electrons are shared

This structure can be shown as H—Cl

136

Double covalent bonds can also be formed between non-metal elements, for example in water (H_2O).

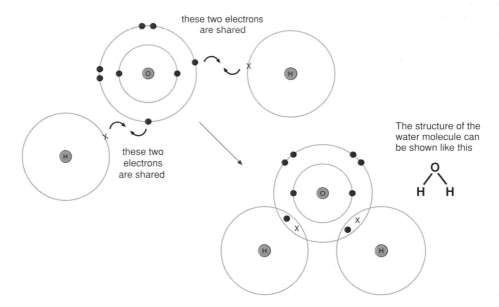

these two electrons
are shared

these two
electrons
are shared

The structure of the
water molecule can
be shown like this

Figure 7.18
How two hydrogen atoms and one oxygen atom form a water molecule

Here are some melting and boiling points for some familiar covalently-bonded materials.

Figure 7.19

Material	Formula	Melting point (°C)	Boiling point (°C)	Normal state at room temperature
carbon dioxide	CO_2	(sublimes)*	−78.5	gas
carbon monoxide	CO	−199	−192	gas
methane	CH_4	−182	−162	gas
ethane	C_2H_6	−183	−88	gas
water	H_2O	0	100	liquid

*The solid turns directly to a gas without melting to a liquid first

Notice how covalent compounds only contain non-metal elements and that they are usually gases, liquids or low melting point solids. The covalent bond in each molecule is a strong bond.

The reason covalently-bonded compounds have such low melting and boiling points is because the forces *between* neighbouring covalent molecules are very weak.

Covalent compounds do not usually dissolve in water nor do they conduct electricity because they do not carry any overall charge.

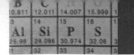
Important points to remember about covalent bonds

- They are formed only between non-metal elements.

- They are formed by sharing pairs of electrons – each atom donating one electron to each shared pair.

- They produce clusters of atoms chemically joined together, known as molecules.

- Whilst the atoms in the molecules are strongly joined together, the forces between molecules are quite weak so covalent compounds are often gases (e.g. oxygen and hydrogen), low boiling point liquids (e.g. water and petrol) or low melting point solids (candle wax).

- They are frequently insoluble in solvents like water.

- They do not conduct electricity in either the solid or molten state.

Giant molecules

Some covalent materials behave in a different way. For example diamond and graphite have high melting points (more than 3500°C) and sand (silicon dioxide) melts at 1610°C. This is because they are not simple little molecules but have giant three-dimensional structures (see Figure 7.20).

Figure 7.20
The arrangement of covalently bonded carbon atoms in (a) diamond and (b) graphite

(a)
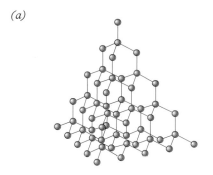

(b)
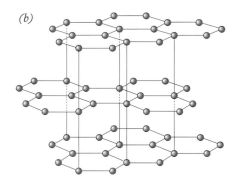

In diamond each carbon atom forms four covalent bonds which produce a rigid, giant covalent structure.

In graphite each carbon atom forms three covalent bonds. The carbon atoms form layers which are free to slide over each other. In graphite there are free electrons which allow the graphite to conduct electricity.

Did you know?

Diamond and graphite are pure forms of the same element, carbon. Diamond is the hardest naturally-occurring substance yet graphite is soft enough to leave a grey mark on paper and is used in pencil lead.

Ionic bonds

Ionic bonds are formed when electrons are transferred (moved) from one atom to another. The electronic structures of non-metals show that they are generally short of electrons, but the opposite is true for metals. Metals have one, two or three electrons outside a full electron shell and they react by losing these electrons to other elements. When metals combine with non-metals the electrons in the outer shell of the metal atoms transfer completely to the outer shell of the non-metal atoms.

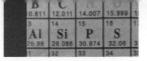
This means that each 'atom' becomes charged. These charged 'atoms' are called **ions**. Positively-charged ions are called **cations** and negatively-charged ions are called **anions**. In atoms there are equal numbers of positively-charged protons and negatively-charged electrons so the overall charge is zero.

In ions this is not the case. A sodium atom (Na = 2,8,1) contains 11 protons and 11 electrons. If it loses its outer electron, it still has 11 protons but now has only 10 electrons so it becomes a positively-charged ion.

$$Na\ (2,8,1) - 1e^- \rightarrow Na^+\ (2,8)$$

Note that the sodium ion (Na$^+$) has the same electronic structure as the noble gas neon, Ne (2,8).

A chlorine atom contains 17 protons and 17 electrons but if it gains an extra electron into its outer shell to make it full, it will have 17 protons and 18 electrons and becomes a negatively-charged ion.

$$Cl\ (2,8,7) + 1e^- \rightarrow Cl^-\ (2,8,8)$$

Note that the chlorine ion (Cl$^-$) has the same electronic structure as the noble gas argon, Ar (2,8,8).

When sodium and chlorine join together, electrons transfer from the outer shells of sodium atoms to the outer shells of chlorine atoms and lots of Na$^+$ and Cl$^-$ ions are formed. Since these ions are of opposite charge they attract strongly and a strong ionic bond is formed (see Figure 7.21).

Figure 7.21
How sodium and chlorine atoms form an ionic bond

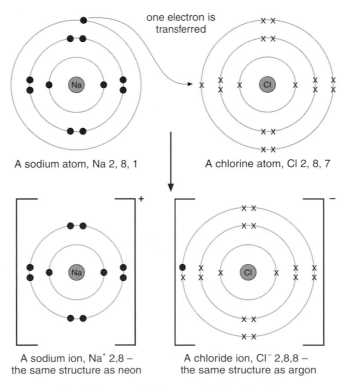

A sodium atom, Na 2, 8, 1   A chlorine atom, Cl 2, 8, 7

A sodium ion, Na$^+$ 2,8 –
the same structure as neon

A chloride ion, Cl$^-$ 2,8,8 –
the same structure as argon

The Na$^+$ ions and Cl$^-$ ions are held together by
electrostatic forces of attraction

139

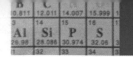

Figure 7.22
How the ionic bond is formed in calcium chloride

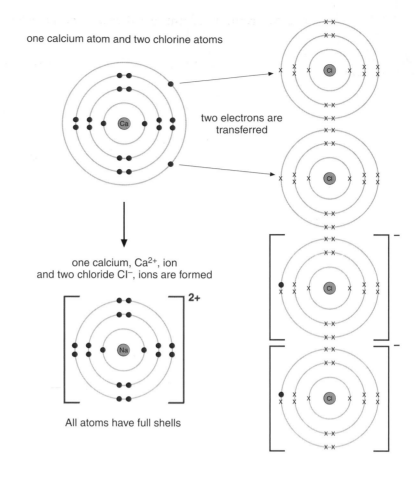

one calcium atom and two chlorine atoms

two electrons are transferred

one calcium, Ca²⁺, ion
and two chloride Cl⁻, ions are formed

All atoms have full shells

Figure 7.23
How the ionic bond is formed in magnesium oxide

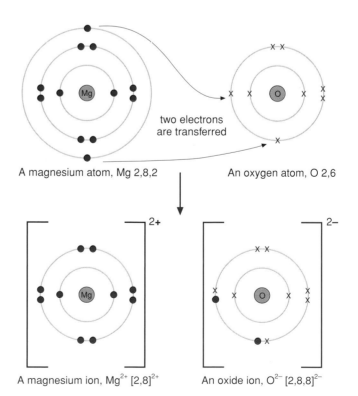

two electrons are transferred

A magnesium atom, Mg 2,8,2

An oxygen atom, O 2,6

A magnesium ion, Mg^{2+} $[2,8]^{2+}$

An oxide ion, O^{2-} $[2,8,8]^{2-}$

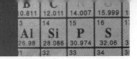
The electronic structures of other common ions are shown in Figure 7.24.

Figure 7.24
The electronic structures of some other ions

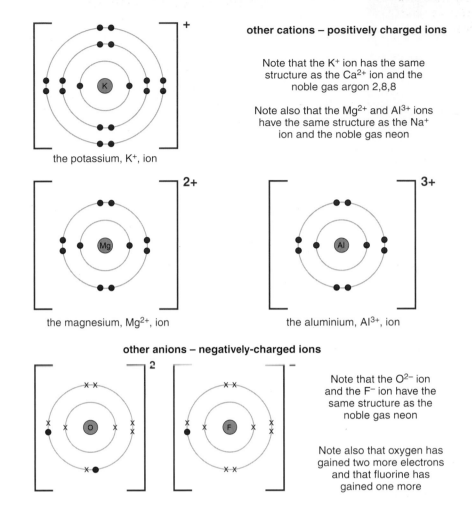

the potassium, K⁺, ion

the magnesium, Mg²⁺, ion

the aluminium, Al³⁺, ion

other cations – positively charged ions

Note that the K⁺ ion has the same structure as the Ca²⁺ ion and the noble gas argon 2,8,8

Note also that the Mg²⁺ and Al³⁺ ions have the same structure as the Na⁺ ion and the noble gas neon

other anions – negatively-charged ions

Note that the O²⁻ ion and the F⁻ ion have the same structure as the noble gas neon

Note also that oxygen has gained two more electrons and that fluorine has gained one more

Here are some melting points and boiling points of familiar ionic compounds. Note that every one contains both a metal and a non-metal element.

Figure 7.25

Material	Formula	Melting point (°C)	Boiling point (°C)	Normal state at room temperature
sodium chloride	NaCl	801	1465	solid
copper oxide	CuO	1326	very high	solid
iron oxide	Fe_2O_3	1565	very high	solid
magnesium oxide	MgO	2800	3600	solid
potassium chloride	KCl	770	1407	solid
lead bromide	$PbBr_2$	373	916	solid
aluminium oxide	Al_2O_3	2045	2980	solid

All ionic compounds are solids with high melting points. This is completely different from covalent compounds that have low melting and boiling points.

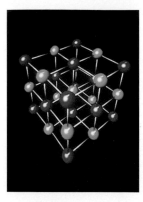

Figure 7.26
The arrangement of sodium (Na⁺) ions and chloride (Cl⁻) ions in sodium chloride

More about ionic bonding

In ionic compounds the charges on the ions will attract any ion that has the opposite charge. Each positive ion becomes surrounded by and attracted to lots of negative ions and each negative ion becomes surrounded by and attracted to lots of positive ions. This can be seen in Figure 7.26 which shows the arrangement of the Na^+ and Cl^- ions in sodium chloride (NaCl). The result is a large three-dimensional structure.

Melting this solid would involve breaking the ionic bonds between all the oppositely-charged ions to release them to become mobile. A lot of energy is required to do this so the melting points of ionic solids are high.

Ionic compounds

Ionic compounds are solids with giant structures but many of them will dissolve in water.

Ionic solids will not conduct electricity when in the solid state. They will, however, conduct electricity when they are molten or dissolved in water. In the solid the ions cannot move (see Figure 7.26) but in the molten state or in solution they can. Mobile ions can carry the current through the liquid. The positively-charged ions move towards the negative electrode and the negatively-charged ions move towards the positive electrode. This two-way flow of ions towards electrodes of opposite charge is known as **electrolysis**. Electrolysis results in some form of chemical change (see Figure 7.27).

Figure 7.27
How the ions move in a liquid ionic compound

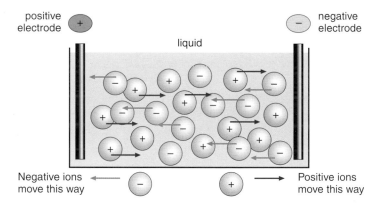

Important points to remember about ionic bonds

- They are formed only between metals and non-metal elements.
- They are formed by the transfer of electrons from metal atoms to non-metal atoms.
- They produce charged particles, known as ions.
- The ionic bond is a very strong bond formed as a result of very strong electrostatic forces of attraction between lots of oppositely-charged ions.
- Ionic compounds are always solids with high melting points.
- Ionic solids are often soluble in water.
- Ionic solids do not conduct electricity but when molten or in aqueous solution they do.

Metals and the metallic bond

With the exception of mercury, all metals are solids at room temperature and most of them have high melting points. Metals are not soluble in water, though some of them react with water and appear to dissolve. Metals are good conductors of electricity.

Metal atoms have either one, two or three electrons in their outer shell. To get a stable noble gas structure the metal atoms need to lose these electrons. When metal atoms are together in a piece of metal, these outer electrons become 'pooled' together to form a 'sea' of **free (mobile) electrons**. The nuclei of the metal atoms and all their completely full electron shells occupy a fixed position within a giant three-dimensional crystal structure and the sea of free, mobile electrons is able to flow through this structure giving metals their characteristic properties. Imagine that a box is filled with marbles to represent the nuclei and full shells of the atoms and then water (representing the sea of outer electrons) is poured into the box to cover the marbles. The marbles will be arranged in a neat ordered pattern but the water will be free to flow between them (see Figure 7.28).

Figure 7.28
How the atoms and outer electrons are arranged in a metal

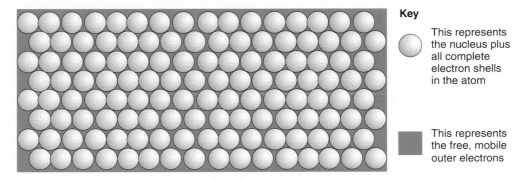

Key

This represents the nucleus plus all complete electron shells in the atom

This represents the free, mobile outer electrons

The free, mobile electrons are responsible for metals being conductors of heat and electricity. The metal atoms are tightly packed together bonded by the sea of free electrons. This means that individual atoms are not easy to separate so metals are strong, hard and have high melting points.

Metals are both malleable and ductile which means that they can be bent into a different shape or can be drawn out into wires. This is because the free electrons allow the tightly packed atoms to move into different positions within the crystal structure (see Figure 7.29).

Figure 7.29
What happens when a metal bends

The atom arrangement in a straight piece of metal

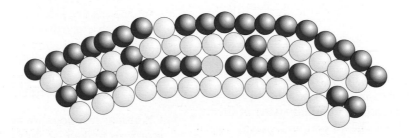

The atom arrangement when the metal is bent – some atoms are forced into different layers

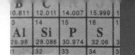

Getting together

The flow chart in Figure 7.30 can be used to work out the type of chemical bond likely to be present in a substance.

Figure 7.30

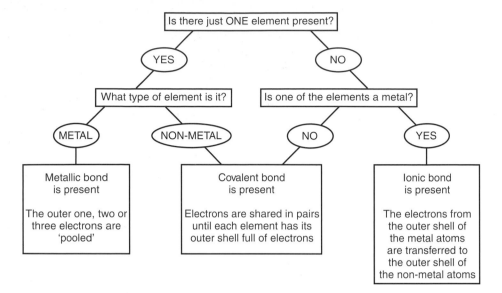

Summary

♦ When non-metal elements join together they do so by sharing electrons from the highest occupied energy level (outermost occupied shell). The strong bonds formed are called **covalent bonds.**

♦ Covalent compounds consist of molecules

♦ Covalent compounds have low melting points, low boiling points and do not conduct electricity.

  ♦ Although the covalent bonds that hold a molecule of a covalent compound together are strong, the intermolecular forces are weak.
  ♦ The molecules of a covalent compound carry no overall electric charge.

♦ Covalent compounds can be represented using either the 'dot and cross' or structural formula conventions.

  ♦ Giant covalent structures have high melting points because of the large number of strong covalent bonds in their structure (lattice). Diamond and graphite and silicon dioxide are examples of giant covalent structures.

♦ Metals and non-metals join together when electrons transfer from the highest occupied energy level (outermost occupied shell) of the metal atom to the highest occupied energy level (outermost occupied shell) of the non-metal atom forming an **ionic bond.**

♦ The gaining or losing of electrons produces ions which have the electronic structure of a noble gas.

♦ Ionic compounds are all solids with high melting points, high boiling points and they conduct electricity when molten or in solution.

♦ Ions and ionic compounds can be represented using either the 'dot and cross' or the 2,8,8,1 conventions.

  ♦ Ionic compounds form giant ionic lattices in which there are strong forces of attraction between the oppositely charged ions.
  ♦ Ionic compounds have high melting points and high boiling points because of the strong forces of attraction.
  ♦ When ionic compounds are melted or dissolved in water the ions become free to move and so the ionic compounds conduct electricity.
  ♦ Metals are made of giant structures in which electrons from the highest occupied energy level (outermost shell) of the metal atom are free to move through the whole structure. These **free electrons** allow the structure to conduct electricity and heat.

♦ Chemical reactions involve the rearrangement of atoms to make new substances.

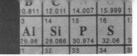

Topic questions

1 Use the copy of the periodic table (on page 208) to identify each of the following elements as either metal or non-metal and then decide the type of bond that would be formed between each pair.
  a) magnesium and fluorine
  b) oxygen and sulphur
  c) zinc and chlorine
  d) phosphorus and oxygen

2 Show how the elements sulphur, $_{16}S$, and hydrogen, $_1H$, can combine together to make the compound hydrogen sulphide, H_2S.

3 Show how the elements lithium, $_3Li$, and fluorine, $_9F$, can combine together to make the compound lithium fluoride, LiF.

4 Show how sodium, $_{11}Na$, and oxygen, $_8O$, can combine to form the compound sodium oxide, Na_2O.

5 The properties of substances often indicate the type of bond present in them. Study the following descriptions of substances A, B, C and D and say what type of chemical bond is likely to be present.

Description of compound	Type of bond present
A is a white solid that melts at 960°C. It dissolves in water and conducts an electric current when in solution and when molten.	
B is a liquid which does not dissolve in water. Its boiling point is 87°C and it burns easily.	
C is a blue crystalline solid that dissolves easily in water. The resulting solution conducts electricity. The solid decomposes before it melts.	
D is a gas with a very strong smell. It is very soluble in water and the resulting solution conducts electricity. If the gas is cooled until it changes into a liquid, this liquid does not conduct electricity.	

6 Use the flow chart in Figure 7.30 to work out the type of chemical bond present in each of the following substances.
  a) magnesium, b) fluorine, c) sodium sulphide, d) lithium oxide, e) nitrogen,
  f) hydrogen sulphide, g) phosphine (a compound of phosphorus and hydrogen)

7.3 Quantitative chemistry

Co-ordinated	Modular
DA 11.8	DA 7
SA n/a	SA n/a

Quantitative chemistry is concerned with the amounts of materials involved in chemical reactions. In some chemical reactions it looks as if the mass has changed. For example when solutions of lead nitrate and potassium iodide are mixed, a dense yellow solid (called a precipitate) of lead iodide is formed. Although it appears that the mass has increased, it hasn't (see Figure 7.31).

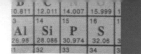

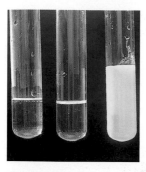

Figure 7.31
The formation of solid lead iodide (right) by mixing solutions of lead nitrate and potassium iodide (left)

Getting together

It is important to remember that **during a chemical change, matter (material) is neither created nor destroyed.** This is a fundamental law in science known as **the Law of Conservation of Mass.** During a chemical reaction atoms are neither created nor destroyed, they are simply rearranged to form new substances. In some reactions gases are produced, for example when magnesium is added to dilute acid in a flask. Gases have mass so the mass of the apparatus will decrease if the gas is allowed to escape into the air.

Chemical calculations are based on the Law of Conservation of Mass. In chemical reactions it is the number of atoms and molecules taking part that is important. In a bank the cashier 'counts' coins by 'weighing' them. The same can be done in chemistry. It is necessary to take into account that different atoms have different masses. This is done using the quantity known as the **relative atomic mass (A_r).**

Did you know?

The element chlorine $_{17}Cl$ has two isotopes – ^{35}Cl and ^{37}Cl. Both forms of chlorine have 17 protons in their nucleus and 17 electrons in shells around the nucleus but ^{35}Cl has 18 neutrons in its nucleus whilst ^{37}Cl has 20. The ratio of ^{35}Cl atoms to ^{37}Cl atoms is 3 : 1. This gives an average value of 35.5 for the relative atomic mass of chlorine.

Relative atomic mass does not have any units because it is a ratio and not a real mass. The relative atomic mass is approximately the number of times one atom of the element is heavier than one atom of hydrogen. One atom of chlorine is (on average) 35.5 times heavier than one atom of hydrogen.

Some commonly used relative atomic masses are shown in Figure 7.32.

Figure 7.32

Element	Relative atomic mass	Element	Relative atomic mass
hydrogen, H	1	sodium, Na	23
chlorine, Cl	35.5	magnesium, Mg	24
carbon, C	12	calcium, Ca	40
oxygen, O	16	aluminium, Al	27
nitrogen, N	14	lead, Pb	207
sulphur, S	32	iron, Fe	56

Did you know?

Relative atomic masses used to be calculated by comparing the mass of an atom to the mass of a hydrogen atom and calling the relative atomic mass of hydrogen 1. For practical reasons carbon is now used as the basis for calculating relative atomic masses. The mass of an atom of an element is compared to the mass of the ^{12}C isotope of carbon on a scale on which the ^{12}C isotope has a mass of 12 atomic mass units.

Calculating relative molecular masses and relative formula masses

Most substances exist either as molecules or ions so it is important to be able to compare the masses of these particles with those of other substances. The **relative formula mass (M_r)** of a simple one element ion (e.g. the Cl^- ion) is the same as the relative formula mass of the atom from which it was formed because the mass of an electron is so small. The relative mass of the Cl^- ion is therefore 35.5.

Where substances consist of molecules, the relative mass of the molecule is obtained by adding together the relative masses of all the atoms within the molecule. For example, water is H_2O, indicating that it consists of two atoms of hydrogen (H = 1) and one atom of oxygen (O = 16).

The relative molecular (or formula) mass of water is $(2 \times 1) + 16 = 18$.

The relative molecular mass of sulphuric acid (H_2SO_4) is $[(2 \times 1) + (32) + (4 \times 16)] = 98$.

Calculating the percentage of an element in a compound

To do this you must first calculate the relative formula mass of the compound and then calculate the percentage of the element.

Example: What is the percentage of carbon (C) in sodium carbonate (Na_2CO_3)?

Na = 23, C = 12, O = 16

The relative formula mass of Na_2CO_3 is $46 + 12 + 48 = 106$.

Carbon makes up just 12 parts of this so:

the percentage of carbon in Na_2CO_3 is $\frac{12}{16} \times 100 = $ **11.43%**

The mole

It is rarely necessary to know the exact number of particles being weighed out. What is needed is the relative number being used. This is important when making new compounds from existing ones so that when the substances are mixed they are all used up and there is no unreacted material left.

Just as shops sell eggs in fractions or multiples of a dozen, so the chemist measures out chemicals in fractions or multiples of a **mole**. The mole is the number of atoms in 1 g of hydrogen (H = 1). It is also the number of atoms in 12 g of carbon (C = 12) or 23 g of sodium (Na = 23) etc.

If the relative mass of a substance is known, then 1 mole of that substance is its relative mass in grams. For example, the relative mass of water is 18 (see earlier) so 1 mole of water is 18 g.

The relative mass of sulphuric acid is 98 so 1 mole of it has a mass of 98 g.

Did you know?

The number of atoms or molecules in a mole is 6×10^{23}. That is 600 000 000 000 000 000 000 000 or six hundred thousand million, million, million!

If hydrogen atoms were laid side by side in a row, there would be 300 million of them in 1 cm!

If one mole of hydrogen atoms were laid side by side, the line would be 20 000 000 000 km long. It would take a ray of light about 18 hours to travel that distance!

Calculating reacting masses

We can use the previous information to calculate the amounts of materials used in chemical reactions.

Example 1: How much copper oxide (CuO) can be obtained by heating 12.4 g of copper carbonate ($CuCO_3$) until there is no further change?

Cu = 64, C = 12, O = 16

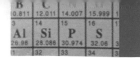

Getting together

Summary

◆ During a chemical reaction matter is neither created nor destroyed.

◆ The **relative atomic mass** of an element is the mass of one atom of the element compared to the mass of one atom of hydrogen.

◆ The **relative molecular mass/relative formula mass** of a substance is found by adding together the relative atomic masses of all the atoms in one molecule of it.

◆ It is possible to calculate the relative formula mass of a compound when the formula is provided.

◆ It is possible to calculate the percentage of an element in a compound when the formula is provided.

◆ It is possible to calculate masses/volumes of reactants or products if the balanced symbol equations or data regarding the volumes or masses of some of the reactants or products are provided.

◆ It is possible to calculate the ratios of elements in a compound and hence the formula if the masses or percentage composition is provided.

Topic questions

Use the following relative atomic masses in any calculations that follow.

H = 1, C = 12, O = 16, N = 14, Cl = 35.5, Na = 23, Mg = 24, S = 32, Pb = 207, Ag = 108, K = 39

1 mole of gas occupies 24 dm³ under room conditions

1 Calculate the relative formula masses of each of the following substances:

$$SO_2 \qquad MgSO_4 \qquad (NH_4)_2SO_4 \qquad NaNO_3$$

2 Which of the following ammonium compounds contains the greater percentage of nitrogen (N)?

Ammonium nitrate, NH_4NO_3, or ammonium carbonate, $(NH_4)_2CO_3$

3 Calculate the mass of each of the following:
   a) 1 mole of nitrogen dioxide (NO_2)
   b) 3 moles of magnesium carbonate ($MgCO_3$)
   c) 0.2 mole of lead nitrate ($Pb(NO_3)_2$)

4 How many moles are present in each of the following?
   a) 24 g of magnesium sulphate ($MgSO_4$)
   b) 2.8 g of potassium hydroxide (KOH)
   c) 48 dm³ of carbon dioxide (under room conditions)

5 When sodium chloride solution (NaCl(aq)) is added to silver nitrate solution ($AgNO_3$(aq)) a white precipitate of silver chloride (AgCl(s)) is formed according to the following equation:

$$NaCl(aq) + AgNO_3(aq) \rightarrow AgCl(s) + NaNO_3(aq)$$

What mass of silver chloride could be obtained from a solution that contained 3.4 g of silver nitrate?

6 What volume of carbon dioxide gas (under room conditions) could be obtained from the action of an excess of dilute sulphuric acid (H_2SO_4(aq)) on 16.8 g of sodium hydrogencarbonate ($NaHCO_3$)?

$$2NaHCO_3(s) + H_2SO_4(aq) \rightarrow Na_2SO_4(aq) + H_2O(l) + 2CO_2(g)$$

Examination questions

1 The diagrams below represent three atoms, **A**, **B** and **C**.

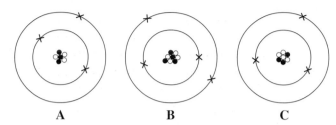

| A | B | C |

a) Two of the atoms are from the **same** element.
   i)   Which of **A**, **B** and **C** is an atom of a different element? *(1 mark)*
   ii) Give one reason for your answer. *(1 mark)*
b) Two of these atoms are isotopes of the same element.
   i)   Which **two** are isotopes of the same element?
   ii) Explain your answer. *(3 marks)*

2 This question is about elements and atoms.
a) About how many different elements are there?

   40   60   80   100   200 *(1 mark)*

b) The following are parts of an atom:

   electron   neutron   nucleus   proton

   Choose from the list the one which:
   i)   has no electrical charge;
   ii) contains two of the other particles;
   iii) has very little (negligible) mass. *(3 marks)*
c) Scientists have been able to make new elements in nuclear reactors. One of these new elements is fermium. An atom of fermium is represented by the symbol below.

$$^{257}_{100}\text{Fm}$$

   i)   How many protons does this atom contain?
   ii) How many neutrons does this atom contain? *(2 marks)*

3 a) Atoms are made of sub-atomic particles. Complete the **six** spaces in the table.

Name of sub-atomic particle	Relative mass	Relative charge
	negligible	
Neutron		
	1	

(3 marks)

b) Complete the spaces in the sentences.
   i)   The atomic number of an atom is the number of _____ in its nucleus and is equal to the number of _____ if the atom is not charged. *(1 mark)*
   ii) The mass number of an atom is the total number of _____ and _____ in its nucleus. *(1 mark)*
c) The table gives information about the atoms of three elements.

Name of element	Chemical symbol	Number of electrons in: 1st shell	2nd shell	3rd shell
Fluorine	F	2	7	0
Neon	Ne	2	8	0
Sodium	Na	2	8	1

Two of these elements can react together to form a chemical compound.
   i)   What is the name and the formula of this compound? *(2 marks)*
   ii) What type of bonding holds this compound together? *(1 mark)*
   iii) Explain, in terms of electron transfer, how the bonding occurs in this compound. *(2 marks)*

4 a) i)   Ammonium nitrate is one type of artificial fertiliser. Calculate the relative formula mass of ammonium nitrate NH_4NO_3. (Relative atomic masses: H = 1, N = 14, O = 16.) *(1 mark)*
   ii) Use your answer to part (a)(i) to help you calculate the percentage by mass of nitrogen present in ammonium nitrate NH_4NO_3. *(2 marks)*

b) One compound of vanadium is vanadium oxide.
   A sample of vanadium oxide contained 10.2 g of vanadium and 8.0 g of oxygen.
   Calculate the formula of this vanadium oxide.
   You must show **all** your working to gain full marks.
   (Relative atomic masses: V = 51, O = 16). *(3 marks)*

151

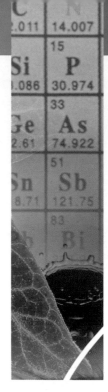

Chapter 8
Representing reactions

8.1	
Co-ordinated	Modular
DA 11.7	DA 7/8
SA Introduction	SA 15/16

Representing chemical symbols, formulae and reactions

Elements

All matter is made up from a limited number of simple substances called **elements**. There are 94 naturally-occurring elements of which 11 are gases, two are liquids and 81 are solids at room temperature.

Substances may contain one element or several different elements. Hydrogen and oxygen are both elements. Water is made up of the two elements hydrogen and oxygen.

The elements may just be mixed together or they can be chemically combined (joined together). Air contains a **mixture** of gases that are not joined together. In water, the two elements are chemically joined together.

The table below shows some of the elements in some common substances.

Figure 8.1
The elements in some common substances

Substance	Elements	Type of substance
aluminium	aluminium	element
iron	iron	element
air	nitrogen, oxygen and argon	mixture
brass	copper and zinc	mixture
water	hydrogen and oxygen	compound
salt	sodium and chlorine	compound
sugar	carbon, hydrogen and oxygen	compound
methane	carbon and hydrogen	compound

Elements are substances made from atoms which contain the same number of protons. Each element has its own individual set of chemical properties that make it react in certain ways. Each element has a fixed position in the **periodic table** (see section 11.2).

Did you know?

The most common element in the Universe is hydrogen, which makes up 90% of all known matter. The most common element in the Earth's crust is oxygen, which makes up 46.6% of the crust by weight.

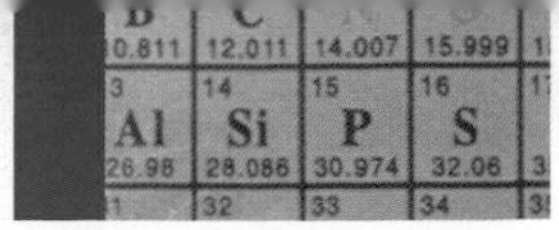

Symbols for elements

Each element is given an abbreviation of one or two letters called a symbol. For example, the symbol for hydrogen is H and the symbol for calcium is Ca.

Some elements have symbols based on Latin names. For example, the symbol for gold is Au. This symbol is derived from the Latin word *aurum*. The symbol for iron is Fe, derived from the Latin word *ferrum*.

Did you know?

Plumbers have their name derived from the Latin word plumbum, which means lead. This is because water pipes used to be made from lead and so the men that fitted and mended them were called plumbers.

The following table gives the names and symbols of some common elements.

Figure 8.2
Some of the common elements and their symbols

Element	Symbol	Element	Symbol
aluminium	Al	iron	Fe
argon	Ar	lead	Pb
barium	Ba	lithium	Li
beryllium	Be	magnesium	Mg
boron	B	manganese	Mn
bromine	Br	neon	Ne
calcium	Ca	nitrogen	N
carbon	C	oxygen	O
chlorine	Cl	phosphorus	P
copper	Cu	potassium	K
fluorine	F	silicon	Si
gold	Au	silver	Ag
helium	He	sodium	Na
hydrogen	H	sulphur	S
iodine	I	zinc	Zn

Did you know?

Every name tells a story. The element helium was discovered in the Sun before it was discovered on the Earth. It was named after the Greek word for the Sun, helios. Copper is named after the Latin word *cuprum*, which in turn was derived from the old name for Cyprus.

Compounds

Compounds are substances which contain two or more elements chemically joined together. Sodium chloride contains the elements sodium and chlorine chemically combined together. Carbon dioxide contains the elements carbon and oxygen chemically combined together.

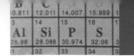

Representing reactions

Most elements are found as compounds in their natural state. Iron in iron ore is combined with oxygen to give iron(III) oxide, aluminium in bauxite is combined with oxygen to give aluminium oxide. Only a few elements are found uncombined in their natural state. For example gold, copper and sulphur are found in the ground as gold, copper and sulphur.

Naming compounds

The names of compounds are usually based on the elements that they are made from. The names of some elements change slightly when they are combined with other elements to form compounds. Usually the combined element has the ending *-ide*. The table below shows some of these names.

Name of element when on its own		Name when joined with another element
chlorine	→	chloride
bromine	→	bromide
iodine	→	iodide
nitrogen	→	nitride
oxygen	→	oxide
sulphur	→	sulphide

Rather than write the full name of a compound each time it is used, it is easier to use an abbreviated version called a formula. The formula of a substance shows:

- which elements have been combined together
- the ratio of the atoms of each of the elements that form it.

For example:

$MgCl_2$ represents 1 atom of magnesium joined with 2 atoms of chlorine
Na_2SO_4 represents 2 atoms of sodium, 1 atom of sulphur and 4 atoms of oxygen joined together.

Some atoms are found grouped together in many compounds. These groups are called radicals and are often formed from acids. For example, SO_4 represents the sulphate radical. It comes from sulphuric acid, H_2SO_4.

Writing the formula of a compound

There are many different compounds, and rather than learn each different formula it is possible to give each element or radical a valency or combining power which can then be used to write the correct formula. The combining power depends on the structure of the atoms making up the compound (see section 7.1).

Some elements can have more than one combining power. In these cases the roman numeral I, II or III in the brackets after their name shows which combining power the element is using in the compound.

Representing chemical symbols, formulae and reactions

Figure 8.3a
Valency or combining power of elements

Combining power 1	Symbol	Combining power 2	Symbol	Combining power 3	Symbol
bromine	Br	barium	Ba	aluminium	Al
chlorine	Cl	calcium	Ca	iron(III)	Fe
hydrogen	H	copper	Cu	nitrogen	N
iodine	I	iron(II)	Fe		
lithium	Li	lead	Pb		
potassium	K	magnesium	Mg		
silver	Ag	oxygen	O		
sodium	Na	sulphur	S		
		zinc	Zn		

Figure 8.3b
Valency or combining power of radicals

Combining power 1	Symbol	Combining power 2	Symbol	Combining power 3	Symbol
ammonium	NH_4	carbonate	CO_3	phosphate	PO_4
hydrogencarbonate	HCO_3	sulphate	SO_4		
hydrogensulphate	HSO_4				
hydroxide	OH				
nitrate	NO_3				
nitrite	NO_2				

The valency or combining power of the element or radical can be used to work out the formula of the substance using the following steps.

Step 1: From the name of the substance write down the symbols of the elements or radicals.
Step 2: Write the valency number above each symbol.
Step 3: Swap the valencies and write them below the other symbol.
Step 4: If the numbers are the same, then just write the symbols.
Step 5: Leave out any that are 1 and put brackets around the radicals if there are two or more.

Examples

Name of compound	Symbols of elements or radicals	Valencies	Formula of compound
Aluminium oxide	Al O	$Al^3 \, O^2$	Al_2O_3
Zinc sulphate	Zn SO_4	$Zn^2 \, SO_4{}^2$	$ZnSO_4$
Sodium sulphide	Na S	$Na^1 \, S^2$	Na_2S
Potassium carbonate	K CO_3	$K^1 \, CO_3{}^2$	K_2CO_3
Iron (III) sulphate	Fe SO_4	$Fe^3 \, SO_4{}^2$	$Fe_2(SO_4)_3$
Ammonium sulphate	NH_4 SO_4	$NH_4{}^1 \, SO_4{}^2$	$(NH_4)_2SO_4$

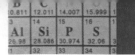

The use of prefixes in naming compounds

Some compounds contain the same elements combined in different positions. The number of atoms of each element can be shown using a prefix.

Figure 8.4
Prefixes used for numbers

Number	Prefix
1	mon
2	di
3	tri
4	tetra

The following compounds have prefixes in their names.

carbon monoxide	CO
carbon dioxide	CO_2
sulphur dioxide	SO_2
sulphur trioxide	SO_3
dinitrogen tetroxide	N_2O_4

Special cases

The correct formula of an element or compound can usually be worked out from its name but there are some special cases where this is not possible.

1 Gases

Most gaseous elements, except the noble gases, form molecules (groups of atoms) rather than remaining as atoms (see section 7.2). Their formulae must show this. For example:

Element	Formula of molecule
hydrogen	H_2
nitrogen	N_2
oxygen	O_2
chlorine	Cl_2
bromine	Br_2
ozone	O_3

The noble gases only have one atom:

Name of element	Formula
helium	He
argon	Ar

2 Acids

Name of acid	Formula
hydrochloric acid	HCl
sulphuric acid	H_2SO_4
nitric acid	HNO_3

3 Other common formulae

water	H_2O
ammonia	NH_3
methane	CH_4

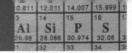

Representing chemical reactions

Many chemicals are harmful in some way or another. The diagrams below show some of the hazard symbols used to identify harmful chemicals.

Figure 8.5

Oxidising
These substances provide oxygen which allows other materials to burn more fiercely.

Highly flammable
These substances easily catch fire.

Toxic
These substances can cause death. They may have their effects when swallowed or breathed in or absorbed through the skin.

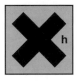

Harmful
These substances are similar to toxic substances but less dangerous.

Corrosive
These substances attack and destroy living tissues, including eyes and skin.

Irritant
These substances are not corrosive but can cause reddening or blistering of the skin.

Chemical change

A chemical change is called a chemical reaction. In a reaction, the substances which react together are called **reactants** and the new substances formed are called **products**. Magnesium reacts with hydrochloric acid to produce hydrogen and magnesium chloride. The reactants are magnesium and hydrochloric acid. The products are hydrogen and magnesium chloride. A chemical change is one in which the products have different chemical properties from the reactants.

The following observations could be an indication that a chemical change is taking place:

- bubbles of gas are formed
- a colour change is seen
- the substances get hotter or colder
- the original substance disappears
- a new substance appears
- the substances burn or glow
- a **precipitate** (an insoluble substance in suspension) forms.

Chemical equations

Chemical equations are a way of describing exactly what happens in a reaction. A word equation only includes the names of the reactants and products. A chemical equation is an abbreviated version of a word equation where the names of the reactants and products are replaced by their formulae.

Chemical equations must balance – there must be the same number of atoms of each element on each side of the equation. An equation does not represent a chemical reaction unless it is completely balanced.

The examples below illustrate the stages involved in writing chemical equations.

Reaction 1: Magnesium ribbon burns in air with a brilliant light to form magnesium oxide.

The magnesium reacts with the oxygen in the air so the equation will only include magnesium and oxygen.

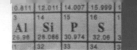

Representing reactions

Word equation: magnesium + oxygen → magnesium oxide

The word equation is first re-written as an unbalanced formula equation

$$Mg + O_2 \rightarrow MgO$$

Note: O_2 is used because oxygen is a molecule made up of two oxygen atoms.

In terms of the atoms: 1'Mg' + 2 'O' → 1 'Mg' + 1 'O'

The numbers of oxygen atoms are not equal on each side. The equation must be balanced by putting numbers in front of each formula. The formulae for each reactant or product must never be changed. It may be helpful to put boxes in front of each formula and then only put numbers in these boxes. So in order to make the number of atoms for each element the same on each side,

$$\boxed{}Mg + \boxed{}O_2 \rightarrow \boxed{}MgO$$
$$\boxed{2}Mg + \boxed{1}O_2 \rightarrow \boxed{2}MgO$$
$$2Mg + O_2 \rightarrow 2MgO$$

In terms of the atoms there are: 2 'Mg' + 2 'O' → 2 'Mg' + 2'O' so the equation is balanced.

Reaction 2: Sulphur burns in oxygen with a blue flame to give sulphur dioxide.

Word equation: sulphur + oxygen → sulphur dioxide

Formula equation: $S + O_2 \rightarrow SO_2$

In terms of the atoms: 1 'S' + 2 'O' → 1 'S' and 2'O'

The equation is already balanced and so the balanced chemical equation is:

$$S + O_2 \rightarrow SO_2$$

Reaction 3: Zinc reacts with hydrochloric acid to give zinc chloride and hydrogen.

Word equation: zinc + hydrochloric acid → zinc chloride + hydrogen

Formula equation: $Zn + HCl \rightarrow ZnCl_2 + H_2$

In terms of the atoms: 1'Zn' + 1 'H' + 1 'Cl' → 1 'Zn' + 2 'Cl' + 2 'H'

The equation is not balanced. The numbers of chlorine and hydrogen atoms are not the same on each side.

$$\boxed{}Zn + \boxed{}HCl \rightarrow \boxed{}ZnCl_2 + \boxed{}H_2$$
$$\boxed{1}Zn + \boxed{2}HCl \rightarrow \boxed{1}ZnCl_2 + \boxed{1}H_2$$
$$Zn + 2HCl \rightarrow ZnCl_2 + H_2$$

In terms of the atoms: 'Zn' + 2 'H' + 2 'Cl' → 'Zn' + 2 'H' + 2 'Cl'

The balanced chemical equation is therefore:

$$Zn + 2HCl \rightarrow ZnCl_2 + H_2$$

Note: 2HCl means that there are two hydrogen atoms *and* two chlorine atoms. The number in front of the formula applies to all the atoms in a compound not just the first element.

Use of state symbols in equations

The states of the reactants and products can be shown in brackets after each formula:

(s) means solid
(l) means liquid
(g) means gas
(aq) means a solution in water (an aqueous solution)

Examples of the use of state symbols are:

$$2Mg(s) + O_2(g) \rightarrow 2MgO(s)$$

$$Zn(s) + 2HCl(aq) \rightarrow ZnCl_2(aq) + H_2(g)$$

Ionic equations

Special equations, called ionic equations, are used to show the **ions** taking part in a reaction (see section 7.2). These are often reactions involving **precipitation**, **neutralisation** and electrolysis.

Reaction 1: When a solution of silver nitrate is added to a solution of magnesium chloride a white precipitate of silver chloride is formed in a solution of magnesium nitrate.

The chemical equation for the reaction is:

$$2AgNO_3(aq) + MgCl_2(aq) \rightarrow 2AgCl(s) + Mg(NO_3)_2(aq)$$

The equation can be re-written in terms of the ions present and their states as:

$$2Ag^+(aq) + 2NO_3^-(aq) + Mg^{2+}(aq) + 2Cl^-(aq) \rightarrow$$
$$2AgCl(s) + Mg^{2+}(aq) + 2NO_3^-(aq)$$

Some of the ions, Mg^{2+} and NO_3^-, appear on both sides of the equation and are not changed during the course of the reaction. These ions are called spectator ions and can be left out of an ionic equation. So the ions remaining are:

$$2Ag^+(aq) + 2Cl^-(aq) \rightarrow 2AgCl(s)$$

which simplifies to:

$$Ag^+(aq) + Cl^-(aq) \rightarrow AgCl(s)$$

An ionic equation must balance in terms of the charge on each side as well as the number of atoms/ions of each element on each side. The net charge on each side must be checked.

Reaction 2: Iron filings will react with copper sulphate solution to give solid copper in a solution of iron(II) sulphate.

The equation for the reaction is:

$$Fe(s) + CuSO_4(aq) \rightarrow Cu(s) + FeSO_4(aq)$$

The equation can be re-written in terms of the ions present and their states as:

$$Fe(s) + Cu^{2+}(aq) + SO_4^{2-}(aq) \rightarrow Cu(s) + Fe^{2+}(aq) + SO_4^{2-}(aq)$$

The spectator ion, SO_4^{2-}, can be left out of the ionic equation. So the ions remaining are:

$$Fe(s) + Cu^{2+}(aq) \rightarrow Cu(s) + Fe^{2+}(aq)$$

The ionic equation is balanced for both charge and atoms/ions of each element.

Topic questions

1 Work out the number of atoms of each element present from the formulae of the following compounds:

   NaCl   CaO   MgBr$_2$ NaNO$_3$ ZnSO$_4$ Al$_2$O$_3$ NH$_4$OH   Fe$_2$(SO$_4$)$_3$

2 Name each of the following compounds:

a) KCl b) $ZnCl_2$ c) MgO d) KNO_3
e) $CuSO_4$ f) $Pb(NO_3)_2$ g) $AgNO_3$
h) Na_2SO_4 i) CH_4 j) CO_2

3 Write the formula of the following compounds:

a) silver chloride   b) calcium sulphate
c) barium chloride   d) sodium carbonate
e) copper nitrate   f) aluminium nitrate
g) iron(III) oxide   h) magnesium nitride
i) calcium phosphate   j) ammonium sulphate

4 Balance the following equations Remember that **the formulae must not be altered**.

a) $Zn + O_2 \rightarrow ZnO$
b) $Mg + HCl \rightarrow MgCl_2 + H_2$
c) $KOH + H_2SO_4 \rightarrow K_2SO_4 + H_2O$
d) $Na_2CO_3 + HNO_3 \rightarrow NaNO_3 + H_2O + CO_2$
e) $FeSO_4 + NaOH \rightarrow Fe(OH)_2 + Na_2SO_4$

5 Write balanced chemical equations for the following word equations:

a) copper + sulphur → copper sulphide
b) carbon + oxygen → carbon monoxide
c) magnesium + nitric acid → magnesium nitrate + hydrogen
d) calcium carbonate + hydrochloric acid →
                 calcium chloride + water + carbon dioxide
e) sodium hydroxide + sulphuric acid → sodium sulphate + water

6 Write balanced chemical equations for the following word equations:

a) iron + oxygen → iron(III) oxide
b) ammonia + sulphuric acid → ammonium sulphate
c) aluminium sulphate + sodium hydroxide →
                 aluminium hydroxide + sodium sulphate
d) magnesium + nitrogen → magnesium nitride

7 Rewrite the following equations showing all the ions present and then write them as ionic equations without the spectator ions. Assume that the solids do not form ions. Show all state symbols.

a) $NaCl(aq) + AgNO_3(aq) \rightarrow AgCl(s) + NaNO_3(aq)$
b) $Mg(s) + CuSO_4(aq) \rightarrow Cu(s) + MgSO_4(aq)$
c) $ZnCl_2(aq) + 2AgNO_3(aq) \rightarrow 2AgCl(s) + Zn(NO_3)_2(aq)$
d) $Na_2SO_4(aq) + BaCl_2(aq) \rightarrow BaSO_4(s) + 2NaCl(aq)$
e) $Cu(NO_3)_2(aq) + 2NaOH(aq) \rightarrow Cu(OH)_2(s) + 2NaNO_3(aq)$

Summary

◆ All matter is made up from a limited range of simple substances called **elements**.

◆ Elements are substances made from atoms that contain the same number of protons.

◆ A **compound** is a substance that contains two or more elements chemically joined together.

◆ The formula of a substance uses chemical symbols to show which elements have been combined together in a compound.

◆ A **mixture** contains two or more different substances that are usually easy to separate.

◆ A chemical change is one in which the products have different chemical properties from the reactants.

◆ A chemical change is called a **reaction**.

◆ In a reaction the substances which react together are called **reactants** and the new substances formed are called **products**.

- Chemical equations are a way of describing exactly what happens in a reaction.

- A word equation only includes the names of the reactants and products and no description of the reaction itself.

- A chemical equation is an abbreviated version of a word equation in which the words are replaced by formulae.

- **Ionic equations** are used to show the ions taking part in a reaction.

8.2	
Co-ordinated	Modular
DA Introduction	DA 7/8
SA Introduction	SA 15/16

Types of chemical reactions

Some common general reactions are given special names to describe the processes taking place.

Thermal decomposition reactions

This happens when a substance is split up into other substances by heat.

Reaction 1: When copper carbonate is heated it thermally decomposes to give copper oxide and carbon dioxide gas. The reaction can be represented as:

word equation: copper(II) carbonate → copper(II) oxide + carbon dioxide

chemical equation: $CuCO_3(s) \rightarrow CuO(s) + CO_2(g)$

Reaction 2: When limestone (calcium carbonate) is heated strongly it thermally decomposes to give calcium oxide and carbon dioxide. The reaction can be represented as:

word equation: calcium carbonate → calcium oxide + carbon dioxide

chemical equation: $CaCO_3(s) \rightarrow CaO(s) + CO_2(g)$

Thermal decomposition is an **endothermic reaction** (see sections 8.4 and 12.1).

Neutralisation reactions

Neutralisation is a reaction between an acid and a base (see section 11.6). A base is a metal oxide or hydroxide. If a base is soluble in water, it is called an alkali. Usually a base is added to an acid in order to neutralise it. Universal indicator can be used to show when all the acid has been neutralised. At that point the pH will equal 7 and the universal indicator will be green.

Reaction 1: When sodium hydroxide solution is added to hydrochloric acid it will neutralise it to form sodium chloride and water. The reaction can be represented as:

word equation: sodium hydroxide + hydrochloric acid → sodium chloride + water

chemical equation: $NaOH(aq) + HCl(aq) \rightarrow NaCl(aq) + H_2O(l)$

Reaction 2: When calcium hydroxide solution is added to nitric acid it will neutralise it to form calcium nitrate and water. The reaction can be represented as:

word equation: calcium hydroxide + nitric acid → calcium nitrate + water

chemical equation: $Ca(OH)_2(aq) + 2HNO_3(aq) \rightarrow Ca(NO_3)_2(aq) + 2H_2O(l)$

Did you know?

Some substances are so thermally unstable that they will decompose explosively at room temperature. Stores of ammonium nitrate have been known to suddenly explode leaving behind a large crater where the buildings were.

Did you know?

Acids and alkalis are used as weapons in the insect world. A bee stings by injecting an acid and an ant 'bite' is caused by a squirt of acid on the skin. Bee stings and ant bites can be treated with a mild alkali to neutralise the acid. A wasp sting is alkaline and can be treated with a mild acid to neutralise the alkali.

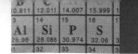

Representing reactions

Displacement reactions

Displacement reactions happen when one element in a substance is replaced or 'pushed out' by another element. Hydrogen can be displaced from acids by metals and a metal can be displaced from a salt by a more reactive metal.

Reaction 1: Magnesium will react with sulphuric acid to form magnesium sulphate and hydrogen gas. The hydrogen in the acid is displaced by magnesium to form a salt called magnesium sulphate.

This displacement reaction can be represented as:

word equation: magnesium + sulphuric acid → magnesium sulphate + hydrogen

chemical equation: $Mg(s) + H_2SO_4(aq) \rightarrow MgSO_4(aq) + H_2(g)$

Reaction 2: Zinc will react with copper(II) sulphate in solution to form zinc sulphate and copper. The copper in copper(II) sulphate has been displaced by the more reactive zinc to form the salt zinc sulphate. This displacement reaction can be represented as:

word equation: zinc + copper(II) sulphate → copper + zinc sulphate

chemical equation: $Zn(s) + CuSO_4(aq) \rightarrow Cu(s) + ZnSO_4(aq)$

The ionic equation is: $Zn(s) + Cu^{2+}(aq) \rightarrow Cu(s) + Zn^{2+}(aq)$

Precipitation

Precipitation is a reaction between two soluble substances in solution that results in an insoluble product. The insoluble product appears as a suspension or precipitate.

In the reaction between solutions of silver nitrate and sodium chloride, silver chloride and sodium nitrate are produced. The silver chloride is insoluble and will be seen as a white precipitate. The sodium nitrate is a colourless solution. The reaction can be represented as:

word equation: silver nitrate + sodium chloride → silver chloride + sodium nitrate

chemical equation: $AgNO_3(aq) + NaCl(aq) \rightarrow AgCl(s) + NaNO_3(aq)$

The ionic equation is: $Ag^+(aq) + Cl^-(aq) \rightarrow AgCl(s)$

Oxidation and reduction

Oxidation occurs when a substance joins with oxygen or *loses electrons*. **Reduction** occurs when a substance loses oxygen or *gains electrons*.

Oxidation

A substance is oxidised if oxygen is added to it or it has gained more oxygen.

Reaction: When magnesium burns in air it forms magnesium oxide. The oxygen in the air has combined with the magnesium. The oxygen has oxidised the magnesium to magnesium oxide. The reaction can be represented as:

word equation: magnesium + oxygen → magnesium oxide

chemical equation: $2Mg(s) + O_2(g) \rightarrow 2MgO(s)$

Reduction

A substance, usually an oxide, is reduced when oxygen is taken away from it.

Reaction: When hydrogen is passed over heated copper(II) oxide, copper and water are formed. The copper(II) oxide has been reduced to copper by losing its oxygen.

The reaction can be represented as:

word equation: copper(II) oxide + hydrogen → copper + water

chemical equation: $CuO(s) + H_2(g) \rightarrow Cu(s) + H_2O(l)$

Oxidation and reduction will always occur together. When one substance is being oxidised another substance is being reduced. In the reaction between copper oxide and hydrogen

- the copper oxide has lost oxygen so has been reduced
- the hydrogen has gained oxygen so has been oxidised
- the hydrogen causes the reduction so is called a reducing agent
- the copper oxide provides the oxygen so is called the oxidising agent.

Reduction and oxidation always occur together in reactions and so these are often called redox reactions from REDuction and OXidation.

The gaining or losing of electrons

When copper metal is oxidised, the copper metal becomes copper ions. For this to happen, the copper metal has to lose electrons to a suitable electron acceptor (in this case oxygen).

$$Cu \rightarrow Cu^{2+} + 2e^-$$

So a broader definition of oxidation is, a substance is oxidised if it loses electrons during a reaction. Transfer of electrons will result in atoms turning into ions, ions turning into atoms, or ions changing their charge.

When a sodium atom loses an electron it forms a positive sodium ion.

$$Na - e^- \rightarrow Na^+$$

Sodium has been oxidised because it has lost an electron.

When a chloride ion loses an electron it forms a chlorine atom.

$$Cl^- - e^- \rightarrow Cl$$

The chloride ion has been oxidised because it has lost an electron.

When copper oxide is reduced by hydrogen, the copper ion becomes copper metal. For this to happen, the copper ion has to gain electrons from a suitable electron donor (in this case hydrogen).

$$Cu^{2+} + 2e^- \rightarrow Cu$$

So a broader definition of reduction is, a substance is reduced if it gains electrons during a reaction.

When a bromine atom gains an electron it forms a negative bromide ion. Bromine has been reduced because it has gained one electron.

$$Br + e^- \rightarrow Br^-$$

When a copper 2+ ion gains two electrons, a copper atom is formed. The copper 2+ ion has been reduced by gaining two electrons.

$$Cu^{2+} + 2e^- \rightarrow Cu$$

Oxidation and reduction will always occur together. When one substance is being oxidised another substance is being reduced. The electrons will be transferred from the substance being oxidised to the substance being reduced.

The ionic version of the reaction of magnesium with oxygen is

$$2Mg(s) + O_2(g) \rightarrow 2Mg^{2+}2O^{2-}(s)$$

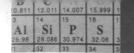

Representing reactions

Each magnesium atom loses two electrons to form a magnesium 2+ ion.

$$Mg - 2e^- \rightarrow Mg^{2+}$$

The magnesium atom has been oxidised to a magnesium ion.

Each oxygen atom gains two electrons to form a negative oxide ion.

$$O + 2e^- \rightarrow O^{2-}$$

Oxygen has been reduced because it has gained two electrons.

Overall, the electrons are transferred from the magnesium atom to the oxygen atom forming new ions.

Reduction and oxidation have taken place in the reaction between magnesium and oxygen. So this reaction is another example of a redox reaction.

Reversible reactions

In most reactions the reactants will change into products and the reaction will stop. In a **reversible reaction** the products can turn back into reactants. Usually a balance is set up between the reactants and products and both are present at the same time. Reversible reactions are also called equilibrium reactions.

When nitrogen and hydrogen react together in the Haber Process (see section 9.2) they form ammonia. The ammonia can then decompose to form nitrogen and hydrogen again.

$$\text{nitrogen} + \text{hydrogen} \rightleftharpoons \text{ammonia}$$
$$N_2(g) + 3H_2(g) \rightleftharpoons 2NH_3(g)$$

The 'double-headed' arrow, $\rightleftharpoons$, shows that the reaction is reversible.

Topic questions

1 For each of the reactions, state whether it is a:

thermal decomposition reaction, neutralisation reaction, displacement reaction, precipitation reaction, reversible reaction, oxidation reaction or reduction reaction.

Give a reason for your choice.

a) When magnesium is burned in oxygen magnesium oxide is formed.
b) Insoluble barium sulphate is produced when a solution of barium nitrate is added to a solution of sodium sulphate.
c) When copper carbonate is heated it splits up into copper oxide and carbon dioxide.
d) Dilute hydrochloric acid reacts with a solution of sodium hydroxide to produce sodium chloride and water.
e) The heating of mercury oxide produces mercury and oxygen.
f) If blue copper(II) sulphate crystals are heated water is produced and the crystals go white. If water is added to the white copper(II) sulphate heat is produced and the copper(II) sulphate goes blue.

g) An iron nail put into a solution of copper(II) sulphate becomes coated in a brown layer of copper.

2 For each of the equations below state whether it is an oxidation or a reduction reaction. Give a reason for your choice. (State symbols have been omitted.)

a) $C + O_2 \rightarrow CO_2$
b) $2Ag_2O \rightarrow 4Ag + O_2$
c) $Cu - 2e^- \rightarrow Cu^{2+}$
d) $Pb^{2+} + 2e^- \rightarrow Pb$
e) $Fe^{2+} - e^- \rightarrow Fe^{3+}$
f) $Br_2 + 2e^- \rightarrow 2Br^-$
g) $K - e^- \rightarrow K^+$

3 Identify the oxidation and reduction reactions in each of the following reactions. Give the reasons for your choices in each case. (State symbols have been omitted.)

a) $PbO + H_2 \rightarrow Pb + H_2O$
b) $CuO + CO \rightarrow Cu + CO_2$
c) $2FeO + C \rightarrow 2Fe + CO_2$

d) $Zn + Fe^{2+} \rightarrow Fe + Zn^{2+}$
e) $Mg + Cu^{2+} \rightarrow Cu + Mg^{2+}$

0.811	12.011	14.007	15.999	1
3	14	15	16	1
Al	Si	P	S	
26.96	28.086	30.974	32.06	3
1	32	33	34	3

Summary

◆ **Thermal decomposition** occurs when a substance is split up into other substances by heat.

◆ **Neutralisation** is a reaction between an acid and a base to form a neutral solution.

◆ **Displacement** occurs when one element in a substance is replaced or 'pushed out' by another element.

◆ **Precipitation** is a reaction between two solutions that results in an insoluble product.

◆ **Oxidation** is the gain of oxygen by a substance.

◆ A substance is oxidised if oxygen is added to it or it has gained more oxygen.

◆ A substance is oxidised if it loses electrons during a reaction.

◆ **Reduction** is the loss of oxygen from a substance.

◆ A substance, usually an oxide, is reduced when oxygen is taken away from it.

◆ A substance is reduced if it gains electrons during a reaction.

◆ In a **reversible reaction**, the products can turn back into reactants.

8.3

Co-ordinated	Modular
DA Introduction	DA 7/8
SA n/a	SA n/a

Exothermic and endothermic reactions

Exothermic reactions

Exothermic reactions *give out* heat. All combustion reactions are exothermic.

When methane is burnt in air, carbon dioxide and water are formed together with a large amount of heat. The reaction can be represented as:

word equation: methane + oxygen → carbon dioxide + water

chemical equation: $CH_4(g) + 2O_2(g) \rightarrow CO_2(g) + 2H_2O(l)$

Endothermic reactions

Endothermic reactions *take in* heat and get colder, or need heat to make them happen. When solutions of magnesium chloride and sodium carbonate are added together a reaction takes place to form magnesium carbonate and sodium chloride. As the products are formed the temperature of the mixture falls and is colder than that of either of the original solutions. The reaction can be represented as:

word equation: magnesium chloride + sodium carbonate → magnesium carbonate + sodium chloride

chemical equation: $MgCl_2(aq) + Na_2CO_3(aq) \rightarrow MgCO_3(aq) + 2NaCl(aq)$

There is more information about exothermic and endothermic reactions in section 12.1.

Summary

◆ **Exothermic reactions** are reactions which give out heat and get warmer.

◆ **Endothermic reactions** are reactions which take in heat and get colder.

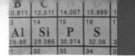

Representing reactions

Examination questions

1 Liquid ammonia is toxic.
Road tankers carrying liquid ammonia must display a hazard warning symbol.

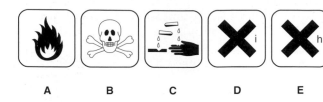

Which hazard warning symbol, A to E, should be displayed on the tanker? *(1 mark)*

2 a) A student studied the effect of temperature on the rate of reaction between hydrochloric acid and sodium thiosulphate.

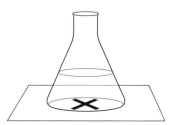

● The student mixed 50 cm^3 of a sodium thiosulphate solution and 5 cm^3 of hydrochloric acid in a flask.
● The flask was placed over a cross.
● The student timed how long after mixing the cross could no longer be seen.

a) i) Balance the chemical equation for this reaction.

$$Na_2S_2O_3(aq) + HCl(aq) \rightarrow$$
$$NaCl(aq) + H_2O(l) + SO_2(g) + S(s)$$

(1 mark)

ii) What causes the cross to be seen no longer? *(1 mark)*

b) The student then tried to make some magnesium sulphate. Excess magnesium was added to dilute sulphuric acid. During this reaction fizzing was observed due to the production of a gas.

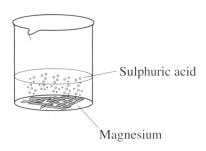

Complete and balance the chemical equation for this reaction.

_____ + H$_2$SO$_4$ →
_____ + _____

(3 marks)

3 Bordeaux Mixture controls some fungal infections on plants. A student wanted to make some Bordeaux Mixture.

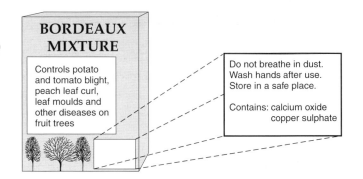

a) The student knew that calcium oxide could be made by heating limestone. Limestone contains calcium carbonate, CaCO$_3$.
i) Write the word equation for this reaction. *(1 mark)*
ii) What type of reaction is this? *(1 mark)*

b) The student knew that copper sulphate, CuSO$_4$, could be made by the following general reaction.

acid + base → salt + water

i) What type of reaction is this? *(1 mark)*
ii) The base used is copper oxide. Name and give the chemical formula of the acid used. *(2 marks)*

4 a) Iron is produced in a blast furnace by the following reaction.

$$Fe_2O_3 + 3CO \rightarrow 2Fe + 3CO_2$$

Both oxidation and reduction take place in this reaction. Explain how. *(2 marks)*
b) Hydrogen is used for the industrial production of ammonia. It is obtained from the reaction between methane and steam. The equation for this reaction is:

$$CH_4 + H_2O \rightarrow 3H_2 + CO$$

Explain how you can tell that this equation is balanced. *(2 marks)*

Chapter 9
The atmosphere

C N
.011 14.007
15
Si P
.086 30.974
33
Ge As
2.61 74.922
51
Sn Sb
8.71 121.75
83
Bi

Key terms	air • atmosphere • catalyst • denitrifying bacteria • endothermic • eutrophication • exothermic • fossilisation • global warming • greenhouse effect • Haber process • nitrifying bacteria • ozone layer • photosynthesis • respiration • reversible reaction • synthesis • ultraviolet radiation

9.1 Changes to the atmosphere

Co-ordinated	Modular
DA 11.9	DA 6
SA n/a	SA n/a

Air

Air is a mixture of gases which for the last 200 million years has had a fairly constant composition. The amount of water vapour in the air varies depending on where you are. In the middle of the Sahara Desert there is very little water vapour; in a tropical rainforest the air contains a lot of water vapour.

Figure 9.1
The composition of air

Gas	Approximate %
nitrogen	80
oxygen	20
small amounts of various other gases, including carbon dioxide, water vapour and noble gases (e.g. argon)	

Did you know?

Although there is only 1% of argon in the air, that is actually quite a large amount. In a normal-sized school laboratory (10 m × 10 m × 3 m) there is about enough argon to fill the passenger compartment of an average sized car (3 m^3).

Evolution of the atmosphere

The Earth's **atmosphere** has not always had the same composition. During the first billion years whilst the Earth was forming there was lots of volcanic activity. This produced an atmosphere that was mainly made up of carbon dioxide, together with some ammonia (NH_3), methane (CH_4) and water vapour. There was little or no oxygen gas, so the Earth's atmosphere would have been similar to the present atmospheres of Mars and Venus. As the Earth gradually cooled the water vapour condensed to form seas and oceans.

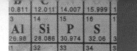

What happened to the carbon dioxide?

Over millions of years the high carbon dioxide content gradually decreased. There were a number of reasons for this.

- Carbon dioxide reacted with sea water to form carbonates (mainly calcium carbonate) which was used to make the shells of many sea creatures. When the creatures died, their shells formed sediments which eventually became rocks such as chalk and limestone ($CaCO_3$).

- Carbon dioxide was removed by green plants through the process of **photosynthesis** (see section 4.1). This process locks carbon dioxide up as glucose and releases oxygen.

- Plants and tiny animals (for example bacteria) became trapped in sediments and decayed. The effect of pressure on the decaying organisms produced fossil fuels, oil and natural gas. **Fossilisation** is a slow process. Most fossil fuels are formed from organisms fossilised about 300 000 000 years ago.

What happened to the ammonia?

As more and more oxygen was released by the process of photosynthesis, the ammonia reacted with the oxygen to produce nitrogen and water.

Some of the earliest forms of life on Earth were bacteria, among which were **nitrifying bacteria** that could convert ammonia into nitrates and **denitrifying bacteria** that could break down some of the nitrogen-containing compounds to form nitrogen gas. The green plants absorbed some of the nitrates from the ground to produce plant protein. The action of the bacteria and the plants reduced the ammonia content of the air and replaced it with nitrogen (see section 6.4).

Other changes to the atmosphere

The action of sunlight on the oxygen in the atmosphere turned some of the oxygen molecules produced by photosynthesis (O_2) into molecules of ozone (O_3). This ozone formed a layer, the **ozone layer** around the Earth at a height varying from about 19 to 48 km. This layer, which was formed many millions of years ago, protects life on Earth by reducing the full effects of the Sun's cancer-causing **ultraviolet radiation** (see section 15.4).

What is happening to the level of carbon dioxide now?

- Some of the carbon dioxide locked up as carbonate rocks is sometimes released back into the atmosphere through the eruption of volcanoes.

- **Respiration** releases the carbon dioxide locked up in carbohydrates.

  - Large amounts of carbon dioxide are released by the burning of fossil fuels. These amounts are so large that the level of carbon dioxide in the atmosphere, which for millions of years had been stable, is now increasing.

  - Because of the increase in atmospheric carbon dioxide there is a increase in the rate at which carbon dioxide reacts with sea water to form insoluble carbonates (mainly calcium carbonate). These carbonates continue to be deposited as sediments or they may form soluble hydrogencarbonates (mainly calcium and magnesium). However, even though there is an increase in the rate at which carbon dioxide is being removed from the atmosphere it is not rapid enough to absorb the additional carbon dioxide now being released into the atmosphere. Increased amounts of carbon dioxide in the atmosphere lead to an increase in the **greenhouse effect** and **global warming** (see section 6.2).

0.811	12.011	14.007	15.999	1
3	14	15	16	1
Al	**Si**	**P**	**S**	3
26.98	28.086	30.974	32.06	3
1	32	33	34	3

Summary

◆ For about 200 million years our atmosphere has been made up of about (80%) nitrogen, (20%) oxygen, together with small amounts of other gases such as carbon dioxide, water vapour and noble gases.

◆ The very early atmosphere is likely to have contained mainly carbon dioxide and water vapour with some methane and ammonia due to volcanic activity.

◆ Green plants removed some of the carbon dioxide during photosynthesis.

◆ Some of the carbon dioxide was locked up as rocks or fossil fuels.

◆ Nitrogen was produced as oxygen reacted with ammonia and the action of denitrifying bacteria.

◆ Some oxygen became ozone to form the ozone layer.

◆ Some carbon dioxide is being released by the action of volcanoes.

◆ Large quantities of carbon dioxide are released by the burning of fossil fuels.

◆ The continued formation of carbonates and hydrogencarbonates in the sea is too slow a process to lock up the extra carbon dioxide being released.

Topic questions

1 In the box are the names of the main gases found in the atmosphere.

> argon   carbon dioxide   nitrogen
> oxygen   water vapour

Use these words to answer the following questions. (You may use each word once.)
a) Which gas makes up most of the atmosphere?
b) Which gas is a noble gas?
c) Which gas is used in photosynthesis?
d) Which gas is made by photosynthesis?
e) Of which gas are there varying amounts depending on the local conditions?

2 In the box are the names of the main gases found in the Earth's early atmosphere.

> ammonia   carbon dioxide
> methane   water vapour

Use these words to answer the following questions. (You may use each word once, more than once or not at all.)
a) Which gas produced the oceans when it cooled down?
b) Which gas was converted into nitrogen?
c) Which gas enabled plants to make protein?
d) Which gas was used to make limestone rocks?

3 Describe why the amount of carbon dioxide in today's atmosphere is less than it was when the Earth was forming.

4 Why is the amount of carbon dioxide in the atmosphere now gradually increasing?

9.2
Co-ordinated	Modular
DA 11.6	DA 7
SA n/a	SA n/a

Useful products from the air

Ammonia

Ammonia is a compound of nitrogen and hydrogen with the formula NH_3. It is made commercially by the Haber process. In this process nitrogen from the air is reacted with hydrogen. The process is widely used to make nitrogenous fertilisers. In fact about 80% of the ammonia produced is used for this purpose. Ammonia is also used to make nitric acid.

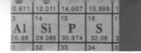

Figure 9.2
The uses of ammonia

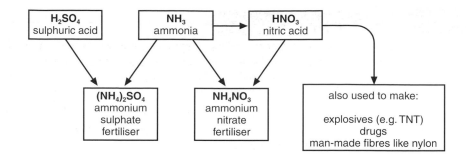

Using ammonia to make nitric acid

Millions of tonnes of nitric acid are made each year, most of which is used in the manufacture of the fertiliser ammonium nitrate. Most nitric acid is made from ammonia. Three stages are involved:

Stage 1   A mixture of air and ammonia is passed over a platinum **catalyst** (see section 12.3) which is heated to a temperature of about 900°C. The oxygen in the air oxidises (see section 8.3) the ammonia into nitrogen monoxide (NO) and water.

Stage 2   The nitrogen monoxide gas is cooled and reacted with more oxygen to form nitrogen dioxide (NO_2).

Stage 3   Further reactions between nitrogen dioxide, oxygen and water produce nitric acid (HNO_3).

The Haber process

In the Haber process ammonia is made by the direct combination of nitrogen and hydrogen using iron as a catalyst to speed up the reaction.

$$\text{nitrogen} + \text{hydrogen} \rightleftharpoons \text{ammonia}$$
$$N_2(g) + 3H_2(g) \rightleftharpoons 2NH_3(g)$$

The raw materials for this process (nitrogen and hydrogen) are inexpensive because they are readily available. Nitrogen is in the air and hydrogen is in the compounds water (H_2O) and methane (natural gas) (CH_4).

Figure 9.3

An industrial plant producing ammonia by the Haber process

The nitrogen can be obtained from the air by burning natural gas in air to use up the oxygen. It is not the only way of getting nitrogen out of the air but it is a fairly inexpensive method.

The hydrogen can be obtained from natural gas (methane) or water. Both substances are relatively inexpensive. The method most frequently used is to react methane with steam in the presence of a suitable catalyst.

$$\text{methane} \quad + \quad \text{steam} \quad \rightarrow \quad \text{carbon dioxide} \quad + \quad \text{hydrogen}$$
$$CH_4(g) \quad + \quad 2H_2O(g) \quad \rightarrow \quad CO_2(g) \quad + \quad 4H_2(g)$$

In the Haber process the nitrogen and hydrogen are passed over a catalyst which is heated to about 450°C, at a pressure of about 200 atmospheres. Some of the hydrogen and nitrogen react to form ammonia. The mixture of gases leaving the catalyst is cooled. The ammonia liquifies and is removed. The unreacted nitrogen and hydrogen mixture is recycled.

The reaction between nitrogen and hydrogen to produce ammonia is a **reversible reaction** (see sections 8.3 and 12.2). This means that some of the ammonia can break down again into hydrogen and nitrogen depending on the temperature and pressure of the reaction.

Increasing the rates of reaction in the Haber process

The chemical industry is geared up to make new substances in a cost effective way. The conditions for any industrial chemical process are often chosen for financial rather than chemical reasons. Several of the above factors are made use of in the manufacture of ammonia by the Haber process.

The chemical industry prefers to produce new materials by a continuous process and at minimal cost. Reaction conditions are carefully selected to achieve both of these aims.

The reaction in which nitrogen and hydrogen are combined together to form ammonia is an exothermic, reversible process as the equation below shows.

$$N_2(g) + 3H_2(g) \rightleftharpoons 2NH_3(g) \quad (+ \text{ heat transferred})$$

Industrial chemists make use of three of the effects already discussed in order to make this reaction as efficient as possible.

1 They alter the pressure of the reactants. (The reaction only involves gases.)

2 They select a suitable temperature for the reaction to take place.

3 They use a catalyst.

Changing the pressure

$$N_2(g) + 3H_2(g) \rightleftharpoons 2NH_3(g)$$

The reaction above shows that one molecule of nitrogen reacts with three molecules of hydrogen to make two molecules of ammonia. Since equal volumes of gases contain the same number of molecules (see section 7.3) it follows that one volume of nitrogen reacts with three volumes of hydrogen to make just two volumes of ammonia. This means that as the reaction proceeds there is a reduction in volume from four volumes (1 + 3) to just two volumes. Increasing the pressure means more ammonia is produced. Increased pressure also speeds up the reaction (see section 12.2). This is because compressing the gas increases its concentration. Figure 9.4 shows how the amount of ammonia produced changes as the pressure inside the reaction vessel changes.

In practice, a moderate pressure of between 150 and 300 atmospheres is chosen. Although high pressures give a higher yield, they are too expensive to maintain.

Figure 9.4
The effect of pressure on the amount of ammonia produced

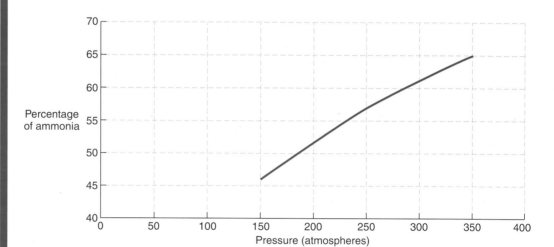

Changing the temperature

$$N_2(g) + 3H_2(g) \rightleftharpoons 2NH_3(g) \quad (+ \text{ heat transferred})$$

The equation shows that the forward reaction is **exothermic**, that is, heat is produced (see section 12.2). We could write the equation in a simplified way:

$$\text{nitrogen} + \text{hydrogen} \rightarrow \text{ammonia} \ (+ \text{ heat transferred})$$

If the reaction mixture is heated, it makes it harder for the reaction to give out heat. This means that less ammonia will be produced. So the cooler the reaction, the more ammonia is produced. The drawback here is that whilst the yield of ammonia will be quite high, the rate of the reaction at the low temperature will be very slow. Figure 9.5 shows how the yield of ammonia changes as the temperature changes.

Figure 9.5 shows that as the temperature is increased at different pressures, the amount of ammonia produced decreases.

Once again a compromise has to be made and temperatures between 400 and 500°C are usually used so that a reasonable yield of ammonia is produced reasonably quickly.

Using a catalyst

Some reactions can be speeded up by the use of a catalyst. The manufacture of ammonia is speeded up by the use of an iron catalyst.

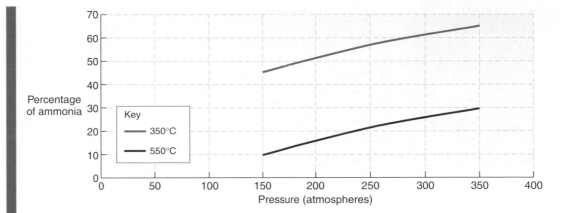

Figure 9.5
The effect of temperature and pressure on the amount of ammonia produced

Summary of the Haber process

In the **synthesis** of ammonia by the Haber process, nitrogen and hydrogen are combined together to form ammonia using a pressure of 150 to 300 atmospheres, a temperature between 400 and 500°C and an iron catalyst.

Nitrogenous fertilisers

Ammonia from the Haber process is used to produce the inexpensive nitrogenous (nitrogen-containing) fertilisers ammonium sulphate ($(NH_4)_2SO_4$) and ammonium nitrate (NH_4NO_3). Ammonium sulphate is made by the neutralisation of ammonia with sulphuric acid and ammonium nitrate is made by the neutralisation of ammonia with nitric acid.

Because these fertilisers are quite inexpensive there is a risk that they will not be used with sufficient care. If too much fertiliser is added to the land the plants cannot use all of it and some gets washed through the soil into rivers. This can cause the following environmental problems.

1 Eutrophication of streams, rivers and lakes (see section 6.2).

- Nitrogenous fertilisers encourage plant growth in water.
- Small plant-like organisms called algae grow rapidly.
- The small animals in the water which live on the algae cannot eat all the extra algae.
- Many of the algae die and are decomposed by bacteria.
- These bacteria take oxygen from the water.
- As the number of bacteria increases, the oxygen content of the water decreases and other larger organisms, such as fish, die of suffocation.

Figure 9.6 ▲
Eutrophication in a river producing large amounts of algae

2 Nitrates in drinking water
Drinking water is taken from rivers. It is usually purified to kill harmful organisms but these processes do not remove any nitrates that have leached into the river from the land. If nitrates are present in drinking water, they can be converted in the body into chemicals called nitrosamines which are carcinogenic and may be a cause of some kinds of cancers of the alimentary canal.

Summary

- Ammonia is manufactured by the Haber process which requires temperatures of about 450°C and pressures of about 200 atmospheres.

- The reaction which uses the raw materials of nitrogen from the air and hydrogen from methane, is reversible.

- In order to maximise the yield of ammonia from the Haber process a number of economic factors associated with the reaction conditions need to be considered.

- Nitrogen is used to manufacture nitric acid, fertilisers and ammonia.

- Nitric acid can be made by the oxidation of ammonia.

- Nitrogenous fertilisers can cause environmental and health problems.

Topic questions

1 What is the formula of ammonia?

2 Answer the following questions about the manufacture of ammonia from nitrogen and hydrogen using the Haber process.
 a) From which source is the nitrogen obtained?
 b) Which two compounds are the source of hydrogen?
 c) What is the name of the catalyst used in the reaction?

3 a) Name the two main nitrogenous fertilisers.
 b) Give the chemical formulae of these substances.

c) i) What is the name given to the polluting of rivers and streams by nitrogenous fertilisers?
 ii) Explain how nitrogenous fertilisers pollute rivers and streams.
 d) Why are nitrates in drinking water a health hazard?

4 In the Haber process for manufacturing ammonia gas from nitrogen and hydrogen, what are the advantages and disadvantages of using
 a) a very high pressure and
 b) a very high temperature?

Examination questions

1 Ammonia is a very important chemical.
 a) The table shows the percentage of ammonia used to make different substances.

Substances made from ammonia	Percentage (%) of ammonia used
fertilisers	75
nitric acid	10
nylon	5
others	10

Copy the pie chart and shade the percentage of ammonia used to make nitric acid.

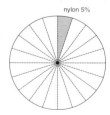

nylon 5%

(1 mark)

b) Ammonia gas is made by the reaction between nitrogen gas and hydrogen gas. Write a word equation to represent this reaction.
(1 mark)

c) Nitrogen is one of the raw materials used to make ammonia. Nitrogen is obtained from air. This pie chart shows the proportion of nitrogen, oxygen and other gases in air. Label the area which represents the proportion of nitrogen in air.

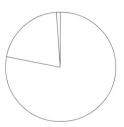

2 The air is a mixture of many gases. Some of these gases are shown in the table.

Name	Chemical formula
nitrogen	N_2
oxygen	O_2
argon	Ar
carbon dioxide	CO_2

a) Which of these gases are:
 i) elements
 ii) compounds? *(2 marks)*
b) Give *one* important use of nitrogen.
  (1 mark)
c) The amount of carbon dioxide in the air varies from place to place.
 The amount of carbon dioxide in the countryside is often lower than in towns and cities. Explain why. *(2 marks)*

3 For 200 million years the proportions of the different gases in the atmosphere have been much the same as today. Over the past 150 years the amount of carbon dioxide in the atmosphere has increased from 0.03% to 0.04%.
a) Describe how carbon dioxide is released into the atmosphere by human and industrial activity. *(2 marks)*
b) Explain how the seas and oceans can decrease the amount of carbon dioxide in the atmosphere. *(3 marks)*
c) i) Give **one** reason why the amount of carbon dioxide in the atmosphere is increasing gradually. *(1 mark)*
 ii) Give **one** effect that increasing levels of carbon dioxide in the atmosphere may have on the environment. *(1 mark)*

4 The Haber process is used to make ammonia NH_3. The table shows the percentage yield of ammonia at different temperatures and pressures.

Pressure (atmospheres)	Percentage (%) yield of ammonia at 350°C	Percentage (%) yield of ammonia at 500°C
50	25	5
100	37	9
200	52	15
300	63	20
400	70	23
500	74	25

a) i) Use the data in the table to draw two graphs. Plot percentage (%) yield of ammonia on the vertical axis and pressures in atmospheres, on the horizontal axis. Draw one graph for a temperature of 350°C and the second graph for a temperature of 500°C. Label each graph with its temperature. *(4 marks)*
 ii) Use your graphs to find the temperature (°C) and pressure (atmospheres) needed to give a yield of 30% ammonia.
  (1 mark)
 iii) On the grid sketch the graph you would expect for a temperature of 450°C.
  (1 mark)
b) i) The equation represents the reaction in which ammonia is formed.

 $N_{2(g)} + 3H_{2(g)} \rightleftharpoons 2NH_{3(g)} + \text{heat}$

 What does the symbol $\rightleftharpoons$ in this equation tell you about the reaction? *(1 mark)*

 ii) Use your graphs and your knowledge of the Haber process to explain why a temperature of 450°C and a pressure of 200 atmospheres are used in industry.
  (5 marks)

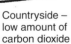

Countryside –
low amount of
carbon dioxide

Towns and cities –
high amount of
carbon dioxide

Chapter 10
The Earth

Key terms

addition polymerisation · alkanes · alkenes · alloys · anode · bauxite · biodegradable · blast furnace · bromine water · carbon dioxide · cathode · coke · combustion · corrosion · cracking · crude oil · crust · cryolite · deposition · electrolysis · erosion · extrusive rock · flammable · fossil fuels · fractional distillation · galvanising · global warming · greenhouse effect · haematite · hydrocarbons · igneous rocks · intrusive rock · lava · limewater · magma · mantle · metamorphic rocks · minerals · monomers · non-renewable (finite) · ores · polymerisation · polymers · reactivity series · resources · rusting · sacrificial protection · saturated hydrocarbons · sedimentary rocks · smelting · structural formula · thermal decomposition · thermit process · transportation · unsaturated hydrocarbons · viscosity · weathering

10.1   The rock record

Co-ordinated	Modular
DA 11.10	DA 6
SA n/a	SA n/a

Igneous rocks

Igneous rocks are formed when **magma** (the liquid rock in the **mantle**) within the Earth cools and solidifies. If the magma solidifies beneath the Earth's surface the igneous rock is known as **intrusive rock**. If the magma erupts from the surface in the form of **lava** from a volcano, the rock formed is an **extrusive rock**.

Intrusive rocks cool much more slowly than extrusive rocks. Slow cooling causes the crystals in the rock to become much larger.

Granite is an intrusive rock with quite large crystals because it cooled slowly; basalt is an extrusive rock with very small crystal because it cooled quickly.

The high ground of Dartmoor and Land's End are made of granite. The Giant's Causeway in Ireland is made of basalt.

Sedimentary rocks

Sedimentary rocks are formed from material that settles on river bottoms or seabeds. In the sea this process of **deposition** (or sedimentation) goes on continually. As the thickness of sediment builds up, the layers underneath become compressed and form sedimentary rocks.

These layers often provide evidence of how and when they were formed. For example, layers of different thickness show periods of discontinuous deposition, ripple marks are often left by water waves, younger rocks usually lie on older rocks. Because the **crust** is unstable and has been subjected to very large forces, sedimentary rock layers are often folded, tilted, fractured (faulted) or even turned upside down.

Sandstone and mudstone are examples of sedimentary rocks.

Limestone is also a sedimentary rock. It is formed by particles of calcium carbonate settling on the sea floor. Often the calcium carbonate is from the shells of dead sea creatures.

Metamorphic rocks

If rocks become buried deep underground they are subjected to very high pressures. If there is magma nearby, the rocks may also get very hot (but not hot enough to melt). High temperatures and/or high pressures can change the structure of the rock. Rocks that have been changed in this way are called **metamorphic rocks**.

Metamorphic rocks are often associated with the violent earth movements – tectonic activity (see section 15.7) that produced present day and ancient mountain ranges.

Slate and marble are metamorphic rocks formed from mudstone and limestone, respectively. Schist is another example of a metamorphic rock.

Large scale movements of the Earth's crust have caused mountain ranges to form very slowly over millions of years. These often replace even older mountain ranges that have been worn down by **weathering**, **erosion** and **transportation**.

The movement of rock from the Earth's mantle to the surface and back again is called **the rock cycle**.

Summary

- **Igneous rocks** are formed by the cooling of molten rock.

- **Sedimentary rocks** are formed by eroded material settling on the sea bed.

- **Metamorphic rocks** are formed by the action of high temperatures and/or high pressures on other types of rock.

- The **rock cycle** is the movement of rock from the Earth's mantle to the surface and back again.

Topic questions

1 Complete the following sentence by choosing the correct word from the alternatives given.

A volcano throws out **gas/lava**. This material cools **rapidly/slowly** to produce **extrusive/intrusive** rocks with **large/small** crystals. An example of this type of rock is **basalt/granite/slate**. If a large mass of underground **lava/magma** cools down the rock formed is **extrusive/intrusive**. **Basalt/granite/slate** formed this way.

2 Metamorphic rocks can be formed when other rocks get hot.

   a) What might cause the rocks to get hot?
   b) Other than heat, what else can produce metamorphic rocks?

3 How are mountain ranges formed?

4 Why are younger sedimentary rocks usually found on top of older sedimentary rocks?

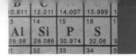

Co-ordinated	Modular
DA 11.4	DA 5
SA 11.4	SA 15

Useful products from metal ores

Rocks and minerals

Most rocks are not pure substances, they are mixtures of different **minerals**. Minerals are usually compounds but some – like gold and sulphur – are elements. Most minerals are unreactive; if they weren't, they would soon be weathered away. Some minerals can be made to react, and useful materials can be extracted from them. These materials are called **ores**.

Figure 10.1
A Venn diagram showing the relationship between rocks, minerals and ores

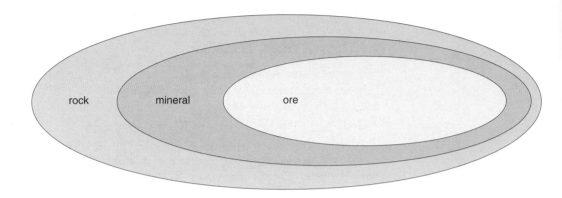

For a mineral to be classed as an ore it must be possible to extract useful amounts of material from it – and to do so economically.

Metal extraction

Most metals are extracted from ores. The exceptions are very unreactive metals like gold and silver which are found as the element. Most ores contain a metal oxide or a substance that can easily be changed into an oxide. There are two main methods of removing the oxygen from the metal oxide. The method used depends on the reactivity of the metal. To decide which method is used it is helpful to include carbon in the **reactivity series** even though this is not a metal.

Figure 10.2
The reactivity series for metals showing the position of carbon

potassium
sodium
calcium
magnesium
aluminium
carbon
zinc
iron
tin
lead
hydrogen – not a metal but included in the series
copper
silver
gold
platinum

Any metal in the reactivity series can displace a metal lower in the series from its compounds. So carbon can displace the metals zinc, iron and copper from their compounds. This process of extracting metals from their ores by heating them with carbon is called **smelting**. Metals above carbon have to be extracted by another method. This process is called **electrolysis**.

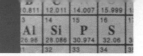

Uses of the reactivity series

Knowing the relative position of one metal to another in the reactivity series can be very helpful in predicting and explaining what happens in many chemical reactions.

- Metals high in the series tend to react or stay as a compound.
- Metals low in the series tend to stay as metals or become metals if they are part of a compound.

Predicting the reaction between a metal oxide and another metal

When a metal oxide is heated with a second metal it is possible for the second metal to remove the oxygen from the metal oxide. A metal higher in the reactivity series will always remove oxygen from the oxide of a metal that is lower than it in the series.

For example:

$$\text{magnesium} + \text{copper(II) oxide} \rightarrow \text{magnesium oxide} + \text{copper}$$
$$\text{Mg(s)} + \text{CuO(s)} \rightarrow \text{MgO(s)} + \text{Cu(s)}$$

but

$$\text{copper} + \text{magnesium oxide} \rightarrow \text{NO REACTION}$$

There is a competition between the two metals for the oxygen. The metal which is higher in the series will gain the oxygen and form an oxide, and the metal which is lower in the reactivity series will lose the oxygen and will form a metal.

The Thermit process

The Thermit process is a method of joining two lengths of railway track together using the greater reactivity of aluminium compared to iron. A mixture of aluminium powder and iron(III) oxide in a crucible is positioned over the gap between the rails and ignited using a magnesium ribbon. A vigorous exothermic reaction takes place forming molten iron and aluminium oxide.

Figure 12.3
The Thermit process for welding together two lengths of railway line

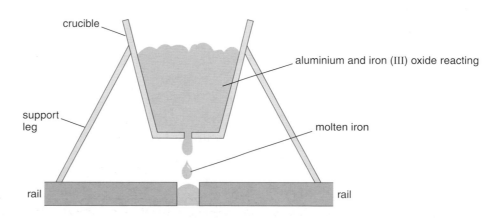

The iron falls through the hole in the crucible and into the gap between the rails. When the iron cools and solidifies, the two lengths of rail will be joined.

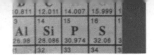

Predicting the method of extraction of a metal from its ore

The position of a metal in the reactivity series can be used to predict the likely extraction process. Aluminium and iron are extracted from their oxides, Al_2O_3 and Fe_2O_3, respectively. Aluminium is high in the series and is a reactive metal. Aluminium is extracted from aluminium oxide by the high energy process of electrolysis. Iron is lower in the series and is much less reactive. Iron is extracted from iron(III) oxide by heating with carbon.

Iron

Iron is obtained from its ore by smelting. This process is carried out in a **blast furnace**.

Figure 10.4

Diagram of the inside of a blast furnace

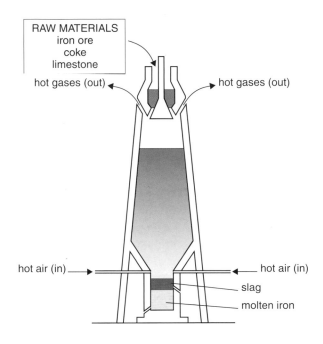

Iron ore (**haematite**, mainly iron(III) oxide, Fe_2O_3), **coke** (fairly pure form of carbon obtained by heating coal) and limestone are put into the top of the blast furnace. Hot air is blown in at the bottom.

In the blast furnace the following reactions take place:

1 Coke burns in the hot air blast producing **carbon dioxide**. (This reaction also helps to keep the furnace hot.)

$$\text{carbon (coke)} + \text{oxygen} \rightarrow \text{carbon dioxide}$$
$$C(s) + O_2(g) \rightarrow CO_2(g)$$

2 The carbon dioxide produced is reduced (see Chapter 7) to carbon monoxide by some more coke.

$$\text{carbon dioxide} + \text{carbon} \rightarrow \text{carbon monoxide}$$
$$CO_2(g) + C(s) \rightarrow 2CO(g)$$

3 The iron(III) oxide is reduced to iron by either carbon or carbon monoxide. The carbon (coke) or carbon monoxide is oxidised to CO_2.

$$\text{iron(III) oxide} + \text{carbon} \rightarrow \text{carbon dioxide} + \text{iron}$$
$$2Fe_2O_3(s) + 3C(s) \rightarrow 3CO_2(g) + 4Fe(l)$$

$$\text{iron(III) oxide} + \text{carbon monoxide} \rightarrow \text{carbon dioxide} + \text{iron}$$
$$Fe_2O_3(s) + 3CO(g) \rightarrow 3CO_2(g) + 2Fe(l)$$

The temperature inside the blast furnace reaches 1400°C. This is hot enough for the iron to melt and the molten iron runs to the bottom of the furnace.

Did you know?

Of the total mass of material produced by the blast furnace only about 25% is iron. About 66% is the gases which the reactions produce. Most of the gas is nitrogen from the hot air blast.

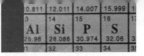

Iron ore is not very pure – it contains a lot of sand (silicon dioxide, SiO_2). Sand does not melt in the blast furnace. If it were left there it would become acidic impurities in the iron. The impurities are removed by the limestone added to the blast furnace.

$$\text{sand} + \text{limestone} \rightarrow \text{calcium silicate (slag)} + \text{carbon dioxide}$$
$$SiO_2(s) + CaCO_3(s) \rightarrow \quad CaSiO_3(l) \quad + \quad CO_2(g)$$

The calcium silicate (slag) melts in the blast furnace. It runs to the bottom and settles on top of the molten iron.

Uses of iron

Iron from the blast furnace is 'cast iron'. It contains about 4% carbon. This amount of carbon makes the iron very brittle. Iron can be converted into steel by passing oxygen through the molten cast iron. The oxygen reacts with some of the carbon to produce carbon dioxide. This lowers the amount of carbon in the iron. Other elements can be added to the iron to make **alloys** with special properties.

Figure 10.3
Some alloys of iron and their uses

Name of alloy	Composition	Properties	Uses
wrought iron	almost pure iron	easy to soften and shape	decorative iron work
mild steel	iron 99.5% carbon 0.5	quite hard but easy to work	buildings, car bodies
hard steel	iron 99% carbon 1%	very hard	ball bearings, blades for cutting tools
'stainless' steel	iron about 74% chromium 18% nickel 8%	resistant to corrosion	containers for corrosive substances, kitchenware
high speed steel	iron about 75% tungsten 18% chromium 4% vanadium 1% carbon 1%	very hard; not easily softened by high temperatures	cutting tools for metal working lathes

The rusting of iron

Rusting is the **corrosion** of iron and occurs in the presence of air (oxygen) and water. Rust is hydrated iron(III) oxide although the exact formula of rust is difficult to give.

Rusting as an oxidation reaction

Rusting is the oxidation of iron – the iron gains oxygen as part of the rusting reaction. If rust is assumed to be iron(III) oxide then the equations are

$$\text{iron} + \text{oxygen} \rightarrow \text{iron(III) oxide}$$
$$4Fe(s) + 3O_2(g) \rightarrow \quad 2Fe_2O_3(s)$$

In terms of electron transfer, the iron has been oxidised because its atoms have lost electrons to form iron 3+ ions (see section 8.3).

$$Fe(s) \rightarrow Fe^{3+}(aq) + 3e^-$$

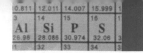

Prevention of rusting

Iron will not rust if it is kept in dry conditions or if it is kept in an atmosphere that does not contain oxygen. Oxygen and water are both needed for the iron to rust.

Rusting can be prevented by any process which makes a barrier between the iron or steel surface and the water and oxygen in the air.

Alloying

Special alloys of steel can be made by adding chromium or vanadium to the iron to reduce its rusting rate. These are called stainless steels and are used to make tools, kitchen utensils and work surfaces. Stainless steels are more expensive than ordinary steels.

Sacrificial protection

This is where a more reactive metal such as magnesium or zinc is attached to the iron or steel object. The more reactive metal will corrode in preference to the iron and steel and thus protect it. The more reactive metal will 'sacrifice' itself for the benefit of the iron or steel. **Sacrificial protection** method is used to protect large steel structures such as pylons and large ships.

Galvanising is a form of sacrificial protection. In this process iron or steel is covered with a layer of zinc. If the zinc coating gets damaged and the iron is exposed, it is the zinc which corrodes first.

Using the reactivity series to explain the prevention of rusting by sacrificial protection using magnesium or zinc

A fairly reactive metal such as magnesium or zinc can be used to prevent iron from rusting as quickly as it would normally. The iron or steel object has the more reactive metal attached to it either directly or connected by a conducting wire.

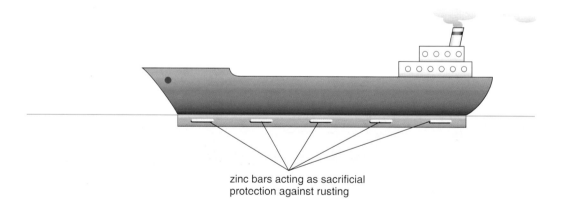

zinc bars acting as sacrificial protection against rusting

Figure 10.6
Sacrificial protection of a ship's hull

The more reactive metal will corrode in preference to the iron – its metal atoms will turn into metal ions.

$$Zn(s) \rightarrow Zn^{2+}(aq) + 2e^-$$

The released electrons will spread onto the iron and help to prevent a similar ionisation process occurring, which would be the first step in the rusting process. The zinc gradually corrodes away and 'sacrifices' itself to protect the iron.

0.811	12.011	14.007	15.999	1
3	14	15	16	1
Al	**Si**	**P**	**S**	
26.98	28.086	30.974	32.06	3
1	32	33	34	3

Aluminium

Aluminium is a reactive metal – it is higher in the reactivity series than carbon. It cannot therefore be produced by smelting so electrolysis has to be used. Electrolysis is a very expensive process because it uses a lot of electricity. It is only possible to extract aluminium economically where there is a cheap source of electricity. The one cheap source of electricity is hydroelectric power (see section 17.3).

The ore of aluminium is called **bauxite**. It is aluminium oxide (Al_2O_3).

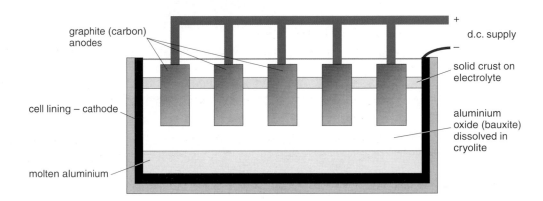

Figure 10.7
Diagram of an aluminium electrolysis cell

Aluminium oxide has a very high melting point (2050°C). To make the process safe to operate, another aluminium ore, **cryolite**, is used. Cryolite melts at a much lower temperature. Bauxite is dissolved in the molten cryolite. The cryolite does not get used up in the electrolysis process.

When an electric current is passed through the molten electrolyte, positively-charged aluminium ions (Al^{3+}) are attracted to the **cathode** (negative electrode). At the cathode, they gain electrons to form aluminium metal.

This is an example of reduction.

$$Al^{3+} + 3e^- \rightarrow Al(l)$$

The molten aluminium settles to the bottom of the tank and is syphoned off from the cell. At the **anode** (positive electrode), oxygen is produced.

The negatively charged ions lose electrons (oxidation).

$$O^{2-} - 2e^- \rightarrow O(g)$$

The carbon anode has to be replaced at regular intervals as it reacts with the oxygen to produce carbon dioxide.

In the electrolytic extraction of aluminium, oxidation and reduction take place. Together these are an example of a redox reaction.

Uses of aluminium

Aluminium has a low density (2.7 g/cm³) compared to iron (7.9 g/cm³). It can be made very strong by alloying with other elements. This makes it ideal for use in the construction of aircraft. Aluminium is also a good conductor of heat so it can be used to make cooking utensils. Because of its good electrical conductivity and low density, it is used to make the high tension cables for the overhead power cables.

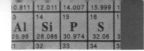

Figure 10.8
Electrolytic cells for the production of aluminium

Using the reactivity series to explain the unexpected behaviour of aluminium due to the formation of an impermeable oxide coating

The high position of aluminium in the reactivity series shows that it should be quite a reactive metal and would be expected to react with water and oxygen in the atmosphere. The normal uses to which it is put, for example window frames and lightweight bodies for railway engines and planes, shows that this is not true.

The surface of the aluminium rapidly reacts with the oxygen in air to form a thin oxide layer, which protects the aluminium below it. This layer is impermeable and prevents any further reaction from taking place with more oxygen or water. The outer oxide layer acts as a protective layer. This layer differs from a rust layer on iron because it stays firmly in place and does not flake off exposing fresh metal surface.

Figure 10.9
The reaction of 'unprotected' aluminium with oxygen in air

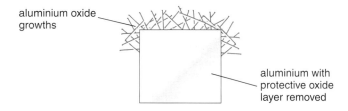

aluminium oxide growths

aluminium with protective oxide layer removed

If the aluminium oxide layer is removed chemically with concentrated hydrochloric acid then a rapid reaction takes place between the newly exposed aluminium surface and oxygen in the air. Delicate, white, fern-like structures of aluminium oxide can be seen growing outwards from the surface of the aluminium. Aluminium is showing its real reactivity.

Using electrolysis to get pure copper

Copper is a good conductor of electricity. Electrolysis is used to make very pure copper.

In this process, copper atoms in the anode lose electrons and become copper ions. These ions pass into the copper sulphate solution. The anode gradually 'dissolves'.

At the cathode, copper ions from the solution gain electrons to become copper atoms. These atoms form on the surface of the cathode as very pure copper.

The electrolyte, a solution of copper(II) sulphate contains the following ions: Cu^{2+} and SO_4^{2-}, from the copper sulphate and H^+ and OH^-, from the water.

The copper **anode** has a positive charge because it has lost electrons to the cathode.

The OH^- and SO_4^{2-} ions are negatively charged so have had electrons added (see section 10.1).

These negative ions are attracted to the positive electrode, the anode, where they lose the extra electrons to form atoms.

However, in this reaction copper atoms in the anode form copper ions which move into the solution. As they do so they lose electrons. The hydroxide ions and sulphate ions remain in the solution. At the anode no substance is produced but copper is lost.

The reaction at the anode can be summarised as:

copper atoms – electrons → copper ions

$$Cu\,(s) \quad - \quad 2e^- \quad \rightarrow Cu^{2+}\,(aq)$$

This is an example of oxidation.

The copper **cathode** has a negative charge because it has gained extra electrons from the anode.

The H^+ and Cu^{2+} ions are positively charged so have had electrons removed (see section 10.1).

These postive ions are attracted to the negative electrode, the cathode, where they collect extra electrons to form atoms. However, because copper is more reactive than hydrogen (see section 10.2) the hydrogen ions remain as ions in the solution and only the copper ions form atoms. The cathode becomes coated with a layer of pure copper.

The reaction at the cathode can be summarised as:

copper ions + electrons → copper atoms

$$Cu^{2+}\,(aq) \quad + \quad 2e^- \quad \rightarrow Cu\,(s)$$

This is an example of reduction.

The Cu^{2+} ions, the OH^- ions, the SO_4^{2-} ions and H^+ ions remain in solution.

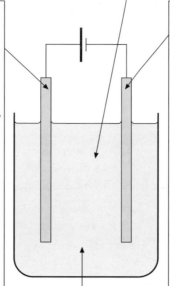

Figure 10.10

Purifying copper by electrolysis

Did you know?

The industrial process of electroplating uses a similar process to the one used to purify copper. To get the metal being deposited to form on the article being plated, the conditions have to be very carefully controlled. Electroplating is used to put a thin layer of gold or silver on cheaper metals to make inexpensive jewellery. It is also used to cover metals like iron with other metals like chromium to prevent corrosion.

Completing and balancing half-equations for the reactions occurring at the electrodes during electrolysis

Half-equations provide information about the charge of the ion and the atomic or molecular nature of the product.

The table gives some information about the product at each electrode.

At the negative electrode (cathode)	At the positive electrode (anode)
Positively charged ions (for example, Cu^{2+}, H^+ and Al^{3+}) gain electrons	Negatively charged ions (for example, Cl^- and O^{2-}) lose electrons
The number of electrons gained equals the size of the charge on the ion	The number of electrons lost equals the size of the charge on the ion
Metallic elements are released as atoms (Cu and Al) Hydrogen released as molecules (H_2)	Gases released as molecules (Cl_2, O_2)

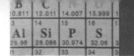

Example 1 Complete and balance the half-equation for the formation of chlorine at an anode.

$$Cl^- \quad - \quad e^- \quad \longrightarrow \quad Cl_2$$

Number of molecules of chlorine (Cl_2) produced = 1
So, number of chloride ions (Cl^-) involved = 2
So, number of electrons to be lost = 2
So, the completed and balanced equation will be

$$2Cl^- \quad - \quad 2e^- \quad \longrightarrow \quad Cl_2$$

Example 2 Complete and balance the half-equation for the formation of copper at a cathode.

$$Cu^{2+} \quad + \quad e^- \quad \longrightarrow$$

Electrons will be gained.
1 atom of Cu will be produced.
So, number of copper ions (Cu^{2+}) involved = 1
So, number of electrons to be gained = 2
So, the completed and balanced equation will be

$$Cu^{2+} \quad + \quad 2e^- \quad \longrightarrow \quad Cu$$

Example 3 Complete and balance the half-equation for the formation of aluminium at a cathode.

$$Al^{3+} \quad + \quad e^- \quad \longrightarrow$$

1 atom of Al will be produced
So, number of aluminium ions (Al^{3+}) involved = 1
So, number of electrons to be gained = 3
So, the completed and balanced equation will be

$$Al^{3+} \quad + \quad 3e^- \quad \longrightarrow \quad Al$$

Example 4 Complete and balance the half-equation for the formation of oxygen at an anode.

$$O^{2-} \quad - \quad e^- \quad \longrightarrow$$

1 molecule of O_2 will be produced
So, number of oxide ions (O^{2-}) involved = 2
So, number of electrons to be lost = 4
So, the completed and balanced equation will be

$$2O^{2-} \quad - \quad 4e^- \quad \longrightarrow \quad O_2$$

Example 5 Complete and balance the half-equation for the formation of hydrogen at a cathode.

$$H^+ \quad + \quad e^- \quad \longrightarrow \quad H_2$$

1 molecule of H_2 is produced
So, number of hydrogen ions (H^+) involved = 2
So, number of electrons to be gained = 2
So, the completed and balanced equation will be

$$2H^+ \quad + \quad 2e^- \quad \longrightarrow \quad H_2$$

Demand for raw materials

The demand for most raw materials is increasing. As the demand goes up, social, economic and environmental problems are created. For example:

- open-cast mining of ore uses up farmland and destroys habitats
- transportation of materials creates noise and air pollution
- mining ore and production processes produce dust and noise
- waste materials produce unsightly tips
- disused underground mines can collapse causing subsidence
- disused open-cast mines look unpleasant
- all disused sites can be dangerous.

Did you know?

The demand for aluminium today is about 1000 times higher than it was 100 years ago but the cost of aluminium is about 20% of what it was 100 years ago.

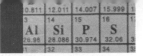

Industrial processes

In this book several industrial processes are mentioned. These processes have been a benefit to people in a number of ways.

- New materials have been developed. A good example of this is the polymer (plastics) industry. This industry has grown as a result of advancements in the oil industry.

- Many industrial products have become easier to get and less expensive. This is because most of these processes take place in very large factories. Large factories can usually make products more efficiently than small factories and this keeps the cost of the product low. If products are made at low cost there will probably be more demand for them. This can mean that more factories will be built to meet that demand.

- The factories have provided employment with good wages.

But there are also pollution problems caused by these industrial processes.

- Big factories need a lot of land. This reduces the land available for growing food. It also affects the habitats of wildlife.

- Transportation – factories need to get their raw materials; they also need to send out the products they make. Both require transportation. Lorries produce noise and air pollution. Heavy lorries can also cause expensive damage to roads and buildings beside the roads. There are similar problems with rail transport, though they are not usually as bad. Ships may be used for transport – they can also pollute. Accidents at sea can be a serious problem because of the huge size of the ships and the amount of material they carry.

- Pollution – many of the processes produce dust and acidic or toxic fumes.

- Waste materials – most of the industrial processes produce substances that are not useful. These substances are often dumped in huge piles (tips), thus using up more land.

People's attitudes to scientific development is an important factor. The Haber process has been used for many years to produce cheap nitrogenous fertiliser (see section 9.2). Once this was considered an important scientific development – it allowed farmers to grow crops every year and increased food production in poor countries. Today, however, many people are unhappy about the use of artificial fertilisers. Because they were inexpensive, these fertilisers were used too much. Some of the fertilisers got into rivers and streams causing pollution (see sections 6.2 and 9.2). Increasingly people are going back to natural fertilisers – they are growing and eating more 'organic' food.

Did you know?

In the last 50 or 60 years many scientific discoveries have been made. In many cases scientists believed these discoveries would be of great benefit to people. Only later was it found that there were problems that no-one has foreseen. Some of these discoveries, like the insecticide DDT and the CFCs that were used as aerosol propellants, harmed the environment. It is now much harder for scientists to convince people that new discoveries will be a benefit. The recent problem with genetically modified (GM) foods is an example of this.

Summary

- Rocks are usually a mixture of **minerals**.

- Minerals are usually chemical compounds.

- **Ores** are minerals that can be used as a source of useful materials.

- The **reactivity series** is a table of metals with the most reactive metals at the top and the least reactive metals at the bottom.

- The reactivity of the metal can be found by reacting it with water, acids and oxygen.

- A metal higher in the reactivity series will always remove oxygen from the oxide of a metal that is lower than it in the series.

- The **Thermit process** is a method of joining two lengths of railway track together using the greater reactivity of aluminium compared to iron.

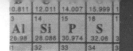

◆ A metal higher in the reactivity series will displace a metal which is lower in the series from a solution of its metal sulphate.

◆ All **corrosion** processes involve a reaction between a metal and substances in the atmosphere.

◆ In **sacrifical protection** a more reactive metal such as magnesium or zinc can be used to prevent iron from rusting as quickly as it would normally.

◆ The surface aluminium rapidly reacts with the oxygen in air to form a thin oxide layer which protects the aluminium below it.

◆ Rusting is the corrosion of iron and happens in the presence of air and water.

◆ Rusting is the oxidation if iron because the iron has gained oxygen and its atoms have lost electrons to form iron 3+ ions.

◆ Rusting can be prevented by an process which makes a barrier between the iron or steel surface and the water and oxygen in the air.

◆ Metals less reactive than carbon can be extracted from their ore by **smelting**.

◆ Metals more reactive than carbon myst be extracted from their ore by **electrolysis**.

◆ Smelting is usually done by heating the ore with carbon.

◆ Smelting is usually done in a **blast furnace**.

◆ In the smelting of iron, iron ore (usually **haematite**), **coke** (fairly pure carbon made from coal) and limestone (calcium carbonate) are added at the top. Hot air is blown in at the bottom.

◆ Extracting a metal by electrolysis means passing an electric current through the molten ore.

◆ Aluminium is extracted by electrolysis.

◆ Half-equations can be used to describe electrolysis.

◆ Aluminium ore, **bauxite**, is dissolved in molten cryolite to lower its melting point.

◆ Pure metals can often be made stronger by adding other elements to make **alloys**.

◆ Industrial processes can cause social, economic and environmental problems.

Topic questions

1 For each of the following, say whether they are *best* described as mineral, ore or rock.
   a) granite
   b) bauxite
   c) haematite

2 Would you use smelting or electrolysis to get the following metals from their ores?
   a) zinc
   b) calcium
   c) iron
   d) potassium

3 a) What three substances are put in the top of a blast furnace?
   b) Which one of these is the ore?
   c) Which one of these is the reducing agent?
   d) What is the purpose of the third substance?
   e) What other substance has to be put into the blast furnace?

4 The gases carbon monoxide and carbon dioxide are made in the blast furnace.
   a) Write a chemical equation for a reaction in a blast furnace that produces carbon dioxide.
   b) Write an equation for a reaction in a blast furnace that produces carbon monoxide.
   c) Which of these gases will reduce iron ore to produce iron?

5 Steel is an alloy of iron and carbon.
   a) What is meant by the term 'alloy'?
   b) What effect does 1% of carbon have on iron?
   c) What effect does 4% of carbon have on iron?
   d) What is the alloy of iron which contains 4% carbon called?

6 a) Bauxite is the main ore of which metal?
   b) What other ore of this metal is mixed with bauxite?
   c) What are the advantages of using a mixture of ores?

7 The cables that are used to carry electricity along overhead power lines are made of pure aluminium. The aluminium is not strong enough by itself and is wrapped round a central core of steel. What properties of aluminium make it a good choice for high tension cables?

8 Complete and balance the following half-equations for products produced at the electrodes during electrolysis.
   a) For chlorine produced at a positive electrode
   $$2Cl^- \quad - \quad e^- \longrightarrow$$
   b) For copper produced at a negative electrode
   $$Cu^{2+} \quad + \quad e^- \longrightarrow \quad Cu$$
   c) For aluminium produced at a cathode
   $$Al^{3+} \quad e^- \longrightarrow$$
   d) For oxygen produced at an anode
   $$O^{2-} \quad e^- \longrightarrow \quad O_2$$
   e) For hydrogen produced at a negative electrode
   $$H^+ \quad e^- \longrightarrow$$

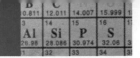

Limestone and its uses

Limestone landscapes are considered by many people to provide the most spectacular landscapes in England. But because limestone is such a useful raw material there are difficult environmental decisions that need to be made when quarrying for limestone.

Figure 10.11
A limestone pavement in Malham, Yorkshire

This diagram shows some uses of limestone.

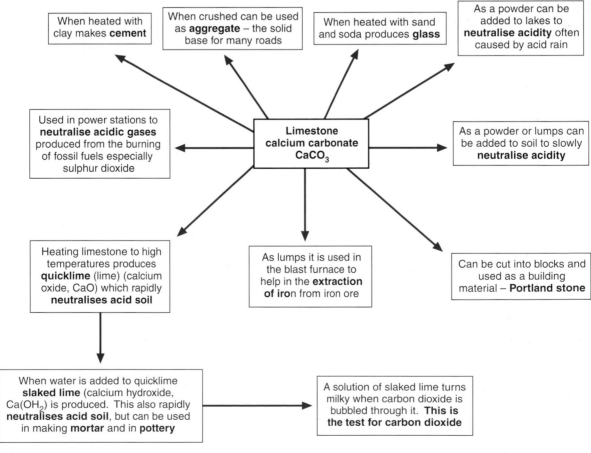

When heated with clay makes **cement**

When crushed can be used as **aggregate** – the solid base for many roads

When heated with sand and soda produces **glass**

As a powder can be added to lakes to **neutralise acidity** often caused by acid rain

Used in power stations to **neutralise acidic gases** produced from the burning of fossil fuels especially sulphur dioxide

As a powder or lumps can be added to soil to slowly **neutralise acidity**

Limestone calcium carbonate CaCO₃

Heating limestone to high temperatures produces **quicklime** (lime) (calcium oxide, CaO) which rapidly **neutralises acid soil**

As lumps it is used in the blast furnace to help in the **extraction of iro**n from iron ore

Can be cut into blocks and used as a building material – **Portland stone**

When water is added to quicklime **slaked lime** (calcium hydroxide, Ca(OH)₂) is produced. This also rapidly **neutralises acid soil**, but can be used in making **mortar** and in **pottery**

A solution of slaked lime turns milky when carbon dioxide is bubbled through it. **This is the test for carbon dioxide**

Figure 10.12
Some uses of limestone

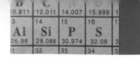

Heating limestone

For many centuries limestone has been heated in lime kilns to provide quicklime and slaked lime which farmers use to reduce soil acidity. The temperature inside these kilns is about 1200°C.

When limestone – calcium carbonate – is heated strongly it breaks down into calcium oxide and carbon dioxide:

calcium carbonate $\rightarrow$ calcium oxide + carbon dioxide
$$CaCO_3 \text{ (s)} \rightarrow CaO \text{ (s)} + CO_2 \text{ (g)}$$
(quicklime/ lime)

This reaction is an example of **thermal decomposition** – the breaking down of a substance by heating (see section 8.3).

Did you know?

If lime is heated strongly it will produce an intense white light. This effect was used in theatres, hence the phrase 'in the limelight'.

Figure 10.13
Limelight being used in theatre spotlights

If water is added to calcium oxide there is a vigorous reaction during which much heat is released and calcium hydroxide is produced.

calcium oxide + water $\rightarrow$ calcium hydroxide
$$CaO \text{ (s)} + H_2O \text{ (l)} \rightarrow Ca(OH)_2 \text{ (s)}$$
(slaked lime)

Large quantities of calcium hydroxide are used in industry as an alkali.

The laboratory test for carbon dioxide

A solution of calcium hydroxide in water is called **limewater**. This is used as a test for carbon dioxide. As carbon dioxide is bubbled through limewater the solution goes milky due to the formation of insoluble calcium carbonate.

calcium hydroxide + carbon dioxide $\rightarrow$ calcium carbonate + water
$$Ca(OH)_2\text{(aq)} + CO_2 \text{ (g)} \rightarrow CaCO_3\text{(s)} + H_2O\text{(l)}$$

Limestone and cement

Cement is made by heating a mixture of crushed limestone, clay and shale at a temperature of about 1600°C. The heating is carried out in a rotary kiln more than 150 m long and 4m in diameter. The resulting clinker is then cooled and ground down into a very fine powder – cement.

0.811	12.011	14.007	15.999	1
3	14	15	16	1
Al	Si	P	S	
26.98	28.086	30.974	32.06	3
1	32	33	34	3

Portland cement mixed with water and sand or gravel forms concrete. The concrete hardens as the water slowly evaporates.

Limestone and glass

Glass was thought to have been first made nearly 4000 years ago. Most of the glass used to make bottles, lamp bulbs, window and plate glass is called soda-lime glass. This glass is made by heating a mixture of lime, soda (sodium carbonate, Na_2CO_3) and sand (silica).

Summary

♦ Limestone (calcium carbonate) is quarried and used as a building material.

♦ Powdered limestone is used to neutralise acidity in lakes and soils.

♦ Limestone heated in kilns breaks down into quicklime (calcium oxide) and carbon dioxide. This is an example of **thermal decomposition**.

♦ Quicklime reacts with water to form slaked lime (calcium hydroxide) which is used to neutralise soil acidity.

♦ Roasting powdered limestone with powdered clay in a rotary kiln produces cement.

♦ When cement is mixed with water, sand and crushed rock, a slow chemical reaction produces concrete.

♦ Glass is made by heating a mixture of limestone, sand and soda (sodium carbonate).

♦ Carbon dioxide is a gas. It will turn limewater milky.

Topic questions

1 Why is limestone such a useful raw material?

2 Write down the chemical names and formulae for each of the following:
   a) limestone
   b) quicklime
   c) slaked lime
   d) soda

3 How is limestone used in power stations?

4 Explain why the heating of limestone is an example of thermal decomposition.

5 Describe the limewater test for carbon dioxide. Use a word equation in your answer.

6 How is cement made?

7 What is concrete?

8 What are the three raw materials needed to make window glass?

Co-ordinated	Modular
DA 11.3	DA 6
SA 11.2	SA 15

Useful products from crude oil

Crude oil

Crude oil and natural gas are **fossil fuels** that were formed from small animals and plants which died and were buried under sediment well over 100 000 000 years ago. The organic matter decomposed (broke down) to form crude oil and natural gas which remained trapped in the rocks.

Crude oil and natural gas are examples of **non-renewable** (or **finite**) **resources**. Coal is a further example. Once non-renewable resources are used up, they cannot be replaced.

Crude oil is a mixture of substances. Most of these substances are **hydrocarbons** (compounds which contain hydrogen and carbon only).

A mixture is made if two or more elements or compounds come together but do not chemically join. In a mixture the chemical properties of each substance remain unchanged. Because their properties are unchanged the substances in a mixture can be separated by physical methods such as distillation.

Fractional distillation

Fractional distillation is used to separate two or more liquids that are mixed together, for example ethanol and water. Fractional distillation is the equivalent of a series of simple distillations and is very efficient. A fractionating column is put in between the boiling liquid mixture and the condenser.

The liquid mixture is heated until it boils and the vapour formed rises up into the fractionating column. The vapour condenses and re-boils many times as it moves up the column. Each time this happens, the composition of the vapour changes to contain more of the most volatile liquid (the one with the lowest boiling point). If the column is long enough, vapour from each of the liquids in the mixture gradually separates and makes its way successively up the column. The temperature of each of the separated vapours is shown by the thermometer before it condenses. As each liquid condenses in turn, it can be collected in separate flasks. The liquid with the lowest boiling point will come out first.

Did you know?

Most of the oil and gas produced by fossilisation has seeped up through the rock and escaped into the atmosphere. It is only when it has been trapped under a layer of non-porous rock that it can be extracted.

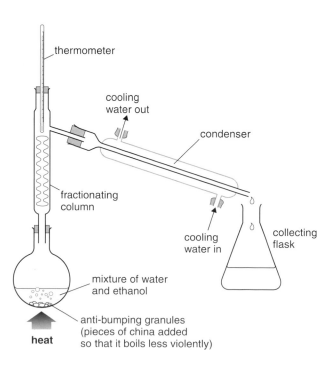

Figure 10.14
Fractional distillation of a liquid mixture

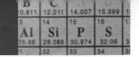

Separating crude oil

Crude oil can be separated into its various useful components by fractional distillation.

The different fractions from crude oil have different properties and therefore different uses.

Figure 10.15
Simplified diagram of the fractional distillation of crude oil

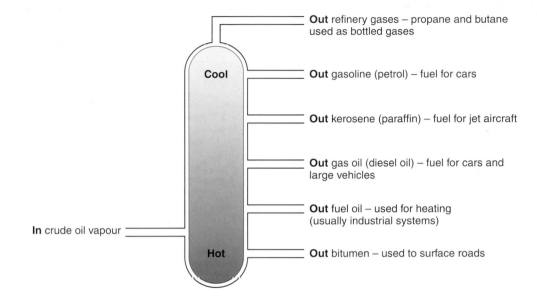

Out refinery gases – propane and butane used as bottled gases

Out gasoline (petrol) – fuel for cars

Out kerosene (paraffin) – fuel for jet aircraft

Out gas oil (diesel oil) – fuel for cars and large vehicles

Out fuel oil – used for heating (usually industrial systems)

Out bitumen – used to surface roads

In crude oil vapour

Cool

Hot

Products that condense lower down the fractional distillation column have a higher boiling point. The higher the boiling point, the more carbon atoms there are in each molecule.

Figure 10.16
Table comparing the properties of some of the fractions of crude oil

Fraction	Approx. no. of carbon atoms	Approx. boiling point/°c	Appearamce	What happens
gasoline (petrol)	4–10	less than 100	almost colourless liquid with low **viscosity***	highly flammable; has an almost colourless, smokeless flame
gas oil (diesel)	16–20	about 200	yellow liquid with a higher viscosity	flammable; burns with a yellow, smoky flame
bitumen	50+	over 350	black substance – so viscous it's almost a solid	can only be made to burn if heated to a high temperature and sprayed as fine droplets

*****Viscosity** is a measure of how runny a liquid is. Water has a low viscosity and treacle has a high viscosity.

Huge quantities of oil are used. Most of it by the industrialised nations of North America and Europe. Much of the oil is transported to these countries in large ocean-going tankers. Oil leaks from these ships – this is a problem if a tanker is involved in an accident. There are many examples where this has happened. Oil does not mix with water, it floats on top producing an 'oil slick'. Oil slicks are a serious environmental hazard. They kill wildlife at sea and can pollute beaches damaging the habitats of many creatures.

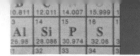

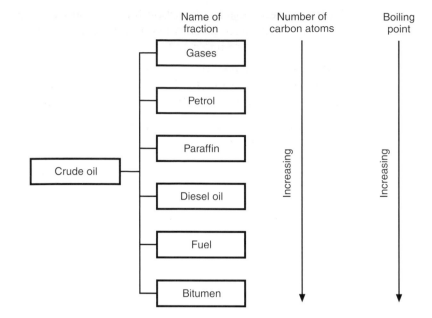

Figure 10.17
*The main fractions
from crude oil*

Did you know?

About 2 tonnes of
North Sea crude oil are
used up each year for
every person in the UK.

Hydrocarbons

Hydrocarbons are compounds containing only HYDROgen and CARBON. Natural gas and most of the substances in crude oil are hydrocarbons. All hydrocarbons are **flammable**. This means that the reaction of these compounds with oxygen is exothermic (see section 12.1). This is why hydrocarbons can be used as fuels. The smaller the size of the molecule, the easier it burns (see Figure 10.16 above). If there is enough air, the products of burning are carbon dioxide and water (vapour).

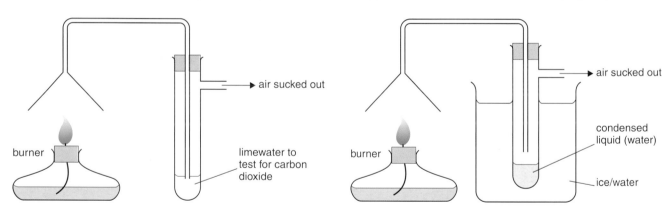

Figure 10.18
*Apparatus to show
that burning
hydrocarbons
produces carbon
dioxide and water
vapour*

The apparatus above can be used to show that burning a hydrocarbon produces carbon dioxide and water. Figure 10.19 gives the tests for carbon dioxide and water (see section 12.2).

Substance	Test for substance	Positive result
carbon dioxide	mix with (or bubble through) limewater	**limewater** goes 'milky'
water	1 add to blue cobalt chloride* 2 add to white (anhydrous) copper(II) sulphate* 3 check the boiling point of the liquid*	cobalt chloride turns pink copper(II) sulphate turns blue water boils at 100°C

*Tests 1 and 2 will show that the liquid contains water. Test 3 will show that the liquid is pure water. (Impurities raise the boiling point above 100°C).

Figure 10.19
*Tests for carbon
dioxide and water*

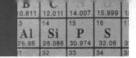

Useful products from crude oil

Unfortunately most hydrocarbon fuels not only contain carbon and hydrogen but often also contain sulphur. When these fuels are burnt, carbon dioxide, water and sulphur dioxide are produced.

If there is not enough air, the **combustion** of the hydrocarbon may be incomplete. If this happens the products of combustion will include carbon monoxide and carbon. If carbon is produced, the flame will be yellow and smoky.

The word equations for burning methane (the main gas present in natural gas) when there is not enough air are:

$$\text{methane} + \text{(not enough) oxygen} \rightarrow \text{carbon} + \text{water vapour}$$

and

$$\text{methane} + \text{(not enough) oxygen} \rightarrow \text{carbon monoxide} + \text{water vapour}$$

If there is enough air the reaction is:

$$\text{methane} + \text{oxygen} \rightarrow \text{carbon dioxide} + \text{water vapour}$$
$$CH_4(g) + 2O_2(g) \rightarrow CO_2(g) + 2H_2O(g)$$

To completely burn one molecule of methane takes two molecules of oxygen (see section 8.4). To completely burn one molecule of the substances in petrol would take about 10 molecules of oxygen. Over 70 molecules of oxygen are needed to burn one of the molecules present in bitumen. This means that the bigger the molecule, the more likely it is that combustion will not be complete. This is why hydrocarbons with large molecules burn with smoky, yellow flames. The black smoke is carbon and the yellow colour of the flame is caused by the carbon particles glowing.

Did you know?

If natural gas is burned in a bunsen burner with the air hole closed, the flame is yellow because there is not enough air. With the air hole open, much more air can get in and the gas is burned completely. When the gas is burned completely, all the available energy of combustion is released. This is why the blue flame is hotter than the yellow flame.

How the burning of hydrocarbon fuels affects the environment

Burning hydrocarbon fuels causes a number of problems.

- The production of large amounts of carbon dioxide has caused the level of carbon dioxide in the atmosphere to rise. The result is an increase in **global warming** – the **greenhouse effect** (see section 6.2).

- If hydrocarbons are burned in a poor supply of air, carbon monoxide is produced. Carbon monoxide is a colourless gas that has no smell. It is highly poisonous.

- Sulphur dioxide is produced because most hydrocarbon fuels contain sulphur. Sulphur dioxide dissolves in the water in the air to form acid rain.

Did you know?

Carbon monoxide kills because it combines with haemoglobin in the blood. This stops the haemoglobin binding to oxygen, so the body's cells get starved of oxygen and stop working. As little as 0.1% of carbon monoxide can be fatal.

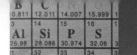

Alkanes

Alkanes are hydrocarbons. The simplest of the alkanes is methane, formula CH_4.

All alkanes have the general formula (C_nH_{2n+2}). In the molecule, all the bonds are covalent (see section 7.2). The **structural formula** of methane is:

$$
\begin{array}{c}
\quad\;\; H \\
\quad\;\; | \\
H - C - H \\
\quad\;\; | \\
\quad\;\; H
\end{array}
$$

Did you know?

The methane molecule isn't really 'flat', it is in the shape of a tetrahedron. A tetrahdera has four faces. All the faces are equilateral triangles. The diagrams below are of two different 'models' of a methane molecule.

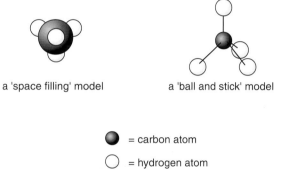

a 'space filling' model   a 'ball and stick' model

● = carbon atom

○ = hydrogen atom

The diagrams above show the shape of a methane molecule.

Figure 10.20 gives the names and formulae of the first three alkanes of the alkane family.

Name of alkane	Molecular formula	Structural formula
methane	CH_4	$\begin{array}{c} H \\ \| \\ H-C-H \\ \| \\ H \end{array}$
ethane	C_2H_6	$\begin{array}{c} H\quad H \\ \|\quad\| \\ H-C-C-H \\ \|\quad\| \\ H\quad H \end{array}$
propane	C_3H_8	$\begin{array}{c} H\quad H\quad H \\ \|\quad\|\quad\| \\ H-C-C-C-H \\ \|\quad\|\quad\| \\ H\quad H\quad H \end{array}$

Figure 10.20
The molecular and structural formulae of the first three alkanes

Alkenes

Alkenes are also hydrocarbons. The simplest of these alkenes is ethene which has the formula C_2H_4. Poly(ethene), common name polythene, is made from ethene.

All alkenes have the general formula C_nH_{2n}. In the molecule all the bonds are covalent (see section 7.2).

Name of alkene	Molecular formula	Structural formula
ethene	C_2H_4	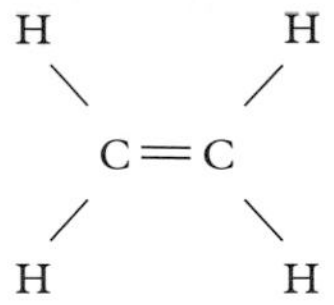
propene	C_3H_6	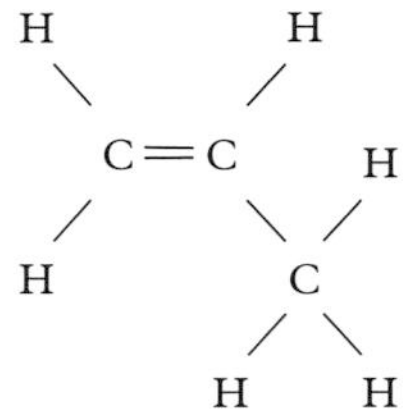

Figure 10.21
The molecular and structural formulae of the first two alkenes

The test for alkenes

Alkenes have a carbon–carbon double covalent bond (see section 7.2). Unlike a carbon–carbon single bond, the double bond is not particularly strong. It can be attacked by some chemicals. This means that alkenes are more reactive than alkanes. For example, **bromine water** will react with ethene but not with ethane. The bromine water loses its colour in the reaction.

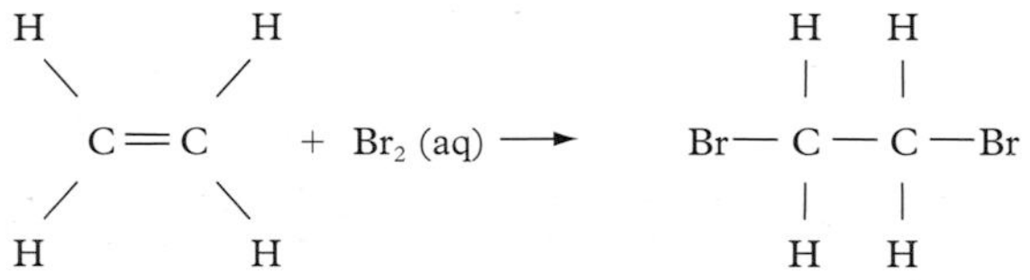

ethene + bromine water (yellow) → dibromoethane (colourless)

Alkenes decolourise bromine water, alkanes do not. This reaction is used as a test to tell the difference between alkanes and alkenes.

Compounds like alkenes which have a $C=C$ double bond are called **unsaturated hydrocarbons**. Alkanes have no $C=C$ double bond, only C—C single bonds. They are called **saturated hydrocarbons**.

Cracking

If the molecules in crude oil are heated strongly, they can break down into smaller molecules, some of which are useful as fuels. Industrially a catalyst is used to speed this process up. This is an example of a thermal decomposition reaction. The thermal decomposition of molecules in crude oil is called **cracking**. It is possible to carry out catalytic cracking on a small scale in the laboratory.

197

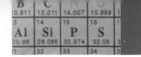
Figure 10.22
Apparatus for producing ethene in the laboratory by cracking petroleum jelly

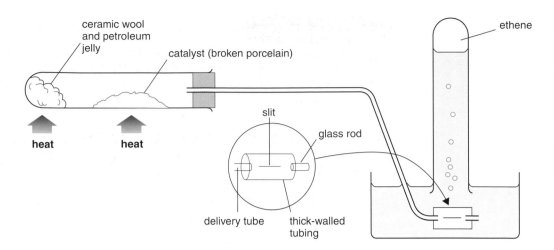

When alkanes are cracked, some of the products are alkenes. In the example below the alkene produced is ethene.

$$alkane \rightarrow alkane + alkene$$
$$C_{10}H_{22} \rightarrow C_8H_{18} + C_2H_4$$
$$decane \rightarrow octane + ethene$$

Cracking is useful because it allows the large molecules of some unwanted products in crude oil to be broken down into more useful substances which have smaller molecules.

Polymerisation

Polymers are large molecules made by joining smaller molecules together. The process is called **polymerisation**. The smaller molecules are called **monomers**. Plastics are polymers. One of the commonest polymers is poly(ethene) or polythene. It is made by polymerising molecules of ethene.

Polymerisation reactions of this type occur when unsaturated molecules (alkenes) join together to produce saturated alkanes with large molecules. Figure 10.23 shows the polymerisation of ethene to produce poly(ethene).

Figure 10.23
The polymerisation of ethene

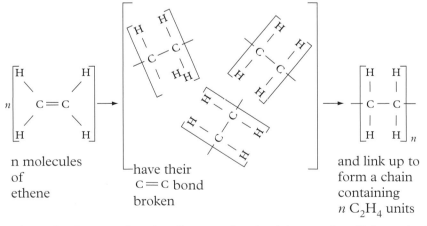

In this polymerisation reaction the alkene molecules join together. Polymerisation reactions of this type are called **addition polymerisation** reactions. This is because the molecules add on to each other.

Some of the products of cracking are used to make plastics. Poly(ethene) is used to make 'plastic' bags and bottles. Poly(propene) tends to be stronger than poly(ethene). It is used to make crates (e.g. milk crates) and ropes.

B	C			
0.811	12.011	14.007	15.999	1
3	14	15	16	1
Al	Si	P	S	
26.98	28.086	30.974	32.06	3
1	32	33	34	3

How the disposal of plastics affects the environment

There are three main ways by which most waste is dealt with. It can be dumped in landfill sites, burnt (incinerated) or recycled.

The demand for polymer materials (plastics) is increasing. Because the plastics are fairly cheap they are often thrown away when finished with. Many landfill sites discourage the disposal of plastics because most plastics are not **biodegradable**. Unlike wood, paper, wool and cotton, plastics do not break down by the action of micro-organisms, and unlike metal, plastics do not corrode over time. Recently some biodegradable plastics have been developed.

Figure 10.24
A landfill site

Figure 10.25
An incinerator

Figure 10.26
Collecting skip for plastic bottles

Incineration involves the burning of waste in a very hot furnace. Some people consider that the burning of many plastics, especially those made of polyvinyl chloride (PVC), produces a harmful group of chemicals called dioxins and heavy metals. The dioxins are thought to be associated with birth defects and some kinds of cancers. The advantage of the incineration of plastics is that it provides a relatively cheap source of energy for the community.

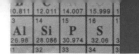

Recycling is now being encouraged in many parts of the world as it seems to be the most practical method to deal with the problem of disposal. For most plastics the process of recycling is fairly cheap and straightforward.

Summary

- **Crude oil** is another useful substance extracted from the Earth.

- Crude oil is the fossilised remains of small animals and plants.

- Crude oil is mainly made of **hydrocarbons**.

- Crude oil can be separated into its various hydrocarbons by **fractional distillation**.

- Hydrocarbons are compounds that contain hydrogen and carbon only.

- Hydrocarbons burn in plenty of air to produce water and carbon dioxide.

- Hydrocarbons burn in a limited air supply to produce water vapour, carbon monoxide and carbon.

- Carbon dioxide is a gas. It will turn limewater milky.

- Hydrocarbons can be **alkanes** or **alkenes**.

- Alkanes have C—C single bond.

- Alkanes are called **saturated hydrocarbons**.

- Alkenes have a C═C double bond.

- Alkenes are called **unsaturated hydrocarbons**.

- Large hydrocarbon molecules can be broken down into smaller ones by a process called **cracking**.

- Small alkene molecules can be joined together to make very large alkane molecules by the process of **polymerisation**.

Topic questions

1 The table opposite lists the properties of some of the hydrocarbons obtained from crude oil. (The hydrocarbons are not listed in any particular order).

Write the letters, A, B or C for each of the following. (You may use each letter once, more than once or not at all.)

Which hydrocarbon:

a) could be petrol?
b) could be bitumen?
c) has the highest boiling point?
d) has seven or eight carbon atoms in its molecule?
e) could be used to power a motor car?

2 Complete the following sentences.

When hydrocarbon fuels are burnt in plenty of air, they produce the gas carbon dioxide and _____. The test for carbon dioxide is to add the gas to _____. If carbon dioxide is present, the liquid goes _____ If there is not enough air present when the hydrocarbon is burnt, the gas _____ can be produced. When another fuel was burned in plenty of air, *only* carbon dioxide was produced. The fuel used was _____.

Hydrocarbon	Appearance	Viscosity	What happens when the hydrocarbon is burned
A	black liquid	very high	very difficult to get it to burn at all
B	colourless gas	very low	burns very easily with a blue, smokeless flame
C	colourless liquid	medium	burns easily with smoky, yellow flame

3 a) For each of the following hydrocarbons, state if the molecule is saturated or unsaturated.

i) methane
ii) ethene
iii) propane

b) Which of the hydrocarbons named in a) is the main substance in natural gas?

c) Which of the hydrocarbons named in a) would decolourise bromine water?

d) Which of the hydrocarbons named in a) would have the general formula C_nH_{2n}?

e) Write a balanced chemical equation for the complete combustion of methane.

f) Draw a structural formula for one alkane and one alkene mentioned in part a).

4 a) What is the name given to the process used to break large hydrocarbon molecules into smaller ones?

b) Why is it useful to be able to break large hydrocarbon molecules into smaller ones?

c) When the process is applied to an alkane are the products:
   A all alkanes?
   B all alkenes?
   C a mixture of alkanes and alkenes?

5 Complete the following sentences using the words or phrases in the box.

> large small poly(ethene) polygon clothes
>
> plastic plastic bottles rope

Polymers are _____ molecules made by joining _____ molecules together. One common polymer is _____ . This polymer is used to make _____ .

Examination questions

1 a) Complete the sentences.
   i) A mineral, or a mixture of minerals, from which a metal can be extracted is called an _____ . *(1 mark)*
   ii) The reactivity series for some metals is:

   potassium
   sodium
   calcium
   magnesium
   aluminium
   zinc
   iron
   copper
   gold

   The name of the method used to extract potassium from the mineral potassium chloride is _____ . *(1 mark)*

   b) The diagram shows the type of furnace used to extract iron.
   i) What is the name of this type of furnace? *(1 mark)*
   ii) Name the main gas, other than nitrogen, in the waste gas, W. *(1 mark)*
   iii) Name the raw materials shown as X, Y and Z. *(3 marks)*
   iv) Carbon monoxide gas forms in the furnace. Balance the chemical equation for a reaction which produces it.

   $$\underline{}\ C(s) + O_2(g) \rightarrow \underline{}\ CO(g)$$

   (1 mark)

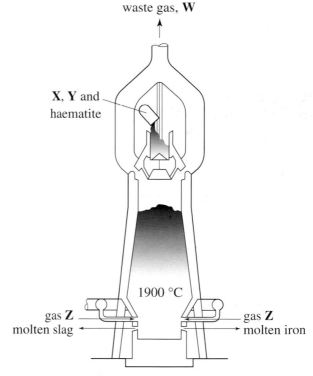

waste gas, **W**

X, **Y** and haematite

1900 °C

gas **Z**
molten slag

gas **Z**
molten iron

c) Carbon monoxide reacts with iron(III) oxide. This is the chemical equation for the reaction which occurs.

$$Fe_2O_3(s) + 3CO(g) \rightarrow 2Fe(l) + 3CO_2(g)$$

   i) What does the symbol (l) mean? *(1 mark)*
   ii) Complete the **two** spaces with the names of the chemicals.
   In this reaction _____ is oxidised to _____ . *(1 mark)*

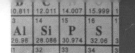

2 Aluminium can be extracted from its ore bauxite. Bauxite contains aluminium oxide, Al_2O_3, which is purified and then processed as shown.

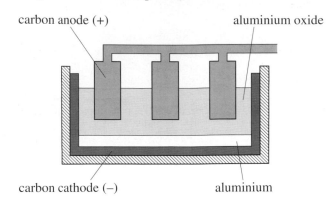

carbon anode (+)   aluminium oxide

carbon cathode (–)   aluminium

a) The word equation for the extraction of aluminium is shown.

aluminium oxide → aluminium + oxygen

i) Write the balanced chemical equation for this reaction. *(3 marks)*
ii) Describe how aluminium is extracted in this process. *(3 marks)*

b) During the extraction carbon dioxide gas is produced. Suggest why. *(2 marks)*

3 The high demand for petrol (octane) can be met by breaking down longer hydrocarbons, such as decane, by a process known as cracking.

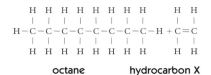

a) Apart from heat, what is used to make the rate of this reaction faster? *(1 mark)*

b) Octane is a **hydrocarbon**.
i) What does **hydrocarbon** mean? *(1 mark)*
ii) Give the molecular formula of octane. *(1 mark)*

c) The hydrocarbon **X** is used to make poly(ethene).
i) What is the name of **X**? *(1 mark)*
ii) What is the name of the process in which **X** is changed into poly(ethene)? *(1 mark)*

4 The table gives information about some alkanes

Name of alkane	Number of carbon atoms in each molecule	boiling point of alkane (°C)
methane	1	−161
propane	3	−42
butane	4	0
pentane	5	
hexane	6	69
octane	8	126
decane	10	

a) Draw a graph of boiling point against number of carbon atoms in the molecule for these alkanes. (Use graph paper.)
Draw your graph so that it can be extended to allow you to find the boiling point of the alkane with ten carbon atoms in each molecule. *(3 marks)*

b) Use your graph to find the boiling point of:
i) pentane
ii) decane. *(2 marks)*

5 The label has been taken from a tube of *Humbrol Polystyrene Cement*, a glue used in model making.

HUMBROL

Polystyrene Cement

X HARMFUL

Paint product contains 1.1.1 TRICHLOROETHANE

• Keep container tightly closed. Harmful by inhalation, in contact with skin and if swallowed.
Avoid contact with eyes.
Keep out of reach of children.

• For use on all polystyrene plastic except expanded or foam. Specially recommended for plastic kits. Thinly coat each surface, press together. To remove cement from fabrics use Humbrol Universal Cleaner.

HUMBROL LTD., HULL, ENGLAND.

a) The solvent used is 1,1,1-trichloroethane. The structural formula of this molecule is:

$$Cl - \overset{\overset{\displaystyle Cl}{|}}{\underset{\underset{\displaystyle Cl}{|}}{C}} - \overset{\overset{\displaystyle H}{|}}{\underset{\underset{\displaystyle H}{|}}{C}} - H$$

i) What do the lines between the atom represent? *(1 mark)*
ii) State whether 1,1,1-trichloroethane is saturated or unsaturated. Give **one** reason for your answer. *(1 mark)*

iii) 1,1,1-trichloroethane is being replaced in favour of a 'better' solvent. Use information on the label to help you to suggest why.   *(1 mark)*

b) Polysytrene is a plastic. Plastics are polymers which are made by the process of polymerisation.

   i)   What is meant by polymerisation?
   (2 mark)

   ii)   The table gives information about monomers and the polymers made from them. Complete the table.   *(3 marks)*

MONOMER		POLYMER	
name	formula	name	formula
ethene	H H \\ / C=C / \\ H H		⎛ H H ⎞ ⎜ | | ⎟ —C — C— ⎜ | | ⎟ ⎝ H H ⎠ₙ
styrene		polystyrene	⎛ H H ⎞ ⎜ | | ⎟ —C — C— ⎜ | | ⎟ ⎝ H C₆H₅ ⎠ₙ
chloroethene	H H \\ / C=C / \\ H Cl	poly(chloro-ethene)	

Chapter 11
Patterns of behaviour

Key terms   acids • alkali • alkali metals • anode • atomic number •
bases • cathode • displacement • electrolysis • electrolyte •
galvanising • group • halide • halogens • hydrogen •
hydrogen ion • hydroxide ions • indicators • inert • ions •
neutral • neutralisation • noble gases • period • periodic table
• pH • pH scale • transition elements • transition metal •
universal indicator

11.1 The development of the periodic table

Co-ordinated	Modular
DA 11.11	DA 8
SA 11.3	SA 16

The story behind the periodic table of the elements

In order to understand why the **periodic table** came into existence it is important to be aware of some of the work on elements that was being carried out between 1770 and 1810.

The work of a French chemist Antoine Laurent Lavoisier (1743–1794) had helped scientists move away from Aristotle's idea of the four elements being air, earth, fire and water by developing the idea of a chemical element. He was the first scientist to define an element as being a substance that could not be broken down by chemical methods.

Lavoisier believed that elements could be divided into four groups according to their chemical behaviour. His table of elements was published in 1789.

Figure 11.1
Table to show Lavoisier's attempt at classifying some simple substances. The modern names are given in brackets

Acid-making elements	Gas-like elements	Metallic elements		Earthy elements
charcoal (carbon)	azote (nitrogen)	cobalt		argilla (aluminium oxide)
phosphorus	caloric (heat)	mercury		barytes (barium sulphate)
sulphur	hydrogen	copper	nickel	lime (calcium oxide)
	light	gold	silver	magnesia (magnesium oxide)
	oxygen	iron	tin	silex (silicon dioxide)
		lead	zinc	

Lavoisier's table includes some substances we know as compounds. He classified them as elements because the chemical method needed to break them down was not known at that time.

In 1808 John Dalton (1766-1844) (see section 7.1), a British chemist and physicist, developed Lavoisier's ideas further when he produced a theory about the structure of elements and compounds. Dalton proposed that:

- all elements were made up of very small solid particles called atoms

- the atoms of a given element are alike and have the same weight

- the atoms of different elements are different

- chemical compounds are formed when atoms combine

- a given compound is always made up of the same number and type of atoms.

Dalton's proposals showed that many of Lavoisier's 'elements' were not in fact elements.

Antoine Lavoisier

During the 19th century chemists were discovering and finding out the behaviour of a large number of new elements. Imagine that as each new element was discovered, the following information was recorded on a card.

```
Name of element ................................

Chemical and physical properties

................................................

................................................

................................................

Atomic weight* ................................
```

(* now called relative atomic mass)

Chemists, just like all good scientists, began to realise that it was necessary to classify this increasing amount of knowledge about the elements into some kind of pattern. The pattern linked together elements with similar properties.

In 1817 Johann Wolfgang Dobereiner, a German chemist, who would have had only a few cards available, showed that the atomic weight of strontium was almost midway between that of calcium and barium and that these three elements had similar properties. Later he showed that the cards of certain other elements with similar properties could also be arranged in sets of three (triads) – for example chlorine, bromine and iodine; iron, cobalt and manganese.

Unfortunately, other chemists failed to grasp the importance of these triads mainly because of the limited number of elements used.

In 1864 John Newlands, a British chemist, proposed and published the idea that if cards were arranged in groups of seven in order of atomic weight, then every eighth element in this grouping shared similar chemical and physical properties. For example lithium and sodium, sodium and potassium, magnesium and calcium.

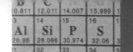

Patterns of behaviour

Newland's arrangement of some of the elements is shown in Figure 11.2.

Figure 11.2
Newland's arrangement of some of the elements

1	2	3	4	5	6	7
H	Li	Be	B	C	N	O
F	Na	Mg	Al	Si	P	S
Cl	K	Ca				

Dmitri Mendeleev

This idea came to be known as the 'law of octaves' because it suggested comparisons with the musical scale. But because of this comparison the law was ridiculed until the work of Mendeleev showed how much truth there was in Newland's idea.

Dmitri Mendeleev (1834–1907) was a Russian chemist who was professor at the University of St Petersburg. At that time there were no good chemistry textbooks, so Mendeleev wrote his own. It was for this book that he started to classify the elements on the basis of their known properties and atomic weights.

In 1869 his first version of a periodic table was published. This was further refined and in 1871 a second version was published. By the time he produced his classification he had information on more than 60 cards. Mendeleev arranged the elements initially in order of atomic weight, but then proceeded to try to make sure that elements with similar properties were in the same vertical column.

	Group 1		Group 2		Group 3		Group 4		Group 5		Group 6		Group 7
Period 1	H (1)												
Period 2	Li (7)		Be (9.4)		B (11)		C (12)		N (14)		O (16)		F (19)
Period 3	Na (23)		Mg (24)		Al (27.3)		Si (28)		P (31)		S (32)		Cl (35.5)
Period 4	K (39)		Ca (40)		?		Ti (48)		V (51)		Cr (52)		Mn (55)
		Cu (63)		Zn (65)		?		?		As (75)		Se (78)		Br (80)
Period 5	Rb (85)		Sr (87)		Y (88)		Zr (90)		Nb (94)		Mo (96)		?
		Ag (108)		Cd (112)		In (113)		Sn (118)		Sb (122)		Te (125)		I (127)

Figure 11.3
Part of Mendeleev's 1869 periodic table. The numbers in brackets are the values for the atomic weights used by Mendeleev

Many scientists did not accept the table at first and treated its contents as no more than an interesting curiosity. But Mendeleev's classification contained many gaps and he believed that the gaps were for elements not yet discovered.

Mendeleev realised that he could prove the value of his classification if it could be used as a tool to predict the missing elements. To do this he set about predicting the properties of the elements likely to occupy the three spaces shaded in the table.

Mendeleev predicted that the missing element in group 4 would have an atomic weight that was likely to be the average of the atomic weights of silicon (Si) and tin (Sn). He also predicted the colour, the density, the melting point and the formula of the oxide of the missing element, which he called ekasilicon. In 1886 germanium was discovered and found to match almost completely all the predictions made for ekasilicon.

Mendeleev made further predictions for the properties of the two elements missing from group 3. The element missing between calcium (Ca) and titanium (Ti) he called ekaboron and the element missing next to zinc (Zn) he called ekaaluminium. In 1875 gallium was discovered and found to match almost completely the predictions made for ekaaluminium and in 1879 scandium was discovered and found to match almost completely the predictions made for ekaboron.

The correctness of Mendeleev's predictions proved to other scientists the importance of the ideas behind his periodic table. Many scientists began to use the table as a working tool that would help them complete the gaps in the rest of the table. Indeed in the 30 years following the publication of the periodic table many more elements were discovered.

The periodic table has undergone two main revisions since being proposed by Mendeleev. The first revision was the inclusion of a new family of elements called the inert or noble gases whose existence was unsuspected for most of the 19th century. The second revision was brought about by the work of Henry Moseley (1887–1915).

Henry Moseley was a British chemist who used X-ray spectra to study atomic structure. His work showed that the **atomic number** and not the atomic weight caused the repeated patterns of behaviour of different elements. In Mendeleev's version of the periodic table based on arranging the elements in order of their relative atomic weights, with elements with similar properties being in the same column most elements are in appropriate groups. However a few are not. Argon atoms, for example, have a greater relative atomic mass (40) than potassium atoms (39). Because of Moseley's work, the elements in the modern periodic table are arranged in order of the atomic number (proton number), and all elements are in appropriate groups.

The contents of the modern periodic table can be used as a tool to predict:

- whether an element is a metal or a non-metal

- the physical properties of an element

- the relative reactivities of the various elements

- the reactions between elements

- the charge on the ions of the elements

- chemical formulae.

All of these predictions are made possible because the modern periodic table is a comprehensive summary of the structure of the atom of each of the known elements.

The modern periodic table is shown in Figure 11.4, on the next page.

Figure 11.4
The periodic table

Mass number A / Atomic number (Proton number) Z: 1 H hydrogen 1

1 (Alkali metals)	2	Transition metals										3	4	5	6	7 (Halogens)	0 (Noble gases)
																	4 He helium 2
7 Li lithium 3	9 Be beryllium 4											11 B boron 5	12 C carbon 6	14 N nitrogen 7	16 O oxygen 8	19 F fluorine 9	20 Ne neon 10
23 Na sodium 11	24 Mg magnesium 12											27 Al aluminium 13	28 Si silicon 14	31 P phosphorus 15	32 S sulphur 16	35 Cl chlorine 17	40 Ar argon 18
39 K potassium 19	40 Ca calcium 20	45 Sc scandium 21	48 Ti titanium 22	51 V vanadium 23	52 Cr chromium 24	55 Mn manganese 25	56 Fe iron 26	59 Co cobalt 27	59 Ni nickel 28	63 Cu copper 29	64 Zn zinc 30	70 Ga gallium 31	73 Ge germanium 32	75 As arsenic 33	79 Se selenium 34	80 Br bromine 35	84 Kr krypton 36
85 Rb rubidium 37	88 Sr strontium 38	89 Y yttrium 39	91 Zr zirconium 40	93 Nb niobium 41	96 Mo molybdenum 42	Tc technetium 43	101 Ru ruthenium 44	103 Rh rhodium 45	106 Pd palladium 46	108 Ag silver 47	112 Cd cadmium 48	115 In indium 49	119 Sn tin 50	122 Sb antimony 51	128 Te tellurium 52	127 I iodine 53	131 Xe xenon 54
133 Cs caesium 55	137 Ba barium 56	139 La lanthanum 57	178 Hf hafnium 72	181 Ta tantalum 73	184 W tungsten 74	186 Re rhenium 75	190 Os osmium 76	192 Ir iridium 77	195 Pt platinum 78	197 Au gold 79	201 Hg mercury 80	204 Tl thallium 81	207 Pb lead 82	209 Bi bismuth 83	Po polonium 84	At astatine 85	Rn radon 86
Fr francium 87	226 Ra radium 88	227 Ac actinium 89															

Elements 58–71 and 90–103 have been omitted.

The value used for mass number is normally that of the commonest isotope, eg ^{35}Cl not ^{37}Cl
Bromine is approximately equal proportions of ^{74}Br and ^{81}Br

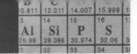

Summary

◆ Early attempts to organise the elements into patterns that reflected their behaviour arranged the elements in order of atomic weight. The modern **periodic table** arranges the elements in increasing order of atomic number.

Topic questions

1 During the 19th century there was a need to develop some form of classification for the elements. Why?

2 It was important that the atomic weight of each element should be measured as accurately as possible. Why?

3 Newland's table was considered to be less useful than that produced by Mendeleev. Why?

4 In what way were the findings of Newlands similar to those of Mendeleev?

5 Use the atomic weights given in Figure 11.3 to predict the likely atomic weight for the element missing from Group 4. Explain how you got to your answer.

6 Explain why Mendeleev's periodic table was finally accepted by other scientists?

7 Which family of elements was missing from Mendeleev's periodic table? Give a reason.

8 Which change to Mendeleev's periodic table resulted from the work of Moseley?

Note: When answering examination questions about the development of the periodic table you will not be expected to remember all the information provided in this section. The examination questions will provide all the background information you might need for your answer.

11.2 Patterns in the periodic table

Co-ordinated	Modular
DA 11.11	DA 8
SA 11.3	SA 16

Period

A **period** is a horizontal row of elements in the periodic table in order of increasing atomic number. Period 2 contains sodium, magnesium, aluminium, silicon, phosphorus, sulphur, chlorine and argon. A new period is started each time a new outer shell of electrons (see section 7.1) is started.

The chemical properties of the elements in a period change gradually across the period as an extra outer electron is added with each new element.

Group

A **group** is a vertical column of elements in the periodic table. The atoms of elements in the same group have the same number of electrons in their outer shell. Because of this, elements in the same group have similar chemical properties.

Each group is given a number which represents the number of electrons in the outer shell of their atoms. (In Group 0 the outer shell is full.)

Metals are found in groups on the left-hand side of the periodic table, e.g. Groups 1 and 2. The non-metals are found in groups on the right-hand side of the periodic table.

There is a gradual change from metal to non-metal as you go across a period. The metals found in the central block are called the **transition metals** or **transition elements**.

Group 0: The noble gases

Group 0 is on the far right-hand side of the periodic table. Each element has atoms with a full outer shell of electrons.

Properties of noble gases

1 They are all gases at room temperature.

2 They are all monatomic and exist as single atoms e.g. He and Ne. (Most elements which are gases form molecules containing more than one atom e.g. H_2, O_2 and O_3.)

3 They are unreactive (**inert**) towards other elements.

Noble gas are unreactive and monatomic because their atoms have a full outer shell of electrons and do not need to gain, lose or share any other electrons to become more stable.

Figure 11.5
Physical properties of helium, neon and argon

Element	Atomic number	Relative atomic mass	Melting point (°C)	Boiling point (°C)
helium	2	4.0	−270	−269
neon	10	20.2	−249	−246
argon	18	40.0	−189	−186

Did you know?

Originally the noble gases were called the inert gases because they were thought to be completely unreactive. However, xenon was found to react with fluorine, which is a very reactive element, and the group name had to be changed.

Uses of noble gases

● Helium is used in airships because it is much less dense than air. It provides the buoyancy needed and is much safer than hydrogen because it is non-flammable.

● Neon is used to give the red coloured light used in advertising signs.

● Argon is used as the inert gas in light bulbs. It is unreactive and, unlike air, does not react with the hot metal filament in the light bulb. This helps the light bulb to last longer.

Group 7: the halogens

Group 7 is near the right-hand side of the periodic table. Each element has atoms with seven electrons in the outer shell of electrons.

The atoms of the **halogens** join together to form diatomic molecules e.g. Cl_2.

Figure 11.6
Physical properties of chlorine, bromine and iodine

Element	Atomic number	Relative atomic mass	Melting point (°C)	Boiling point (°C)
chlorine	17	35.5	−101	−34
bromine	35	80	−7	58
iodine	53	127	114	183

B	C	N	O	
0.811	12.011	14.007	15.999	1
3	14	15	16	1
Al	**Si**	**P**	**S**	3
26.98	28.086	30.974	32.06	3
1	32	33	34	3

Properties of chlorine, bromine and iodine

1 Melting points and boiling points are low, but increase down the group as the atomic number increases.

2 They are brittle/crumbly when solid.

3 They are poor conductors of heat and electricity when solid or liquid.

4 Reactivity decreases down the group as the atomic number increases.

In a reaction with a metal, a halogen atom will gain an electron to form a halide ion, X^-. Adding an extra electron to the outer shell of electrons becomes more difficult as the size of the atom increases down the group. This is because the outer electrons are further from the nucleus and less strongly attracted. The extra inner electron shells also shield the outer electrons from the full attractive force from the nucleus. Both of these effects mean that the extra electron is more difficult to add to the outer shell of electrons as part of a reaction.

5 They are all non-metals with coloured vapours.
- Fluorine is a straw coloured gas
- Chlorine is a yellow-green gas
- Bromine is a red liquid with a red-brown vapour
- Iodine is a grey shiny solid and gives off a violet vapour on heating.

6 They react with hydrogen forming covalent hydrogen **halides**.
Hydrogen reacts with chlorine to form hydrogen chloride.
$$H_2 + Cl_2 \rightarrow 2HCl$$
Hydrogen reacts with bromine to form hydrogen bromide.
$$H_2 + Br_2 \rightarrow 2HBr$$
Hydrogen reacts with iodine to form hydrogen iodide.
$$H_2 + I_2 \rightarrow 2HI$$

7 **Displacement** reactions.
Each halogen will displace a halogen lower in the group to form a solution of the lower halide.
Chlorine will displace bromine and iodine from bromide and iodide solutions.
$$Cl_2 + 2KBr \rightarrow Br_2 + KCl$$
$$Cl_2 + 2KI \rightarrow I_2 + KCl$$
Bromine will displace iodine from iodide solutions but not chlorine from chloride solutions.
$$Br_2 + 2KI \rightarrow I_2 + KBr$$

8 They react with alkali metals. Ionic solids are formed with the halogen ion, Cl^-, Br^- and I^-, having a single negative charge.

9 They react with non-metallic elements to form covalent molecular compounds e.g. CCl_4.

Summary

- A **period** is a horizontal row of elements in the periodic table arranged in order of increasing atomic number.

- A **group** is a vertical column of elements whose atoms have the same number of electrons in their outer shell.

- Similarities and differences in the properties of elements can be explained in terms of their electronic structure.

- More than three quarters of the elements are metals found in Groups 1 and 2 and in the block of transition metals.

- The elements in Groups 7 and 0 are non-metals.

- The elements in Group 7 are called the **halogens**, their ions carry a single negative charge.

- The reactivity of the halogens decreases going down the group.

- The elements in Group 0 are called the **noble gases**.

- The noble gases are unreactive because their atoms have a complete outer shell.

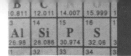

Topic questions

1 What is the name given to:
   a) each horizontal row of elements in the periodic table?
   b) each column of elements in the periodic table?

2 What happens to the atomic numbers of the elements in a particular period as you move from left to right across the periodic table?

3 What can you say about the number of electrons in the outer shell of the elements in a particular group?

4 What can you say about the outer shell of the elements in Group 0?

5 In which groups are the non-metals found?

6 a) Give three properties of the noble gases.
   b) What happens to the boiling points of the noble gases as their atomic number increases?

7 a) What is the name of the elements in Group 7?
   b) How many electrons are in the outer shell of the elements in Group 7?
   c) What happens to the reactivity of the Group 7 elements as their atomic number increases?

8 Use your periodic table to work out the number of electrons in the outer electron shells of the following elements:
   a) sodium
   b) chlorine
   c) oxygen
   d) phosphorus
   e) magnesium
   f) carbon
   g) aluminium
   h) neon
   i) fluorine
   j) boron

11.3   Metals and the periodic table

Co-ordinated	Modular
DA 11.11	DA 8
SA 11.3	SA 15

Group 1: The alkali metals

Group 1 is on the left-hand side of the periodic table. Each element has atoms with one electron in the outer shell of electrons.

Figure 11.7
Physical properties of lithium and sodium and potassium

Element	Atomic number	Relative atomic mass	Melting point (°C)	Boiling point (°C)	Density (g/cm³)
lithium	3	6.9	180	1330	0.53
sodium	11	23.0	98	883	0.97
potassium	19	39.1	64	760	0.86

Properties of lithium, sodium and potassium

1 They are metals with a low melting point and a low density (they float on water).

2 Their melting and boiling points decrease as you go down the group with increasing *atomic number* (see section 7.1).

3 They react readily with cold water to form **hydrogen** and metal hydroxides which dissolve in water to give **alkaline** solutions.

$$\text{lithium} + \text{water} \rightarrow \text{lithium hydroxide} + \text{hydrogen}$$
$$2\text{Li(s)} + 2\text{H}_2\text{O(l)} \rightarrow 2\text{LiOH(aq)} + \text{H}_2\text{(g)}$$

$$\text{sodium} + \text{water} \rightarrow \text{sodium hydroxide} + \text{hydrogen}$$
$$2\text{Na(s)} + 2\text{H}_2\text{O(l)} \rightarrow 2\text{NaOH(aq)} + \text{H}_2\text{(g)}$$

$$\text{potassium} + \text{water} \rightarrow \text{potassium hydroxide} + \text{hydrogen}$$
$$2\text{K(s)} + 2\text{H}_2\text{O(l)} \rightarrow 2\text{KOH(aq)} + \text{H}_2\text{(g)}$$

The reactivity increases down the group as the atomic number increases. This is shown by the increasingly vigorous reactions of lithium, sodium and potassium with water. A test for hydrogen is described on the next page.

Figure 11.8
a) lithium, b) sodium and c) potassium reacting with water

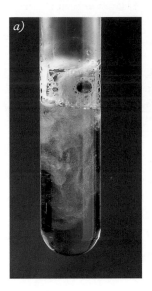  

In a reaction, the metal atom will lose an electron to form a metal ion, M^+. The outer electron on the atom becomes easier to remove as the size of the atom increases down the group. This is because the outer electron is further from the nucleus and less strongly attracted by the positively-charged protons. The extra inner electron shells also shield the outer electron from the full attractive force from the nucleus. Both of these effects mean that the outer electron is less firmly held by the nucleus and is easier to lose as part of a reaction.

4 They react with halogens to give ionic halides.
The resulting metal ion has a single positive charge, M^+.
Lithium reacts with chlorine to give lithium chloride, Li^+Cl^-.

$$2Li(s) + Cl_2(g) \rightarrow 2LiCl(s)$$

They react with bromine to give ionic bromides.
Sodium reacts with bromine to give sodium bromide, Na^+Br^-.

$$2Na(s) + Br_2(l) \rightarrow 2NaBr(s)$$

They react with iodine to give ionic iodides.
Potassium reacts with iodine to give potassium iodide, K^+I^-.

$$2K(s) + I_2(s) \rightarrow 2KI(s)$$

5 They react with oxygen to form ionic oxides $(M^+)_2O^{2-}$.
The resulting metal ion has a single positive charge, M^+.
Lithium reacts with oxygen to form lithium oxide, $(Li^+)_2O^{2-}$.

$$4Li(s) + O_2(g) \rightarrow 2Li_2O(s)$$

Sodium reacts with oxygen to form sodium oxide, $(Na^+)_2O^{2-}$.

$$4Na(s) + O_2(g) \rightarrow 2Na_2O(s)$$

Potassium reacts with oxygen to form potassium oxide, $(K^+)_2O^{2-}$.

$$4K(s) + O_2(g) \rightarrow 2K_2O(s)$$

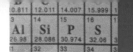
A laboratory test for hydrogen

If a lighted splint is held over the top of a test tube of hydrogen, the hydrogen burns with a squeaky explosion.

Alkali metal compounds

Alkali metal halides are ionic compounds, which dissolve in water to form separate aqueous ions. Each ion is surrounded by water molecules in solution.

$$Na^+Cl^-(s) \xrightarrow{(H_2O)} Na^+(aq) + Cl^-(aq)$$

Patterns of behaviour

Alkali metal hydroxides are ionic compounds that dissolve in water to form alkaline solutions with a pH of >7.

These alkaline solutions contain **hydroxide ions**, $OH^-(aq)$. Sodium hydroxide dissolves in water to form $Na^+(aq)$ and $OH^-(aq)$ ions.

$$Na^+OH^-(s) \xrightarrow{(H_2O)} Na^+(aq) + OH^-(aq)$$

Industrially, sodium hydroxide solution is formed by the electrolysis of sodium chloride solution. Chlorine and hydrogen are also formed at the same time. The simultaneous production of three useful industrial chemicals helps to make the whole process more economical.

Summary

◆ The elements in Group one are called the **alkali metals,** their metal ion carries a single positive charge.

◆ The reactivity of the alkali metals increases going down the group.

◆ The alkali metals react with water to form hydrogen and metal hydroxides which dissolve in water to give **alkaline** solutions.

◆ The test for hydrogen is that it causes a lighted splint to burn with a squeaky explosion.

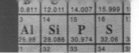

Topic questions

1 a) What is the name given to the elements in Group 1?
   b) How many electrons are there in the outer shell of these elements?

2 What happens to the reactivity of the Group 1 elements as their atomic number increases?

3 Write down the word equation for the reaction between sodium and water.

4 What is the laboratory test for hydrogen?

5 a) Write down the word equation for the reaction between sodium and bromine.
   b) Write down the balanced symbol equation for the reaction between potassium and iodine. (Ignore state symbols)

11.4

Co-ordinated	Modular
DA 11.12	DA 5
SA n/a	SA n/a

Patterns in the transition elements

The **transition elements** are found between Groups 2 and 3 in the periodic table.

Figure 11.9
Physical properties of manganese, iron, copper, zinc and mercury

Element	Atomic number	Relative atomic mass	Melting point (°C)	Density (g/cm³)
manganese	25	54.9	1244	7.2
iron	26	55.8	1535	7.9
copper	29	63.5	1083	8.9
zinc	30	65.4	420	7.1
mercury	80	201.0	−39	13.6

Properties of the transition elements

1 They are metals with high melting points and high densities (except for mercury).

2 They can be used as catalysts to speed up reactions. For example, iron is used as a catalyst in the Haber process to make ammonia from nitrogen and hydrogen (see section 9.2).

3 They are hard, tough and strong.

4 They are much less reactive than the alkali metals and so do not corrode so quickly with oxygen and/or water.

Did you know?

Manganese, iron, copper and zinc are needed by plants and animals in very small amounts for healthy growth – they are known as trace elements. Iron is found in red blood cells as part of a haemoglobin molecule, which carries oxygen from the lungs to all body cells. A lack of iron in the body will result in anaemia.

Properties of compounds

1 They form coloured compounds. These can often be seen in pottery glazes of various colours or in weathered copper (green).

2 They have catalytic properties and speed up some reactions e.g. manganese (IV) oxide, MnO_2, helps to decompose hydrogen peroxide into water and oxygen.

Uses

- Cast iron is used to make manhole covers.
- Iron is used as a catalyst to manufacture ammonia.
- Copper is used in electrical wiring and domestic hot water pipes.
- Zinc is used in **galvanising** iron to prevent it from rusting.

Summary

◆ The **transition elements** all have metallic properties and form coloured compounds.

◆ Many transition elements are used as catalysts.

Topic questions

1 Identify the following elements from their descriptions:
   a) a metal that is used in household electrical wiring.
   b) a metal that is used as a catalyst in the manufacture of ammonia.
   c) a metal that is used to galvanise iron.
   d) a metal oxide that acts as a catalyst.

2 Between which groups in the periodic table are the transition elements found?

3 Why are these elements often called the transition metals?

4 Give four properties of transition elements.

5 Give two properties of the compounds formed from transition elements.

11.5	Patterns in the reactions of metal halides (halogens)

Co-ordinated	Modular
DA 11.12	DA 8
SA n/a	SA n/a

Common salt (sodium chloride, NaCl)

Most of the salt used on our foods comes not from the sea but from mines in Cheshire. Because salt is soluble in water it is extracted from deep underground by a process called solution mining. In hot countries salt is obtained from sea water by evaporation.

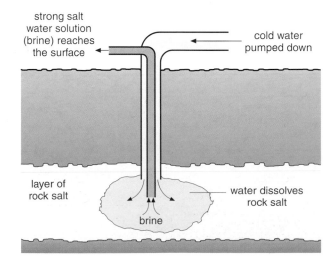

Figure 11.10
Solution mining

Figure 11.11
Salt pans – the water evaporates and the salt remains

0.811	12.011	14.007	15.999	1
Al	**Si**	**P**	**S**	**3**
26.98	28.086	30.974	32.06	3

Sodium chloride is a compound made from sodium (a very reactive alkali metal) and chlorine (a very poisonous and reactive halogen). This illustrates the fact that the chemical properties of the separate elements in a compound are completely different from the properties of the compound they form.

The electrolysis of sodium chloride solution (brine)

The **electrolysis** of brine is an important industrial process. It is important because of the three very useful chemicals produced, chlorine, sodium hydroxide and hydrogen.

Electrolysis of sodium chloride solution

Hydrogen (H₂)

Used in the making of margarine, paper and ceramics (pottery)

Chlorine (Cl)

Used to kill bacteria in drinking water and in swimming baths. Used in the making of disinfectants, hydrochloric acid, bleach and the plastic polymer PVC

Sodium hydroxide (NaOH)

Used in the making of ammonia and margarine

Figure 11.12
Products of electrolysis of sodium chloride

What happens during the electrolysis of sodium chloride solution?

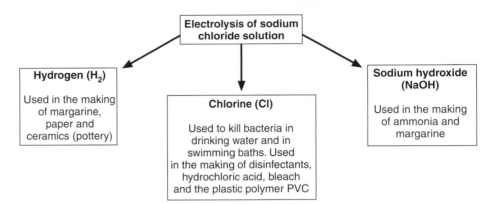

Because sodium chloride solution is the conducting solution (**electrolyte**) the ions present in it are H^+ ions and OH^- ions from the water and Na^+ ions and Cl^- ions from the sodium chloride.

The **anode** has a positive charge because it has lost electrons to the cathode. The anode therefore can gain electrons.

The OH^- and Cl^- ions are negatively charged so have gained electrons.

These negative ions are attracted to the positive electrode, the anode, where they can lose the extra electrons to make them into atoms. However, only the chloride ions become atoms. These atoms pair up to form Cl_2 molecules. The hydroxide ions remain in the solution. So, at the anode chlorine gas is released.

The reaction at the anode can be summarised as:
chloride ions − electrons → chlorine

$$2Cl^- (l) - 2e^- \rightarrow Cl_2 (g)$$

This is an example of oxidation.

The **cathode** has a negative charge because it has gained extra electrons from the anode. The cathode has electrons to lose.

The H^+ and Na^+ ions are positively charged so have lost electrons (see section 10.2).

These positive ions are attracted to the negative electrode, the cathode, where they can collect the extra electrons to make them into atoms.

However, because sodium is more reactive than hydrogen the sodium ions remain as ions in the solution and only the hydrogen ions form atoms. These atoms pair up to form H_2 molecules. So, at the cathode hydrogen gas is released.

The reaction at the cathode can be summarised as:
hydrogen ions + electrons → hydrogen

$$2H^+ (aq) + 2e^- \rightarrow H_2 (g)$$

This is an example of reduction.

The Na^+ ions and the OH^- ions remain in solution to form a solution of sodium hydroxide

Figure 11.13
Electrolysis of sodium chloride solution

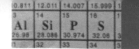

A health warning about chlorine

Chlorine gas even in small amounts is poisonous. If breathed in, it affects the cells lining the lungs, making them produce large quantities of fluid. If a lot of chlorine gas is breathed in then the large amounts of fluid produced can fill the lungs and can cause permanent lung damage or even death.

Did you know?

Chlorine was used as a poison gas in the first World War. When the wind was in the right direction, it was released towards the opposing army. It is denser than air and rolled along the ground until it reached the enemy trenches, filling them with the choking gas. However, if the wind changed direction, the chlorine gas could blow back towards the attacking army.

The laboratory test for chlorine gas

When chlorine is used as a bleach it combines with the coloured dyes in a cloth and turns them into colourless compounds. This property is made use of in the laboratory test for chlorine. In the test a piece of damp litmus paper is held above a small amount of the gas. If the gas is chlorine the litmus paper loses its colour.

The effect of light on silver halides (halogens)

Almost all types of photograph rely on the light-sensitive properties of the silver halides silver chloride, silver bromide and silver iodide. Photographic film consists of a thin layer of gelatin on a base of transparent plastic. The emulsion contains crystals of silver halides. When the film is exposed to light the silver halide crystals undergo a chemical change. As the film is developed the silver halide crystals in those areas exposed to most light become reduced to particles of metallic silver. Photographic paper works in much the same way as the film, but fewer silver halide crystals are used.

X-rays and radiation from radioactive substances produce the same changes to silver halide crystals as light.

Reactions of the hydrogen halides

Hydrogen halides form acidic solutions when added to water.

Did you know?

Hydrofluoric acid is so corrosive that it will corrode glass. It is stored in lead, steel or plastic containers.

This is due to the formation of $H^+(aq)$ ions.

Hydrogen chloride gas forms hydrochloric acid when it is dissolved in water. Hydrogen chloride in water gives H^+ and Cl^- ions.

$$HCl(g) \xrightarrow{(H_2O)} H^+(aq) + Cl^-(aq)$$

Hydrogen reacts with the halogens in ways that reflect the relative reactivities of the Group 7 elements.

	0.811	12.011	14.007	15.999	1
3		14	15	16	1
Al	**Si**	**P**	**S**		3
26.98	28.086	30.974	32.06		3
1	32	33	34		3

Figure 11.4
Reactions of hydrogen and halogens

Rate of reaction	Action with water
A mixture of hydrogen and fluorine will react explosively to form hydrogen fluoride (HF)	Forms hydrofluoric acid
Chlorine and hydrogen can be mixed together without a reaction if they are kept in the dark. If the mixture is exposed to sunlight they react together explosively to form hydrogen chloride (HCl)	Forms hydrochloric acid (a commonly used laboratory acid)
A mixture of hydrogen and bromine needs to be heated if they are to react and form hydrogen bromide (HBr)	Forms hydrobromic acid
Very little reaction between hydrogen and iodine even if they are heated	

Summary

♦ The electrolysis of sodium chloride solution produces chlorine gas at the anode, hydrogen gas at the cathode and a solution of sodium hydroxide.

♦ The test for chlorine is that it will decolorise moist litmus paper.

♦ Silver halides react in light to form particles of silver.

Topic questions

1 Explain why salt can be mined using a process called 'solution mining' but coal cannot.

2 What three important products are formed when a solution of sodium chloride is electrolysed? Give two uses for each product.

3 During the electrolysis of sodium chloride solution.
   a) What substance is produced at the cathode?
   b) What substance is produced at the anode?

   Explain in terms of ions and electrons why this substance is formed.

4 How would you test a gas to find out if it was chlorine?

5 What happens to crystals of silver bromide when they are exposed to the light?

Co-ordinated	Modular
DA 11.12	DA 5
SA 11.4	SA 15

Patterns in making metal compounds

Acids

Pure **acids** are covalent molecules (see section 7.2). They only behave as acids in water.

● A dilute acid is formed when an acid is dissolved in a lot of water.

● A concentrated acid is formed when an acid is dissolved in only a small amount of water.

Common acids used in the laboratory are shown in Figure 11.15.

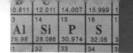

Figure 11.15
Common laboratory acids

Name	Formula
hydrochloric acid	HCl
nitric acid	HNO_3
sulphuric acid	H_2SO_4
ethanoic acid	CH_3COOH

Did you know?

There are may acids that occur naturally. Citric acid is found in oranges and lemons. Malic acid is found in apples. Ethanoic acid (acetic acid) is the main constituent of vinegar. Lactic acid is found in sour milk, yoghurt and some cheeses. Ants defend themselves when attacked using a spray of methanoic acid (formic acid). Stinging nettles have hollow hairs containing methanoic acid that break off in the skin and cause irritation.

Bases

Bases are usually oxides or hydroxides of metals. **Alkalis** are soluble bases.

Common bases used in the laboratory are shown in Figure 11.16.

Figure 11.16
Common laboratory bases

Slightly soluble bases		Soluble bases (alkalis)	
calcium hydroxide	$Ca(OH)_2$	sodium hydroxide	NaOH
calcium oxide	CaO	potassium hydroxide	KOH
		aqueous ammonia	$NH_3(aq)$

The pH scale

The **pH scale** is used to measure the acidity or alkalinity of an aqueous solution. It is a set of numbers which run from 0 to 14.

Indicators and the pH scale

Indicators are dyes which change colour when mixed with acids and alkalis. **Universal indicator** is a mixture of dyes which have been chosen to give the same order of colours as the visible spectrum (rainbow).

- An acid solution has a pH less than 7, i.e. pH 1 → 6.
- A neutral solution has a pH of exactly 7.
- An alkaline solution has a pH greater than 7, i.e. pH 8 → 14.

Universal indicator is used to measure the pH of a solution to show whether the solution is acidic, **neutral** or alkaline.

Figure 11.17
Universal indicator colours at different pH values

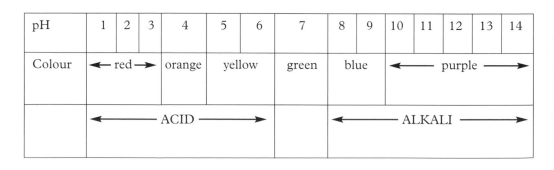

pH	1	2	3	4	5	6	7	8	9	10	11	12	13	14
Colour	← red →			orange	yellow		green	blue		← purple →				
	← ACID →							← ALKALI →						

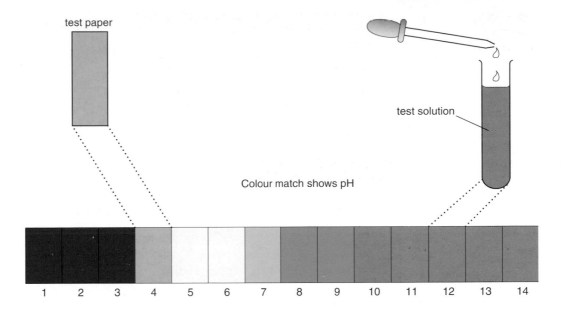

Figure 11.18
Colour matching to the pH chart

The solution can be tested by adding a few drops of the indicator liquid or by using indicator paper. In both cases the colour seen is compared to a coloured chart and the pH is read off.

The pH scale is a measure of the concentration of hydrogen, H^+, ions in the solution.

> ## Did you know?
>
> Different types of water have slightly different pH values. Rainwater has a pH of 6 due to dissolved acidic gases such as carbon dioxide and sulphur dioxide. Pure distilled water has a pH of 7. Mains water has a pH of 7.8 so that it does not react with the water pipes in a house. Sea water has a pH of 8 due to the many dissolved substances that it contains.

Acids and alkalis and ions

Acids

The formula of all acids contains hydrogen (H). An acid is a substance, which forms **hydrogen ions,** H^+(aq) **ions,** when added to water.

So, HCl in water gives H^+ and Cl^-

$$HCl(g) \xrightarrow{(H_2O)} H^+(aq) + Cl^-(aq)$$

HNO_3 in water gives H^+ and NO_3^-

$$HNO_3(l) \xrightarrow{(H_2O)} H^+(aq) + NO_3^-(aq)$$

H_2SO_4 ionises into $2H^+$ and SO_4^{2-}

$$H_2SO_4(l) \xrightarrow{(H_2O)} 2H^+(aq) + SO_4^{2-}(aq)$$

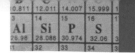
Alkalis (soluble bases)

The formula of all alkalis contains hydroxide (OH). Soluble bases (alkalis) are substances which form **hydroxide ions**, OH^-(aq), when added to water. Both NaOH and KOH form hydroxide ions when dissolved in water.

NaOH in water gives

$$NaOH(s) \xrightarrow{(H_2O)} Na^+(aq) + OH^-(aq)$$

KOH in water gives

$$KOH(s) \xrightarrow{(H_2O)} K^+(aq) + OH^-(aq)$$

Aqueous ammonia reacts with water to give NH_4^+ and OH^-.

$$NH_3(g) + H_2O(l) \rightleftharpoons NH_4^+(aq) + OH^-(aq)$$

The 'double-headed' arrow, $\rightleftharpoons$, shows that the reaction is reversible (see section 8.3).

Aqueous ammonia is a base because as it dissolves in water, it also reacts with water to form OH^- ions.

Neutralisation

Neutralisation is a reaction between an acidic solution and an alkaline solution. A typical neutralisation reaction would be the reaction between the acid, hydrochloric acid, and the alkali, sodium hydroxide (see section 8.3).

The process of neutralisation can be followed using the change in colour of Universal Indicator.

$$
\begin{array}{ccccccc}
\text{sodium hydroxide} & + & \text{hydrochloric acid} & \rightarrow & \text{sodium chloride} & + & \text{water} \\
NaOH(aq) & + & HCl(aq) & \rightarrow & NaCl(aq) & + & H_2O(l)
\end{array}
$$

The solution of the sodium hydroxide is carefully added in small amounts to the hydrochloric acid and the resulting solution is tested with universal indicator paper. When the universal indicator paper turns green, the solution is neutral and no more sodium hydroxide solution need be added.

Neutralisation can be summarised as:

$$alkali + acid \rightarrow salt + water$$

The name of the salt depends upon the metal in the alkali and the acid used as can be seen from the table:

Figure 11.19
Names of salts produced by neutralisation

Name of alkali	Name of salt produced with	
	Hydrochloric acid	Sulphuric acid
sodium hydroxide	sodium chloride	sodium sulphate
calcium hydroxide	calcium chloride	calcium sulphate
magnesium hydroxide	magnesium chloride	magnesium sulphate

Neutralising nitric acid produces salts called nitrates. So if ammonia solution is neutralised by nitric acid a salt called ammonium nitrate is produced.

0.811	12.011	14.007	15.999	1
3	14	15	16	1
Al	Si	P	S	
26.98	28.086	30.974	32.06	3
1	32	33	34	3

Examples of neutralisation

- Occasionally the stomach may produce too much acid and the painful problem of acid indigestion may be felt. Indigestion tablets are a base and will neutralise the excess acid being formed by the stomach.

- Farmers add calcium hydroxide or calcium oxide to their soils in order to make them less acidic. This allows a greater range of crops to be cultivated because some plants will not grow properly on soils that are too acidic.

Neutralisation as an ionic process

When neutralisation reactions are looked at in detail it is found that all the different reactions involving acids and alkalis have the same essential ionic reaction.

The reaction between hydrochloric acid and sodium hydroxide involves the following ions.

$$H^+(aq) + Cl^-(aq) + Na^+(aq) + OH^-(aq) \rightarrow Na^+(aq) + Cl^-(aq) + H_2O(l)$$

When the spectator ions (see Section 8.2) that are common to each side, $Na^+(aq)$ and $Cl^-(aq)$, are removed, only the ionic reaction between $H^+(aq)$ and $OH^-(aq)$ to make $H_2O(l)$ remains.

The essential ionic reaction for all reactions between acids and alkalis is

$$H^+(aq) + OH^-(aq) \rightarrow H_2O(l)$$

Making salts of transition metals

The oxides and hydroxides of transition metals do not dissolve in water – they are examples of insoluble bases. To make a solution of a transition metal salt, the metal oxide (or metal hydroxide) is added to an acid until no more dissolves. The excess metal oxide (or hydroxide) is removed by filtering.

Figure 11.20
Making copper(II) sulphate

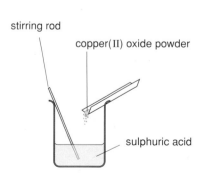

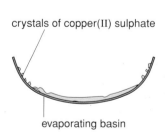

stirring rod
copper(II) oxide powder
sulphuric acid

unreacted copper(II) oxide

crystals of copper(II) sulphate
evaporating basin

Stage 1
The copper(II) oxide is added to the acid until no more dissolves.

Stage 2
The mixture from the beaker is filtered. Collecting in the empty beaker will be a solution of copper(II) sulphate.

Stage 3
The copper(II) sulphate solution is left so that the water from the solution can evaporate to leave crystals of copper(II) sulphate.

223

The reaction can be summarised as :

copper(II) oxide	+	sulphuric acid	→	copper(II) sulphate	+	water
$CuO(s)$	+	$H_2SO_4(aq)$	→	$CuSO_4(aq)$	+	$H_2O(l)$

Reactions of hydrochloric acid

Hydrochloric acid and some metal hydroxides

Metal hydroxides react with hydrochloric acid to give metal chlorides and water.

sodium hydroxide	+	hydrochloric acid	→	sodium chloride	+	water
$NaOH(aq)$	+	$HCl(aq)$	→	$NaCl(aq)$	+	$H_2O(l)$

calcium hydroxide	+	hydrochloric acid	→	calcium chloride	+	water
$Ca(OH)_2(aq)$	+	$2HCl(aq)$	→	$2CaCl_2(aq)$	+	$2H_2O(l)$

Hydrochloric acid and ammonia

Ammonia reacts with hydrochloric acid to give a salt called ammonium chloride.

ammonia	+	hydrochloric acid	→	ammonium chloride
NH_3	+	HCl	→	NH_4Cl

Reactions of sulphuric acid

Sulphuric acid and some metal hydroxides

Metal hydroxides react with sulphuric acid to form metal sulphates and water.

sodium hydroxide	+	sulphuric acid	→	sodium sulphate	+	water
$2NaOH(aq)$	+	$H_2SO_4(aq)$	→	$Na_2SO_4(aq)$	+	$2H_2O(l)$

magnesium hydroxide	+	sulphuric acid	→	magnesium sulphate	+	water
$Mg(OH)_2(aq)$	+	$H_2SO_4(aq)$	→	$MgSO_4(aq)$	+	$2H_2O(l)$

Sulphuric acid and ammonia

Ammonia reacts with sulphuric acid to form the salt ammonium sulphate.

ammonia	+	sulphuric acid	→	ammonium sulphate
$2NH_3$	+	H_2SO_4	→	$(NH_4)_2SO_4$

Summary

- Hydrogen halides dissolve in water to produce acidic solutions.

- The **neutralisation** reaction between an acid and an alkaline hydroxide solution produces a salt and water.

- Neutralisation can be represented as:

  $$H^+ (aq) + OH^-(aq) \rightarrow H_2O (l)$$

- Hydrochloric acid produces chlorides.

- Sulphuric acid produces sulphates.

- Nitric acid produces nitrates.

- Hydrogen ions H^+ make solutions acidic.

- Hydroxide ions OH^- make solutions alkaline.

Topic questions

1 What type of solution would have:
   a) a pH range of 1 to 6
   b) a pH of 7
   c) a pH range of 8 to 14?

2 What element do all acids contain?

3 a) What are the oxides or hydroxides of metals called?
   b) What are soluble bases called?

4 a) Write down the word equation for the reaction between sodium hydroxide solution and hydrochloric acid.

b) Write down the balanced symbol equation for the reaction between sodium hydroxide solution and hydrochloric acid. Include state symbols.

c) What is the chemical name of the salt produced in this reaction?

5 a) Write down the word equation for the reaction between copper(II) oxide and sulphuric acid.

b) Write down the balanced symbol equation for this reaction. Include state symbols.

c) What is the chemical name of the salt produced in this reaction?

6 a) Write down the word equation for the reaction between calcium hydroxide solution and hydrochloric acid.

b) Write down the balanced symbol equation for this reaction. Include state symbols.

c) What is the chemical name of the salt produced in this reaction?

7 a) Write down the word equation for the reaction between ammonia and hydrochloric acid.

b) Write down the balanced symbol equation for this reaction. Ignore state symbols.

c) What is the chemical name of the salt produced in this reaction?

8 a) Write down the ion present in all acids.

b) Write down the ion present in all alkalis.

c) Write down the ionic equation that represents neutralisation. Include state symbols.

Examination questions

1 Part of the periodic table which Mendeleev published in 1869 is shown below.

	Group 1	Group 2	Group 3	Group 4	Group 5	Group 6	Group 7
Period 1	H						
Period 2	Li	Be	B	C	N	O	F
Period 3	Na	Mg	Al	Si	P	S	Cl
Period 4	K	Ca	★	Ti	V	Cr	Mn
	Cu	Zn	★	★	As	Se	Br
Period 5	Rb	Sr	Y	Zr	Nb	Mo	★
	Ag	Cd	In	Sn	Sb	Te	I

a) Name **two** elements in Group 1 of Mendeleev's periodic table which are **not** found in Group 1 of the modern periodic table. *(2 marks)*

b) Which group of elements in the modern Periodic Table is missing on Mendeleev's table? *(1 mark)*

c) Mendeleev left several gaps on his periodic table. These gaps are shown as asterisks(★) on the table above.
Suggest why Mendeleev left these gaps. *(1 mark)*

d) Complete the following sentence.
In the **modern** periodic table the elements are arranged in the order of their
_____ numbers. *(1 mark)*

2 Part of the periodic table is shown below. Use the information to help you answer the questions which follow.

H							He
Li	Be	B	C	N	O	F	Ne
Na	Mg	Al	Si	P	S	Cl	Ar

a) Write the symbol for;
   i) chlorine; *(1 mark)*
   ii) sodium; *(1 mark)*

b) i) What is the symbol of the element which is in Group 2 and Period 3? *(1 mark)*
   ii) What name is given to Group 7? *(2 marks)*

c) The arrangement of electrons in sulphur (S) is 2.8.6. Write the arrangement of electrons for:
   i) neon (Ne) *(1 mark)*
   ii) aluminium (Al) *(1 mark)*

d) The periodic table is an arrangement of elements in order of increasing atomic number. What is the atomic number of an element? *(1 mark)*

e) What is the name of the uncharged particle in the nucleus of an atom? *(1 mark)*

3 Chlorine, hydrogen and sodium hydroxide are produced by the electrolysis of sodium chloride solution.

A student passed electricity through sodium chloride solution using the apparatus shown in the diagram.

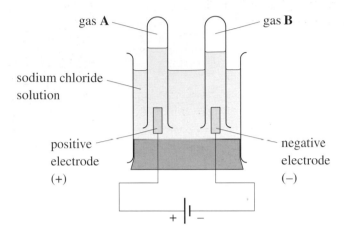

a) Name gas A and gas B *(1 mark)*
b) Describe and give the result of a test you could do in a school laboratory to find out which gas is chlorine. *(2 marks)*
c) Chlorine is used for treating water for drinking and in swimming pools. Why? *(1 mark)*

d) i) Balance the half equation for the production of hydrogen at the electrode.

_____H^+ + _____e^- → H_2 *(1 mark)*

ii) Which word, from the list, best describes the reaction in part (d)(i)?

**decomposition   cracking   neutralisation
oxidation   reduction**

4 Acids and bases are commonly found around the home.
a) Baking powder contains sodium hydrogencarbonate mixed with an acid. When an acid is added, the baking powder releases carbon dioxide. How could you test the gas to show that it is carbon dioxide?
(2 marks)

b) Indigestion tablets contain bases which cure indigestion by neutralising excess stomach acids.

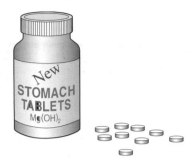

i) One type of indigestion tablet contains magnesium hydroxide. This base neutralises stomach acid as shown by the balanced chemical equation.

$$Mg(OH)_2 + 2HCl \rightarrow MgCl_2 + 2H_2O$$

Write a balanced **ionic** equation for the neutralisation reaction. *(2 marks)*
ii) How does the pH in the stomach change after taking the tablets? *(1 mark)*

5 The alkali metals, in Group 1, are the most reactive metals. The symbols and atomic numbers for the first three alkali metals are shown here.

$$_3Li \quad _{11}Na \quad _{19}K$$

a) Describe what you see when sodium reacts with water. *(3 marks)*
b) Write a balanced chemical equation for the reaction of sodium with water. *(3 marks)*
c) If the reaction is repeated using lithium it is much slower, but with potassium it is much faster. Give reasons for the similarities and the differences in the reactions of lithium, potassium and sodium with water. *(4 marks)*

Chapter 12
Chemistry in action

C N
.011 14.007
15
Si P
.086 30.974
33
Ge As
2.61 74.922
51
Sn Sb
8.71 121.75
83
Bi

Key terms activation energy · anhydrous · catalyst · denaturing · endothermic reaction · energy level diagram · enzyme · equilibrium · exothermic reaction · fermentation · products · reactants · reversible reaction · substrate

12.1   Energy transfers in chemical reactions

Co-ordinated	Modular
DA 11.16	DA 7
SA n/a	SA n/a

Chemical reactions make new substances. In a chemical reaction the starting materials – called **reactants** – are changed into new substances – called **products**. The atoms in the reactants are re-arranged when they change into products (see section 8.2). This only happens when the reactant particles collide with each other. Sometimes when reactants are mixed there is an instant reaction, for example when magnesium is added to dilute acid the fizzing starts immediately. In other reactions the mixture of reactants has to be 'encouraged' to react by applying energy, often in the form of heat. An example of this is when natural gas (methane) and air are mixed as a gas tap is turned on. At first nothing happens – the reaction only begins when a flame is applied.

Each of these reactions get hot. This is because the reaction gives out energy. Reactions that give out energy are called **exothermic reactions**. Exothermic reactions give out energy as heat (see section 8.4).

Some reactions need energy to make them work. These reactions absorb energy. Reactions that absorb energy are called **endothermic reactions** (see section 8.4). An example of an endothermic reaction is the heating of limestone (calcium carbonate) to produce quicklime (calcium oxide) and carbon dioxide (see section 10.3).

$$CaCO_3(s) \rightarrow CaO(s) + CO_2(g)$$

calcium carbonate   calcium oxide
(limestone)   (quicklime)

In all chemical reactions two major changes take place:

1   The starting materials have to be decomposed by breaking the chemical bonds in them to release the elements so that they can recombine in a different way to make the products (see section 7.2). This process absorbs energy from the surroundings and is called an endothermic process.

2   Once the reactants have been decomposed, new chemical bonds are formed between the elements that have been released so that new chemical products can be produced. This releases energy to the surroundings and is called an exothermic process.

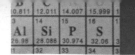

Chemistry in action

The energy changes that take place during a chemical reaction can be shown in **energy level diagrams**.

Figure 12.1 is an energy level diagram showing how the energy of the system changes during an exothermic reaction.

Figure 12.1
An energy level diagram for an exothermic reaction

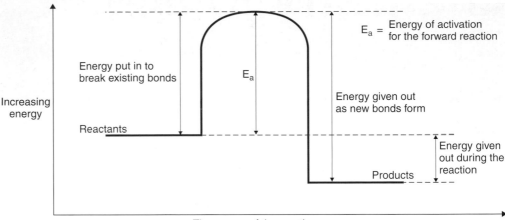

In this case the amount of energy released as the new chemical bonds are formed is greater than the energy that is put in to break the existing bonds and start the reaction. This means there is an energy bonus and the reaction mixture will become warmer. Energy will be given out to the surroundings. Most of the energy released in this way will be heat energy but sometimes light energy is also emitted.

Figure 12.2 shows how the energy of the system changes during an endothermic reaction.

Figure 12.2
An energy level diagram for an endothermic reaction

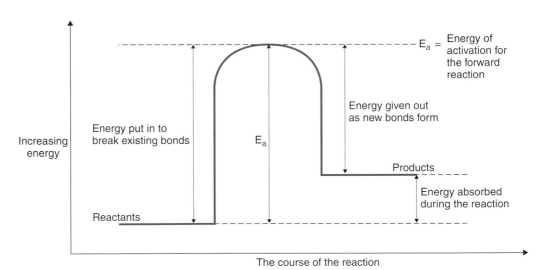

In this case the amount of energy needed to break the existing chemical bonds and start the reaction is greater than that released when new bonds are formed so there is an energy shortage. This means the reaction mixture will need to absorb energy. In some reactions this will cause the reaction mixture to become cooler. Usually endothermic reactions only work if they are heated. Endothermic reactions are less common than exothermic reactions.

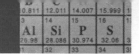

Using bond energy values to find out if a reaction is exothermic or endothermic

The table shows the energy transfers when some common chemical bonds are made or broken.

Chemical bond	Energy transferred in kJ/mol*
C—H	413
O=O	497
C=O	745
O—H	464
(*See section 7.2)	

Calculating the energy transferred when methane burns in air.

The chemical equation for the reaction is: $CH_4(g) + 2O_2(g) \rightarrow CO_2(g) + 2H_2O(l)$

The bonds involved are:

$$\begin{array}{c} H \\ | \\ H—C—H \\ | \\ H \end{array} + 2 \times (O{=}O) \rightarrow O{=}C{=}O + 2 \times (H—O—H)$$

In terms of bonds there are $4 \times (C—H) + 2 \times (O{=}O) \rightarrow 2 \times (C{=}O) + 4 \times (O—H)$

The contents of the table show the bond energies involved (in kJ/mol).

	Bond breaking			Bond formation	
No.	Type	Energy in	No.	Type	Energy out
4	C—H	$4 \times 413 = 1652$	2	C=O	$2 \times 745 = 1490$
2	O=O	$2 \times 497 = 994$	4	O—H	$4 \times 464 = 1856$
Total energy in = 2646 kJ/mol			Total energy out = 3346 kJ/mol		

This shows that there is a net energy transfer of heat energy to the surroundings of $(3346 - 2646)$ kJ. The reaction is therefore exothermic, releasing 700 kJ of energy.

Activation energy

During any chemical reaction the reacting particles must first collide with each other, however, collision alone does not guarantee that the reaction will start. Reaction will only occur if the colliding particles have sufficient combined energy to get over the energy barrier between reactants and products. This minimum amount of energy that is required is called the **activation energy** – shown as E_a in Figures 12.1 and 12.2.

The energy level diagrams are similar to a fairground roller coaster – if the car carrying the passengers does not have enough energy to get up and over the first hill, the ride will not even start. This explains why some reactions occur spontaneously (i.e. as soon as the reactants are mixed), such as the reaction between magnesium and acid, and others require 'help'. A spontaneous reaction gets enough energy from the reactants and their surroundings to overcome the energy barrier between reactants and products. Reactions that require 'help' do not.

Summary

◆ Chemical reactions involve the re-arrangement of existing atoms to make new substances.

◆ Matter is neither created nor destroyed.

◆ Chemical bonds have to be broken before new substances can be made and this is an **endothermic** process.

◆ Endothermic processes require an input of energy.

◆ When new bonds are made, energy is released and this is an **exothermic** process.

◆ Exothermic processes release energy to the surroundings, usually as heat.

◆ The **activation energy** is the minimum amount of energy required by colliding particles before reaction can take place.

◆ **Energy level diagrams** link the making and breaking of bonds together to show whether the whole reaction is endothermic or exothermic.

Topic questions

1 What energy transfers take place in:
   a) exothermic reactions
   b) endothermic reactions
   c) bond breaking
   d) bond formation?

2 Explain in terms of bond breaking and bond formation the energy transfers in an exothermic reaction.

3 In a reaction the energy needed to break bonds is 300 kJ/mol and the energy transferred during bond formation is 200 kJ/mol. Is the reaction exothermic or endothermic? Give a reason for your answer.

4 What is activation energy?

12.2 Reversible reactions

Co-ordinated	Modular
DA 11.15	DA 7
SA n/a	SA n/a

Some chemical changes are permanent whilst others can be easily reversed (see section 8.3). An example of a permanent chemical change is seen when paper is burned. The materials that make up the piece of paper combine with oxygen from the air in an exothermic reaction producing carbon dioxide and water vapour and leaving an ash. To reverse this process in the laboratory would be impossible. This is called an irreversible reaction.

Some reactions are fairly easy to reverse, for example the action of heat on solid ammonium chloride (NH_4Cl).

When ammonium chloride is heated as shown in Figure 12.3, the open end of the test tube remains cold whilst the solid itself becomes very hot and decomposes into two new products, ammonia (NH_3) and hydrogen chloride (HCl). These are both gases and they diffuse along the tube towards the cold end where they recombine to form solid ammonium chloride again. This is an example of a **reversible reaction**.

A reversible reaction must not be confused with similar physical changes that occur. For example, when paraffin wax is heated it first melts and then boils. If the paraffin wax vapour is cooled it condenses and then solidifies. This is quite different because when wax is heated it does not change chemically into a new substance. At all stages in the process the wax is present as wax molecules in either solid, liquid or gas state. With ammonium chloride, thermal decomposition (see section 8.3) of the solid into completely new compounds takes place and this is followed on cooling by the reverse chemical change taking place.

Figure 12.3
The action of heat on solid ammonium chloride

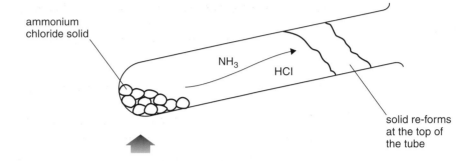

A reversible reaction is one that can take place in either direction depending on the conditions. Chemical reactions that are reversible have a special sign ($\rightleftharpoons$) in the equations that represents the change.

$$\xrightarrow{\text{heat}}$$
$$NH_4Cl(s) \rightleftharpoons NH_3(g) + HCl(g)$$
$$\xleftarrow{\text{cool}}$$

The laboratory test for water

Another example of a reversible reaction is the action of heat on blue copper sulphate crystals (see section 10.4). These have the formula $CuSO_4.5H_2O$ and when heated, the water of crystallisation separates and a white powder with the formula $CuSO_4$ is produced. This powder is called **anhydrous** copper sulphate. It is often used to detect the presence of water. When water is added to it, the colour changes from white back to the original blue. The reversible reaction can be summarised in the following equation:

$$\underset{\text{blue}}{CuSO_4.5H_2O} \rightleftharpoons \underset{\text{white}}{CuSO_4} + 5H_2O$$

Reversible reactions in a closed system

If the reversible reaction:

$$A + B \rightleftharpoons C + D$$

were to take place in a closed container so that none of the products could escape, after a while it would seem that nothing was happening. This would be because the reaction $A + B \rightarrow C + D$ was happening at exactly the same rate as the reaction $C + D \rightarrow A + B$. The reversible reaction would now be described as being in a state of **equilibrium**.

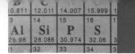

What can affect equilibrium?

1 Changing the temperature

- If the forward reaction is *endothermic* and the temperature is *increased*, the yield of products is *increased*.

- If the forward reaction is *endothermic* and the temperature is *decreased*, the yield of products is *decreased*.

- If the forward reaction is *exothermic* and the temperature is *increased*, the yield of products is *decreased*.

- If the forward reaction is *exothermic* and the temperature is *decreased*, the yield is *increased*.

2 Changing the pressure on reactions that involve reacting gases

- If there are more molecules on the left of the balanced symbol equation, then increasing the pressure will increase the rate of the reaction.

- If there are more molecules on the right of the balanced symbol equation, then increasing the pressure will decrease the rate of the reaction.

- If there are equal numbers of molecules on each side of the balanced symbol equation, then increasing the pressure will have no affect on the rate of the reaction.

These changes in conditions and reaction rates are important in determining the optimum conditions in many industrial processes, including the Haber process (see section 9.2).

Summary

♦ Some reactions can proceed in either direction depending on the conditions. These are **reversible reactions**.

♦ When a reversible reaction occurs in a closed system, an **equilibrium** is reached when the reactions occur at exactly the same rate in each direction.

Topic questions

1 What is the difference between a permanent chemical change and a reversible reaction?

2 Describe what happens when ammonium chloride is placed at the bottom of a long glass tube and heated?

3 a) What is the chemical test for water?
   b) How can this reaction be made to reverse?

Rates of reaction

The rate of a chemical reaction is the speed at which it takes place. It can vary from a fraction of a second to several weeks, months or even years and it is not always possible to predict how fast a given reaction will take place.

- Some reactions take place very quickly e.g. an explosion.
- Some take place quite quickly e.g. the action of water on potassium metal.
- Some take place quite slowly e.g. the rusting of iron.
- Some take place very slowly e.g. the fermentation of grapes into wine.

The only real way to find out how fast a chemical reaction takes place is to carry it out.

Do not confuse how *fast* a reaction takes place with how *much* reaction takes place. The amount of reaction taking place depends on the amounts of chemicals used. If marble chippings (calcium carbonate) are added to dilute hydrochloric acid, the gas carbon dioxide is given off. The actual volume of gas produced depends on how much marble and how much acid was used.

- If a lot of marble and acid are used, a lot of carbon dioxide is produced.

- If a small piece of marble and a lot of acid are used, the marble is likely to be used up. Some of the acid will remain unreacted. The amount of gas produced here depends on the amount of marble used.

- If a huge lump of marble and just a small amount of acid are used, the acid will be used up and some of the marble will remain. The amount of gas produced this time depends on the amount of acid used.

The amount of reaction is determined by the material that is present in *least* quantity.

Factors affecting the rate of a reaction

The rate at which a reaction takes place depends on a number of factors. Several things can be done to either speed up or slow down a chemical reaction. These are:

- changing the *surface area* of any solids involved in the reaction
- changing the *concentration* of any reactants that are in solution
- changing the *pressure* in reactions where gases are involved
- changing the *temperature* of the reaction mixture
- adding a *catalyst*.

For each of these factors it is important to realise exactly how and why the reaction rate changes.

Factors that can have their values changed are known as variables. If the value of one variable (the independent variable) is changed, then the value of a second variable (the dependent variable) will also change. To find out how changing independent variables affects a dependent variable it is essential that we change only one variable at a time.

In rates of reaction experiments, if more than one independent variable is altered at a time it is impossible to decide which one is causing the change in the reaction rate. The idea of changing one independent variable at a time is used in each of the experiments described in this chapter.

Changing the surface area

This factor only affects reactions where a solid is involved. Figure 12.5 shows what happens on the surface of a piece of marble when it is dropped into dilute hydrochloric acid.

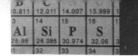

Chemistry in action

The particles that really cause the reaction are the hydrogen ions ($H^+(aq)$) that come from the acid. These ions are able to move freely through the solution and will eventually bump into the surface of the piece of marble, reacting with it. The marble in the centre of the lump has to wait until all the marble around it has been reacted before it has a chance to react with the H^+ ions.

Figure 12.4
Hydrogen ions colliding with a lump of marble

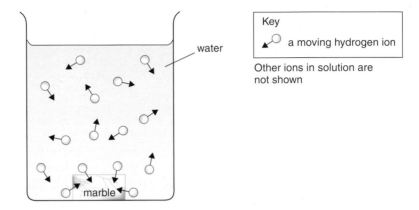

Each time a hydrogen ion collides with the marble it reacts and is neutralised. The number of hydrogen ions steadily decreases reducing the concentration of the acid. As this happens the rate of the reaction slowly decreases. If the marble is in excess, eventually all the hydrogen ions are used up and the reaction stops altogether. If the experiment is repeated with the same amount of marble broken into several smaller pieces, more of the marble that was in the centre of the large lump will become exposed to the acid and will be able to react immediately and not have to wait for the marble around it to be dissolved first (see Figure 12.5).

Figure 12.5
Hydrogen ions colliding with smaller pieces of marble

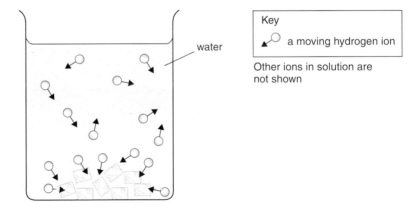

There are several applications where changing the particle size can alter the rate of a reaction.

- The burning of coal in power stations can be speeded up by converting it into a very fine powder. It can then be blown through a pipe where it burns as if it were a gas. Using a powder speeds up the combustion process.

- Large pieces of graphite are used as anodes in the extraction of aluminium by electrolysis (see section 10.2). This slows down the rate at which the graphite burns in the atmosphere of oxygen that is generated around the anodes during the process. Using large lumps slows down the combustion process.

For a given amount of material, the smaller the size of the pieces, the larger the surface area and the faster reactions will take place.

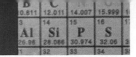

The effect of particle size/surface area on the rate of a chemical reaction can be demonstrated using the reaction between marble and dilute hydrochloric acid (see Figure 12.6).

Figure 12.6
The reaction between acid and marble

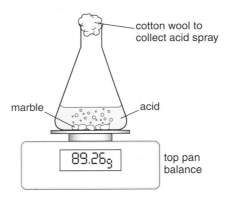

cotton wool to collect acid spray

marble

acid

89.26g

top pan balance

A known mass of large pieces of marble is dropped into a known volume of dilute hydrochloric acid in a conical flask. A plug of cotton wool is placed in the neck of the flask to trap any acid spray but allow any gas to escape. As the gas escapes, the mass of the apparatus decreases. The decrease in mass is recorded each minute until there is no further change, i.e. until the reaction has stopped.

The experiment is then repeated using the same volume of acid of the same concentration and at the same temperature. The same mass of marble pieces is used but the size of the pieces is changed. In one repeat, medium-sized pieces are used; in the next, small pieces are used. (Note that the only variable changed here is the size of the particles being used.) Figure 12.7 shows a typical set of results obtained and Figure 12.8 shows these results in the form of three graphs.

Time (mins)	Total mass loss (g)		
	Large pieces	Medium pieces	Small pieces
0	0.00	0.00	0.00
1	1.00	1.48	2.96
2	1.80	2.52	3.68
3	2.52	3.20	3.85
4	3.00	3.58	3.93
5	3.29	3.73	3.98
6	3.50	3.85	3.99
7	3.66	3.92	4.00
8	3.76	3.96	4.00
9	3.82	3.99	4.00
10	3.88	4.00	4.00
11	3.93	4.00	4.00
12	3.97	4.00	4.00
13	3.99	4.00	4.00
14	4.00	4.00	4.00
15	4.00	4.00	4.00

Figure 12.7
Table of results

235

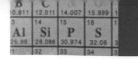

Figure 12.8
The effect of chip size on reaction rate

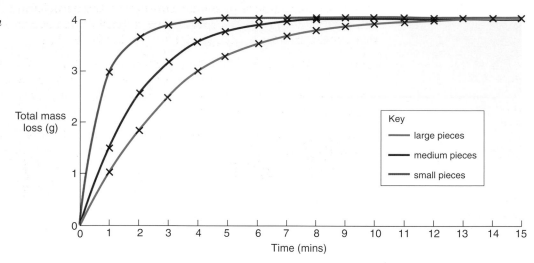

Notice how in Figure 12.8 each graph eventually reaches the same height but at a different time. The same overall mass loss is obtained because equal amounts of reactants have been used in all three experiments. The slope of the initial part of each graph is an indication of the rate of the reaction. The steeper the slope, the faster the reaction. So the smaller the marble pieces, the faster the reaction. This is because smaller pieces have a larger surface area.

Did you know?

Smoking is not allowed in flour mills partly because of the danger of an explosion caused by the extremely fine particles of flammable material (flour) floating in the air. If these particles caught fire they would burn like a gas and the flame could spread very quickly! Workers wear rubber-soled shoes and sometimes the mills even have rubber floors to prevent the risk of a spark.

Changing the concentration of reactants in solution

The rate of a reaction can also be altered by using solutions of different concentrations. Figure 12.9 shows the effect of doubling the concentration of the acid. There are now twice as many particles in the same volume of solution. This means there is twice the chance of these acid particles colliding with the marble and causing a chemical reaction to occur.

Figure 12.9
Changing the concentration of a substance in solution

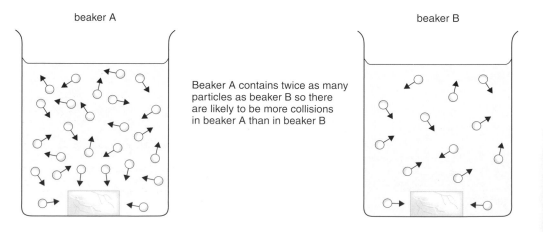

Beaker A contains twice as many particles as beaker B so there are likely to be more collisions in beaker A than in beaker B

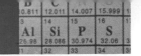

In very dilute solutions of soluble substances, the particles are a long way apart and will have to travel greater distances before they can collide and react. This is going to take time, so reactions involving dilute solutions are slow.

The effect of changing the concentration of a solution on the reaction rate can be followed by measuring the volume of gas given off over a period of time. Equal volumes of solutions of hydrochloric acid of different concentrations are added to the same mass of small marble chips.

A typical set of apparatus for doing this is shown in Figure 12.10. (The graduated measuring cylinder shown could be replaced by a gas syringe or burette.)

Figure 12.10
Collecting and measuring the volume of gas evolved in a reaction

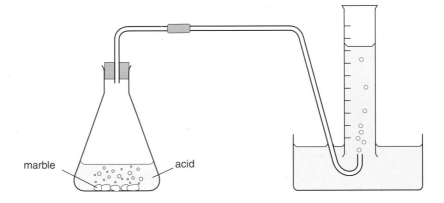

marble   acid

Readings of the volume of gas collected after each minute are made until the reaction is complete. The reaction stops in each case when the marble chip has completely dissolved.

Typical readings using solutions of three different concentrations are shown in Figure 12.11.

Figure 12.11
Table of results

Time (mins)	Concentration of acid used		
	Low concentration	Medium concentration	High concentration
0	0.0	0.0	0.0
1	3.5	5.5	10.0
2	7.0	10.5	16.0
3	10.0	14.5	20.5
4	13.0	18.0	23.0
5	15.5	20.5	24.7
6	18.0	22.5	25.7
7	20.1	24.0	26.3
8	21.9	25.0	26.5
9	23.5	25.5	26.5
10	25.0	26.0	26.5
11	26.0	26.5	26.5
12	26.5	26.5	26.5

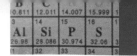

Chemistry in action

The graphs of these results (shown in Figure 12.12) are similar in shape to those for changing the particle size. Once again the steeper the slope of the graph, the faster the reaction is taking place. The graphs show that as the concentration of the acid increases, the rate of reaction also increases.

Figure 12.12
The effect of concentration on reaction rate

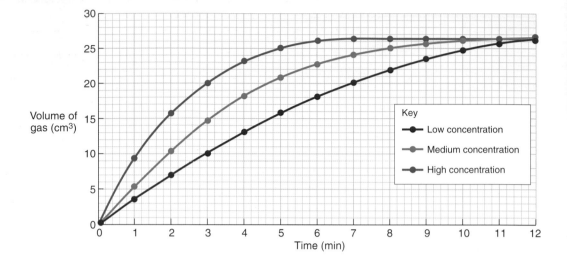

Another useful reaction to use to study the effect of changing the concentration of a solution on the reaction rate is that between sodium thiosulphate solution ($Na_2S_2O_3(aq)$) and dilute hydrochloric acid ($HCl(aq)$).

The equation for the reaction is:

$$Na_2S_2O_3(aq) + 2HCl(aq) \rightarrow 2NaCl(aq) + H_2O(l) + SO_2(g) + S(s)$$

During the reaction solid sulphur is produced but it appears very slowly and at first there appears to be no reaction. After a few moments, however, the mixture slowly starts to turn cloudy and if the reaction vessel is placed over a pencilled cross drawn on a piece of paper (see Figure 12.13) the cloudiness gradually obscures the cross until it can no longer be seen. The time taken for this to happen can be used as the basis for following the reaction.

Figure 12.13
Sodium thiosulphate and hydrochloric acid

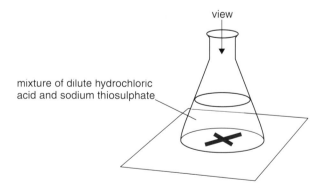

In the experiment, as the sodium thiosulphate is diluted, the reaction particles become more spread out so they will have to travel further before they can react. Figure 12.14 lists some typical results obtained. The longer a reaction takes, the slower it is. The rate of reaction is 'inversely proportional' to the time taken. This means that:

$$\text{rate of reaction is proportional to } \frac{1}{\text{time}}$$

Figure 12.14

Experiment number	Volume of sodium thiosulphate used (cm³)	Volume of water used (cm³)	Volume of hydrochloric acid used (cm³)	Time (sec)	1/time (sec⁻¹)
	Concentration				Rate
1	50	0	5	38	0.0263
2	40	10	5	47	0.0213
3	30	20	5	62	0.0161
4	25	25	5	74	0.0135
5	20	35	5	95	0.0105
6	10	40	5	182	0.0055

The last column in the table shows the value of 1/time which is a measure of the rate of the reaction. Figures 12.15 and 12.16 show the graphs produced when the 'concentration' of the thiosulphate solution is plotted against 'time' and also against '1/time'.

Figure 12.15
The effect of concentration on the time of a reaction

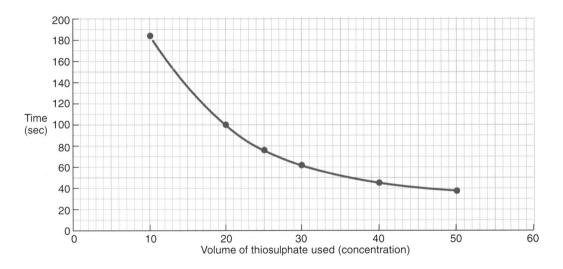

Note that in Figure 12.16, as the concentration of the thiosulphate solution increases, the rate of reaction increases.

Figure 12.16
The effect of concentration on rate of reaction

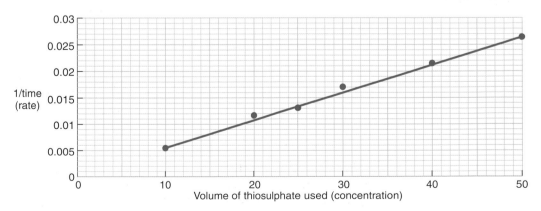

Changing the temperature

The kinetic theory says that when the temperature of a substance is increased, the particles in a solid vibrate more and the particles in liquids and gases move faster.

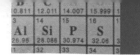

Chemistry in action

When a reaction mixture is heated up, each particle in it acquires more energy. The particles move around faster so that collisions between them are likely to happen more often. This increases the reaction rate for two reasons:

- the number of collisions taking place will increase

- the number of collisions where the energy of the colliding particles exceeds the energy of activation for the reaction will increase.

The effect of changing the temperature on the reaction rate can be shown using the reaction between sodium thiosulphate solution and dilute hydrochloric acid. The equipment and method of following the reaction used would be similar to that described earlier in this chapter (see Figure 12.13). A number of reactions would be carried out in which the concentrations and volumes of sodium thiosulphate solution and dilute hydrochloric would be kept the same but the temperature of each reaction would be different. The time taken for the cross to disappear when viewed from above through the solution would be recorded. Some typical results are shown in Figure 12.17.

Figure 12.17

Temperature at start of reaction (°C)	Time for cross to disappear (sec)
47	33
42	38
38	49
35	66
30	100
21	181

These results are shown in the form of a graph in Figure 12.18. Note how the time for the reaction to take place decreases as the temperature increases.

Figure 12.18
The effect of temperature on reaction time

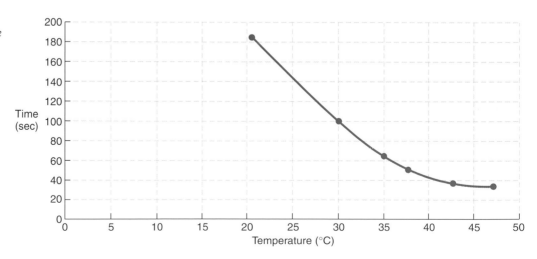

Adding a catalyst

A **catalyst** is a substance that can alter the rate of a chemical reaction but is not used up and remains chemically unchanged at the end of the process. Catalysts are used extensively in industry and most modern cars now incorporate them in catalytic converters to convert harmful exhaust gases to comparatively harmless ones before they are released into the atmosphere.

Some reactions are very slow, for example the gradual decomposition of hydrogen peroxide (H_2O_2) into water and oxygen.

$$2H_2O_2(aq) \rightarrow 2H_2O(l) + O_2(g)$$

This reaction can be speeded up by adding manganese(IV) oxide ($MnO_2(s)$) as a catalyst. When the two substances are mixed there is a much more vigorous reaction and oxygen is readily given off. Tiny amounts of catalyst can bring about very large changes in reaction rates. Not all reactions can be speeded up in this way. As the amount of catalyst used increases, the rate of reaction also increases until a point is reached where the reaction will go no faster.

If increasing amounts of catalyst are added to hydrogen peroxide solution, the time to produce 50 cm³ of oxygen decreases as shown in Figure 12.19. To test a gas to see if it is oxygen, put a glowing spill into the gas. If it is oxygen, the spill will relight.

Figure 12.19
The effect of amount of catalyst on reaction rate

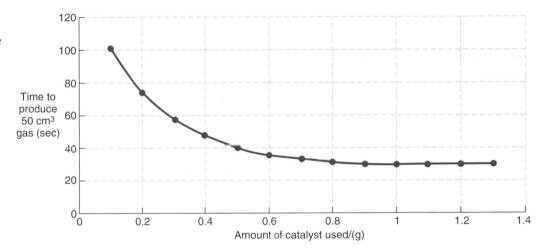

Note that the addition of more than 0.9 g of catalyst does not produce any further increase in the reaction rate.

Catalysts work by providing an easier route from reactants to products and this is done by lowering the energy barrier – i.e. reducing the energy of activation for the reaction.

Figure 12.20
An energy level diagram for a reaction involving a catalyst

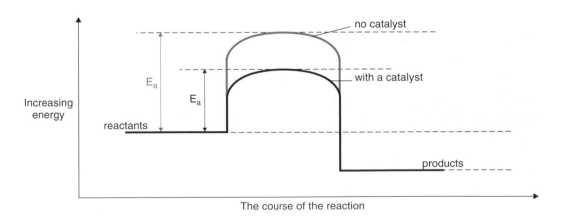

Figure 12.20 shows the energy level diagram for a reaction both with and without a catalyst. The catalyst has provided a lower energy barrier and therefore it is easier for the particles to change into products.

241

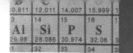

Chemistry in action

Did you know?

Whilst catalysts are not used up during chemical reactions they can be easily 'poisoned'. This happens if they come in contact with substances that block the active sites on their surface. If, for example, leaded petrol is used in a car that has a catalytic converter, the surface of the catalyst can become covered in lead compounds and this will prevent the gases getting absorbed on to the surface and therefore the catalytic activity is destroyed.

Summary

◆ The rate of a chemical reaction can be altered by:
  – changing the surface area of any solids involved
  – changing the concentration of any solutions involved
  – changing the temperature at which it is carried out
  – changing the pressure, if gases are involved
  – the use of a **catalyst**.

◆ A catalyst is a substance that can alter the rate of a reaction but is not used up in the process.

Topic questions

1 a) When a piece of fresh liver is added to hydrogen peroxide solution (H_2O_2) there is a vigorous reaction and oxygen gas is given off. If in a second similar experiment a piece of liver that has been boiled in water for 5 minutes is used, no gas is produced. Explain this reaction.
   b) Explain why food keeps for longer periods of time if stored in a refrigerator.

2 Explain each of the following:
   a) Sugar dissolves faster in hot water than in cold water.
   b) When producing carbon dioxide gas by adding marble to an acid, the reaction is quickest when the marble is in powdered form.

3 Why does increasing each of these variables increase the rate of a reaction?
   a) surface area
   b) concentration
   c) temperature
   d) pressure

4 When the temperature increases the rate of reaction increases. What happens to the amount of products produced? Give a reason for your answer.

5 a) What does adding a catalyst do to the rate of a reaction?

   b) How does the addition of a catalyst affect the activation energy level for the reaction?

12.4   Reactions involving enzymes

Co-ordinated	Modular
DA 11.14	DA 7
SA 11.6	SA 16

Enzymes are complicated molecules made up of proteins. They are often described as biological catalysts. Most catalysts contain metals – often transition metals from the centre of the periodic table. Enzymes usually do not contain metals.

Enzymes act as catalysts in a different way to metal and metal compound catalysts. Each enzyme has a particular shape, so each enzyme can only combine with molecules that have a matching shape. The enzyme and its matching molecule lock together like the pieces of a jigsaw. This is called the 'lock and key' model of enzyme action – because only the matching molecule (the key) will fit the shape of the enzyme (the lock).

The fact that enzymes will only react with molecules of a particular shape explains why each enzyme is specific to a particular reaction.

Once the reactant molecule (also called the **substrate**) combines with the enzyme, it can change to a product molecule. This product will fall away from the enzyme. Therefore enzymes can be used again and again without being used up. This can be seen in Figure 12.21.

Figure 12.21
How enzymes catalyse reactions

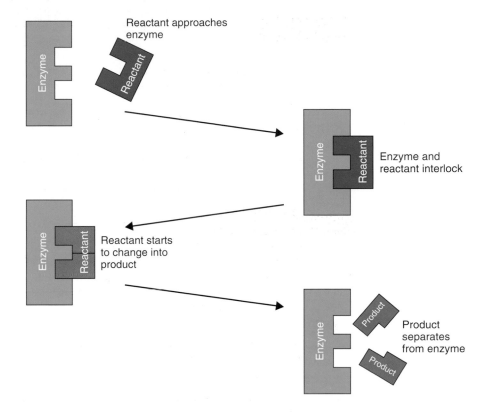

Examples of enzymes can be found in the digestive system where they break down large molecules of starch, protein and fats into smaller molecules which can be absorbed by the bloodstream (section 2.1). Amylase is the enzyme, which breaks down starch into glucose, proteases break down proteins into amino acids and lipases break down fats to fatty acids. In these examples, the enzymes (amylase, protease, and lipase) are the 'locks' and glucose, protein and fats are the 'keys'.

The chemical composition of enzymes, and hence their shape and catalytic activity, can be easily destroyed by heating. This is called **denaturing**. Enzyme activity is poor at low and high temperatures but is quite vigorous at temperatures around 25 to 35°C. Enzymes are therefore only able to catalyse reactions below about 50°C.

Different enzymes work best at different pH values.

Reactions involving enzymes

Three important processes that involve enzymes are making bread, brewing beer and wine and making yoghurt.

Making bread

The ingredients for making bread include flour, sugar, salt and yeast. Yeast is a naturally-occurring material that lives off decaying plant or animal material. Yeast does not contain chlorophyll so cannot make its own food but it does contain enzymes.

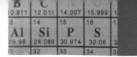

Chemistry in action

If yeast is mixed with the other bread-making ingredients and left in a warm place, the mixture (the dough) begins to rise because carbon dioxide is given off and the warmth makes the gas expand. This process is called **fermentation**. The reaction that takes place is:

$$C_6H_{12}O_6 \rightarrow 2C_2H_5OH + 2CO_2$$
$$\text{glucose} \qquad \text{ethanol} \qquad \text{carbon dioxide}$$

The risen dough is then baked in an oven where the high temperature destroys the enzymes in the yeast, makes the carbon dioxide gas bubbles expand even further and causes the bread to rise.

Making beer and wine

Another common process which uses enzymes is the conversion of sugar into ethanol (alcohol). This process is also a fermentation reaction. It requires warm conditions – high temperatures destroy the enzymes present. The reaction is the same as the one for making bread.

$$C_6H_{12}O_6 \rightarrow 2C_2H_5OH + 2CO_2$$
$$\text{glucose} \qquad \text{ethanol} \qquad \text{carbon dioxide}$$

Many of the starting materials for making wine or beer contain starch. Enzymes in the yeast are responsible for catalysing the breakdown of the complex organic starch molecules into simpler sugars such as sucrose and glucose. These are then converted into ethanol (alcohol) and carbon dioxide. As the fermentation process proceeds, bubbles of carbon dioxide gas are evolved and if a still wine is required, all of this gas is allowed to escape. Where a fizzy wine or beer is required, a secondary fermentation process is allowed to take place. This time the gas is not allowed to escape until the bottle is opened.

Making yoghurt

This process involves allowing the enzymes in the bacteria in unpasteurised milk to convert the natural sugar present in the milk – lactose – into the acid lactic acid. This makes the milk taste sour but it also helps to preserve it. If pasteurised milk is used – and this is more common nowadays – then yoghurt makers have to add their own strains of bacteria in order to make the process work. The process is carried out under warm conditions because high temperatures destroy the enzymes.

 Figure 12.22
Some uses of enzymes

Uses of enzymes in the home and industry

Use	Enzyme	Action	Comments
Biological washing powders	lipases proteases	remove fat-based stains remove protein-based stains	will work at low temperatures
Baby foods	proteases	pre-digest protein	the baby's digestive system has less work to do
Sweeteners	isomerase	converts glucose into fructose	fructose is very much sweeter than glucose, so less needs to be used
Production of sugar syrup	carbohydrases	converts starch syrup obtained from plants into maltose and glucose both of these sugars are sweet	different amounts of enzyme produce either a syrup containing more maltose than glucose – used in the brewing industry – or a syrup containing more glucose than maltose. Further enzyme action on the latter turns the glucose into fructose and produces a syrup used in foods and drinks

Enzyme technology

Figure 12.23
Fermentation tanks

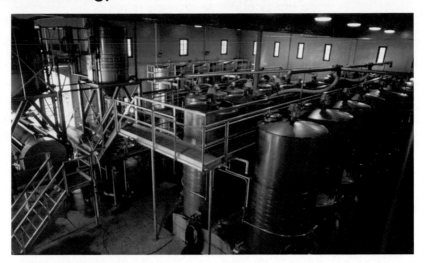

The enzymes used in industry are usually extracted from microbes (bacteria, moulds and fungi). It is now possible for scientists to 'programme' bacteria to make certain enzyme drugs. Human insulin (see section 5.4) can now be manufactured in this way.

The use of enzymes in industry has the benefit of allowing chemical reactions that would otherwise require the use of high temperatures and/or high pressures to take place at normal temperatures and pressures. Their use also means that the chemical reactions are likely to be less costly and that the reactions can be allowed to take place as a continuous process.

However, a successful product will only be produced if certain conditions can be guaranteed. These include:

● That the bacteria, which are grown in large quantities from which the enzymes are to be extracted, are all identical. This is necessary to ensure that the extracted enzymes are identical.

● That the enzymes must be identical to the strain required. This is necessary to ensure that the correct chemical reaction is controlled and that unwanted products are not produced.

● That there is close monitoring of the reaction conditions to ensure that a safe and uncontaminated product results.

● That all modified bacteria are contained safely until it is proven that they are no risk to any other organisms.

Getting the best from the enzymes

Any enzyme used in industry must be pure and not contaminated in any way that will affect its action as a biological catalyst, and must be able to be made in large quantities.

The required enzyme is extracted and purified and allowed to reproduce in sterile conditions in a fermenter. (A large tank where pH, temperature and food source can be monitored and regulated so that very large quantities of the identical enzyme can be produced.)

Whilst the enzyme is carrying out its intended function the main problem is to make sure that it stays working for as long as possible by controlling and monitoring the immediate working environment. This is called stabilisation.

In order for the reactants involved in the process to come into contact with the enzyme in a continuous process the enzyme needs to be immobilised. This can be achieved by trapping the enzyme in an inert solid support or a carrier made of alginate beads. (Alginate is an extract of seaweed, similar to agar.)

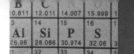

Chemistry in action

Summary

◆ **Enzymes** are complex organic molecules that can catalyse specific reactions. They are biological catalysts that can be destroyed by heating.

◆ Enzymes are used in the manufacture of alcohol, bread and yoghurt.

◆ Enzymes are used in industry to control reactions that would otherwise require high temperatures and/or pressures.

Topic questions

1 When magnesium ribbon is added to dilute sulphuric acid it starts to dissolve and give off hydrogen gas. Suggest three ways in which the rate at which the magnesium dissolves could be increased.

2 What are enzymes?

3 Use the 'lock and key' model to explain how enzymes work.

Examination questions

1 Some types of filler go hard after a catalyst is added from a tube. A manufacturer tested this reaction to see what effect the amount of catalyst had on the time for the filler to harden. The results are shown in the table.

Volume of catalyst added to filler (cm³)	Time for the filler to harden (minutes)
1	30
2	15
3	10
4	7
6	4

a) Draw a graph of these results. Plot 'time for the filler to harden', in minutes, on the vertical axis and 'volume of catalyst added to filler', in cm³, on the horizontal axis.
(3 marks)

b) Use your graph to suggest the time taken for the filler to harden using 5 cm³ of catalyst.
(1 mark)

c) What is the effect of the catalyst on the rate of this reaction? *(1 mark)*

2 Hydrogen peroxide, H_2O_2, is often used as a bleach. It decomposes forming water and oxygen.

a) Write the balanced chemical equation for the decomposition of hydrogen peroxide.
(3 marks)

b) The rate of decomposition of hydrogen peroxide at room temperature is very slow. Manganese oxide is a catalyst which can be used to speed up the decomposition. Complete the sentence.
A catalyst is a substance which speeds up a chemical reaction. At the end of the reaction, the catalyst is _____ . *(1 mark)*

c) Two experiments were carried out to test if the amount of manganese oxide, MnO_2, affected the rate at which the hydrogen peroxide decomposed.

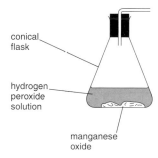

conical flask

hydrogen peroxide solution

manganese oxide

i) Complete the diagram to show how you could measure the volume of oxygen formed during the decomposition.
(2 marks)

ii) The results are shown in the table below. Draw a graph of these results. The graph for 0.25 g MnO_2 has been drawn for you.
(3 marks)

iii) Explain why the slopes of the graphs become less steep during the reaction.
(2 marks)

iv) The same volume and concentration of hydrogen peroxide solution was used for both experiments. What *two* other factors must be kept the same to make it a fair test? *(2 marks)*

Time in minutes	0	0.5	1	1.5	2	2.5	3	3.5
Volume of gas in cm³ using 0.25 g MnO₂	0	29	55	77	98	116	132	144
Volume of gas in cm³ using 2.5 g MnO₂	0	45	84	118	145	162	174	182

Volume of gas in cm³ (graph showing 0.25g MnO₂ curve, y-axis 0–200, x-axis Time in minutes 0–4)

3 a) i) A student added a few drops of Universal Indicator to some sodium hydroxide solution. The indicator turned purple. What was the pH of the sodium hydroxide solution? *(1 mark)*

ii) The student added an acid until the solution was neutral. What is the colour of Universal Indicator when the solution is neutral? *(1 mark)*

b) i) Which acid should the student add to sodium hydroxide solution to make sodium sulphate? *(1 mark)*

ii) Write the formula of sodium sulphate. *(1 mark)*

c) The student noticed that the solution in the beaker got warm when the acid reacted with the alkali. The energy diagram below represents this reaction.

i) In terms of **energy**, what type of reaction is this? *(1 mark)*

ii) Use the energy diagram to calculate a value for the amount of energy released during this reaction. *(1 mark)*

iii) Explain, in terms of bond breaking and bond forming, why energy is released during this reaction. *(3 marks)*

iv) The reaction takes place very quickly, without the help of a catalyst. What does this suggest about the activation energy for this reaction? *(1 mark)*

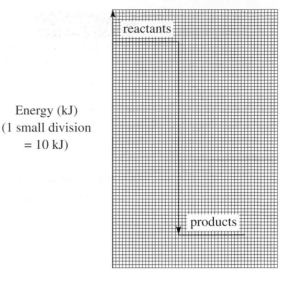

Energy (kJ)
(1 small division = 10 kJ)

4 This item appeared in the Wolverhampton *Express and Star* on October 31st, 1997. Read the passage and answer the questions that follow.

> ## Fumes scare at factory
>
> Workers were forced to flee a factory after a chemical alert. The building was evacuated when a toxic gas filled the factory.
> It happened when nitric acid spilled on to the floor and mixed with magnesium metal powder.

a) The equation which represents the reaction between magnesium and nitric acid is:

$$Mg(s) + 4HNO_3(aq) \rightarrow Mg(NO_3)_2(aq) + 2H_2O(l) + 2NO_2(g)$$

Give the formula of the toxic gas that was produced. *(1 mark)*

b) What does toxic mean? *(1 mark)*

c) The reaction of nitric acid with magnesium metal powder is more dangerous than if the acid had fallen on to the same mass of magnesium bars. Explain why. *(1 mark)*

d) i) Water was sprayed on to the magnesium and nitric acid to slow down the reaction. Explain, in terms of particles, why the reaction would slow down. *(2 marks)*

ii) Explain why it is better to add alkali, rather than just add water to the spillage. *(1 mark)*

Chapter 13
Electricity and magnetism

13.1	Electric charge

Co-ordinated	Modular
DA 12.5	DA 10
SA n/a	SA n/a

Electrostatics

If a balloon is rubbed on a cloth, the balloon will stick to the wall. The attraction between the balloon and the wall is caused by **electrostatic forces**.

Figure 13.1 ▶

Figure 13.2
Each hair has the same charge, so they repel each other

Friction between the balloon and the cloth moves **electrons** (see section 7.1) from some atoms in the cloth. These electrons transfer to the balloon from the cloth and so the balloon is negatively **charged**. The cloth is positively charged as it has lost electrons. Charged objects affect each other. The rules are:

- Objects with the *same* charge *repel*.
- Objects with *opposite* charges *attract*.

248

The negative charge on the balloon repels electrons on the wall surface. Close to the balloon the wall becomes positively charged. As the balloon is light it is attracted to the wall. The balloon stays on the wall until its negative charge leaks away to the air or to the wall.

Figure 13.3 ▶
Positive charges stay on the wall surface

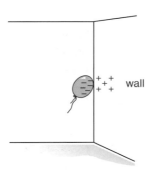

Charges at work

The sparks seen when clothing is removed quickly are due to electrostatic charges caused by friction. Clothes in a tumble dryer often become charged by friction causing 'static' clicks as the clothes are separated. Walking on a carpet can cause a person to get an electrostatic charge. This can result in the person getting an electric shock when they touch a metal door handle.

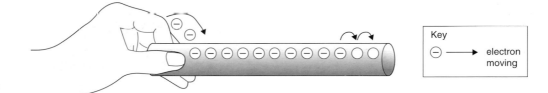

Figure 13.4 ▶
A conducting metal rod

This is because metals have atoms whose outer electrons are very free to move (see section 7.2). Should some of these **free electrons** be rubbed off then other electrons easily take their place, producing a flow of electrons. Substances, such as metals, in which electrons can flow easily are good **conductors** of charge. Plastic, polythene and rubber conduct badly and are called **insulators**. If electrons are rubbed off an insulator, then other electrons do not flow to take their place. When electrons are rubbed onto an insulator they do not move away. A static charge builds up.

Figure 13.5 ▶
A large Van der Graaf generator

Insulators and insulated objects can become very highly charged. The very high charge is not able to flow.

Problems created by a large charge
Lightning

Figure 13.6 ▶
Lightning between oppositely charged clouds

If a cloud with a large charge passes close to tall objects on the ground, then the charge can run to earth causing a lightning strike.

Figure 13.7▶
A lightning strike between a charged cloud and a tree

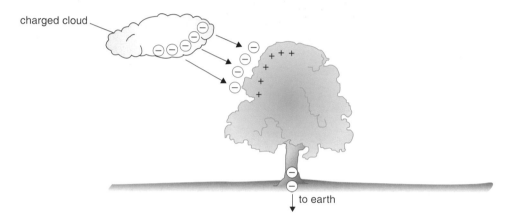

Many trees are damaged by lightning as they are often the tallest objects on the ground. The air particles between the charged cloud and the tree are **ionised**. Some air particles are stripped of electrons to become positive ions; others gain electrons to become negative ions. These ions allow the electrons to move quickly from the cloud to the tree during the discharge. The flash occurs as the air particles rejoin to their missing charge.

High voltage cables

High voltage cables are bare metal wires conducting large amounts of charge. They are held far above the ground by insulated pylons. Getting too close to such a cable is dangerous. An electric discharge can flow to the close object when the air stops insulating the cable. As the charge moves, a large **electric current** flows to earth.

Figure 13.8
The insulated supports are clearly visible on these pylons

A kite touching the cables will complete an electric circuit. There is now a path for the current to **earth** through the person. The large flow of charge can kill the person holding the kite. Even a small current can stop the heart muscle working.

Figure 13.9

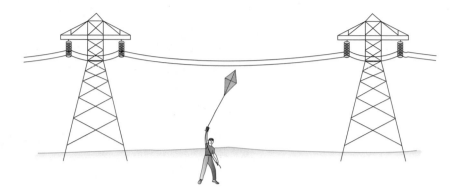

Did you know?

The lightning conductor was invented by Benjamin Franklin. He flew a kite during a thunderstorm and saw the line react as charge ran down it to earth. This gave him the idea for the lightning conductor. A Russian scientist copying Franklin's experiment died as the charge was conducted through him to the ground when he held a similar kite.

Delivering fuel

Figure 13.10
Tanker delivering fuel safely

As large tankers deliver petrol to a garage or aviation fuel to an aircraft, the movement of liquid in the pipe causes the pipe to be charged by friction. A wire connecting the pipe nozzle to the ground prevents a large charge forming on the pipe, because any charges leak away along the wire to the earth. If a large charge is formed on the pipe then a discharge spark to an uncharged object could ignite the petrol vapour.

Fine powders

In some industries, such as flour making, the friction between fast moving powders can cause the build up of electric charges. If the charges are not discharged safely, they can cause the powders to explode.

Making use of electric charge

An electrostatic paint spray gun

An electrostatic paint spray gun like the one in Figure 13.11 charges the paint droplets as they leave the gun. The droplets repel each other forming a fine spray. This coats the objects evenly with paint. Less paint is wasted if the object is given an opposite charge, as all the paint is attracted to the object.

Figure 13.11
An electrostatic paint spray gun

Electrostatic smoke precipitator

The burning of fossil fuels, such as coal, in power stations and factories pollutes the atmosphere not only with waste gases but also with smoke. Smoke consists of tiny particles of solid material. The smoke can be removed from the waste gases before they pass into the atmosphere by using the principles of electrostatics in a device called a smoke precipitator (Figure 3.12).

On their way up through the chimney the waste gases pass through a negatively charged metal grid. The smoke particles become charged as they pass by the grid and are repelled by the similar charge on the grid. The large collecting plates on the lining of the chimney are given an opposite (positive) charge to that on the grid. The smoke particles are attracted to the oppositely charged metal plates connected to earth. The smoke particles lose their charge and drop to the bottom where the plates are tapped (they are precipitated). The waste gases are then free of smoke particles.

Figure 13.12
An electrostatic smoke precipitator

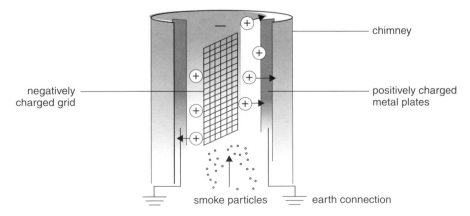

The photocopier

Substances such as selenium, arsenic and tellurium are called photoconductors because they are electrical insulators in the dark but electrical conductors in the light.

Figure 13.13 ▶
A photocopier

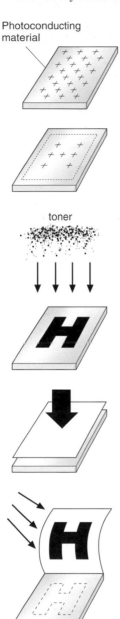

Photoconducting material

toner

1. In a photocopier there is a plate that consists of a layer of photoconducting material on a thin metal backing sheet. At the start of a copy cycle this plate is given a positive charge.

2. The page to be copied is lit with a strong light and an image of the page forms on the surface of the charged plate. The white areas of the page light up the photoconducting layer so it becomes conducting and the charge leaks away to the metal backing. The black areas of the page are left as an image in the positive electric charges on the belt.

3. Negatively charged toner powder is spread over the plate and is attracted to the parts of the plate still positively charged (the dark parts of the original copy).

4. A blank sheet of paper is given a positive charge and rolled over the belt where it attracts the negatively charged toner.

5. Rollers then apply heat and pressure to make the toner stick to the paper.

Electric current

An **electric current** is the movement of charged particles. In a metal wire, the electric current is the flow of electrons. Electrons always move from a place where there are too many electrons to a place lacking electrons. When a wire is connected to the terminals of an electrical source, the electrons travel from the negative end of the source, along the wire to the positive end or terminal of the source. This is the same as thinking of a flow of positive charges from the positive terminal. Remember, positive charges do not move in a solid.

Only the outer electrons of metal atoms are free to move through a solid.

The electric current in a gas or liquid is a movement of positively and negatively charged particles called **ions** (see section 7.2).

Figure 13.14
Flowing electrons make an opposite electric current

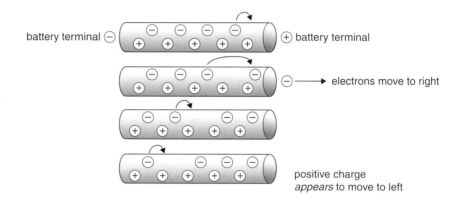

Did you know?

Scientists who carried out early work on electric current did not know about electrons. They thought a positive electric current flowed from the positive to the negative terminal. This idea is still used today. This idea is called conventional current.

Electrolysis

Certain chemical compounds conduct electricity when they are melted or dissolved in water. The electric current is formed by negatively charged ions moving to the positive **electrode** (**anode**), and positively charged ions moving to the negative electrode (**cathode**). The conducting liquid is called the **electrolyte** and the process is called **electrolysis**. When the ions reach the electrodes, simpler substances are given off at the electrode as gases, or deposited there as solids.

During the electrolysis of copper sulphate solution (see section 10.2), the negative electrode becomes coated in a layer of copper, whilst the positive electrode loses copper (Figure 13.15).

In the electrolysis of slightly acidified water, hydrogen gas bubbles off at the negative electrode, and oxygen gas at the positive electrode.

The results of electrolysis reactions like these show that the mass or volume of a substance freed during electrolysis increases in proportion with:

- the current – doubling the current doubles the mass or volume
- the time for which current flows – doubling the time doubles the mass or volume.

Figure 13.15
*How liquids conduct
electric current*

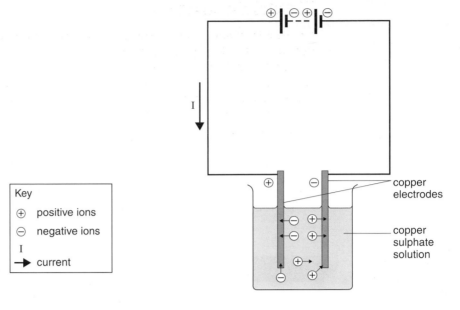

Key
⊕ positive ions
⊖ negative ions
I
→ current

copper
electrodes

copper
sulphate
solution

Example: In the electrolysis of copper sulphate solution, 0.9 g of copper were deposited when a current of 1.5 A flowed for 30 minutes.

How much copper would there be for a current of 1.5 A for 90 minutes?

30 minutes	give	0.9 g
so 3 × 30 minutes	give	3 × 0.9 g
New mass	=	3 × 0.9 g
	=	2.7 g

Did you know?

Electroplating uses electrolysis. Metal spoons have an even coating of silver.

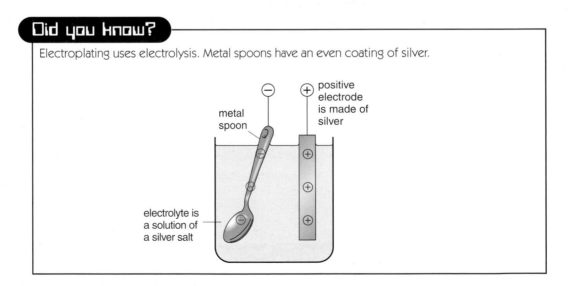

metal
spoon

positive
electrode
is made of
silver

electrolyte is
a solution of
a silver salt

Summary

◆ There are two types of electrical charge: negative and positive.

◆ An object becomes negatively charged when it gains **electrons**, and positively charged when it loses electrons.

◆ Two objects with a similar charge repel; two objects with opposite charges attract.

◆ **Electrostatic forces** can be used to explain the working of paint spray guns, photocopiers and smoke precipitators.

◆ A build up of charge can be dangerous.

◆ A large build up of charge on an object can cause a spark to jump between the object and any earthed conductor brought near to it.

◆ A charged object can discharge when it is connected by a conductor to earth (earthing).

◆ An electric current in a wire is due to the movement of charged particles. In a solid the charged particles are electrons. In a gas or liquid the charged particles are **ions**.

◆ Metals are good conductors of electricity because they contain free electrons that can move easily through the metal.

◆ **Electrolysis** is the process by which an electric current flows in a liquid to release simpler substances at the anode and cathode.

◆ The mass/volume of a substance deposited or released at the electrodes during electrolysis increases in proportion to the time and the current.

Topic questions

1 Fill the gaps in the following passage using these words:

attract	charged	electrons	friction	induces	positively
		repel	rubbed	small	

A polythene rod is negatively _____ when it is _____ with a cloth. The cloth loses _____ to the rod. The cloth is _____ charged. The rod can attract _____ objects. The rod _____ opposite charges on the objects. Opposite charges _____ and like charges _____.

2 Match the term to each explanation.

discharge   earthing wire   electrolysis   electrons

a) Transfer a large charge safely to the ground.
b) Tiny particles with negative charge.
c) Rapid movement of electrons to a place where electrons are missing.
d) Electric current as ions move in a liquid.

3 A small polystyrene bead hangs on a nylon thread and is given a positive charge. Different rods charged by friction are brought near the bead. The movement of the bead is recorded in the table. Tick each charge present on each rod.

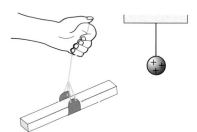

Material of rod	Bead movement	Charge on rod		
		Positive	Negative	Uncharged
Cellulose acetate	Repelled			
Ebonite	Attracted			
Perspex	Repelled			
Polythene	Attracted			
Steel	None			

4 Why are the nozzles of fuel pipes earthed?

5 A current flows through a solution of copper sulphate.

Which are the positive and negative ions?
Explain how you work out your answer.

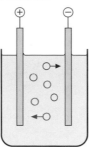

6 A lightning conductor is connected to this church
tower, to prevent damage during a lightning discharge.
The conductor is a metal spike joined by thick copper
wire to a metal bar in the earth.

a) What happens to the air between the cloud and
the tower before a lightning discharge occurs?

b) What can you say about the size of the current
and the time it flows during a discharge?

c) Give reasons why the lightning conductor must:
i) be made of metal
ii) be made of a thick copper connecting wire
iii) have its end buried in the ground.

7 In an electroplating process, 2 g of silver are deposited on a spoon when a current
of 1 A flows for 30 minutes.

What would be the mass deposited if
a) 1 A of current flows for 15 minutes
b) 2 A of current flows for 30 minutes.

13.2   Circuits

Co-ordinated	Modular
DA 12.1	DA 10
SA 12.1/12.2	SA 17

Figure 13.16▼
*Electrical components
and their symbols*

A circuit is a closed pathway of wires and components which conduct the electrons
from and back to the electrical source. Circuits are represented by circuit diagrams.
They show all the components and wires using symbols.

Figure 13.16 shows the symbols with the components they represent.

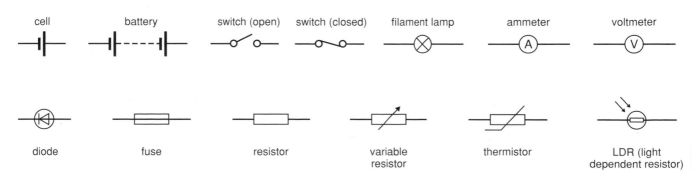

Series and parallel circuits

A series circuit is a single pathway for an electric current through a set of wires and components. A parallel circuit gives a number of pathways for the electric current.

In a series circuit with several lamps, removing one lamp breaks the circuit. All the other lamps go out. This also happens if a lamp 'blows'. Some sets of Christmas tree lights are wired in series. When one lamp burns out, all the lamps go out – with a long search to find the lamp that made all the others turn off.

In a parallel circuit with several lamps, removing one lamp does not stop the others working. Most lighting circuits are wired in parallel.

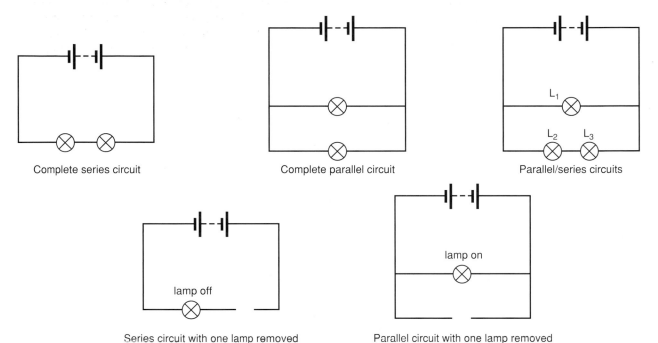

Complete series circuit   Complete parallel circuit   Parallel/series circuits

lamp off   lamp on

Series circuit with one lamp removed   Parallel circuit with one lamp removed

Figure 13.17

Electric current in series and parallel circuits

An electric current is the flow of electrons in a wire or the movement of ions in a fluid. Electric current, I is measured in units called **amperes** (A), often called amps. **Ammeters** are special instruments that measure the amount of electric current flowing in a circuit. Because ammeters measure the current in a circuit, they are always connected in series in the circuit.

The current in a series circuit is the same at all points in the path. The current in a parallel circuit divides between the different paths. It rejoins where the parallel paths meet. In a parallel circuit, the total current $= I_1 + I_2$.

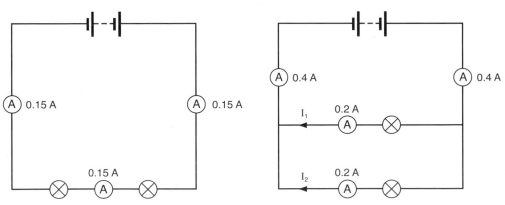

Figure 13.18
Current in series and parallel circuits

Potential difference (voltage)

The **potential difference** (p.d.) between two points in a circuit measures the difference in electrical potential energy for a charge moved between those two points. It is also called **voltage**. A **voltmeter** measures this energy drop in units of **volts** (V). Because potential difference is the energy difference between two points, a voltmeter is always placed across the component in the circuit. So voltmeters are always connected in parallel with the component they are monitoring.

Potential difference – lamps in series

Lamps in series have the same current flow through each of them. This is because the same charge moves through each lamp. The charge gives up its energy in stages. Each stage sees a fall in the electrical energy of the charge. The greater the energy transferred to each lamp by each unit of charge, the larger the potential difference across the lamp.

Figure 13.19
Voltage drop and lamps in series

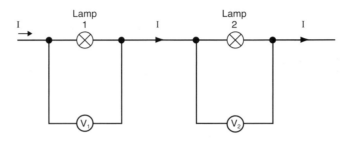

In a series circuit the total potential difference $= V_1 + V_2$.

Potential difference – lamps in parallel

Lamps in parallel have separate currents in each path. But the charges passing through each different path all have the same energy transfers. The potential difference across the set of parallel lamps is the same.

Figure 13.20
Current and lamps in parallel

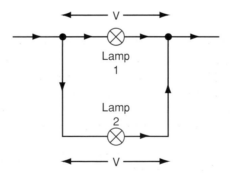

Potential difference – cells in series

The potential difference provided by cells connected in series is the sum of the potential differences of each **cell** separately – as long as they are connected together correctly, + to −.

Sometimes a single cell is wrongly called a **battery**. A battery is really a set of cells in series. A typical car battery has six cells, each 2V, with a total of 12V.

Figure 13.21
Potential difference of cells in series

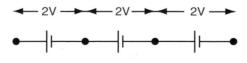

Total potential difference = 6 V

Resistance

Some devices let a large current flow through them. These devices have a low **resistance** to the flow of current. A device with a high resistance allows less current to flow through it when the same potential difference is applied. The amount of resistance is measured by seeing how much current flows for a certain potential difference. A typical circuit shows a component being tested for its resistance (Figure 13.22). The ammeter is always in series with the component, and the voltmeter in parallel with it.

The current is changed by the variable resistor for a set of readings.

Current, in A	Potential difference, in V
0	0
0.5	2
1.0	4
1.5	6
2.0	8

By inspecting each pair of readings in the results table, it can be seen that $\frac{\text{potential difference}}{\text{current}}$ is a constant value (= 4).

A graph of potential difference against current shows a straight line through the origin. The slope of the graph is 4.

These results show that $\frac{\text{potential difference}}{\text{current}} = \text{constant}$

The constant is the resistance for that object and so

$$\begin{array}{ccccc} \text{potential difference} & = & \text{current} & \times & \text{resistance} \\ (\text{volt, V}) & & (\text{ampere, A}) & & (\text{ohm, } \Omega) \\ \text{V} & = & \text{I} & \times & \text{R} \end{array}$$

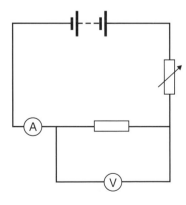

Figure 13.22

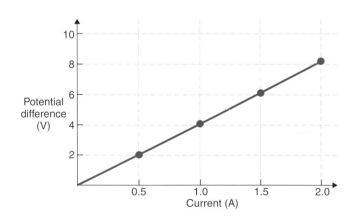

Figure 13.23

This resistance stays constant provided the component does not change its physical conditions (for example, the wire does not get hot). This relationship was first investigated by a scientist called George Ohm and the unit of resistance is named after him, the **ohm** (Ω).

Example: What is the potential difference across the 10 ohm wire in Figure 13.24?

$$\begin{aligned}
\text{potential difference} &= \text{current} \times \text{resistance} \\
&= 3 \times 10 \\
&= 30 \text{ volt}
\end{aligned}$$

Figure 13.24

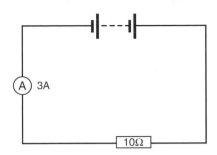

Resistances in series and parallel

Resistances in series all have the same current. As another resistance is connected in series the current decreases because the total resistance increases. When components are connected in series, the total resistance is the sum of their separate resistances.

Example:

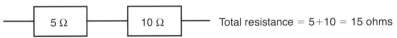

Total resistance = 5+10 = 15 ohms

For resistances in parallel, each additional resistor connected in parallel allows another pathway for the current. So the total current increases, and the total resistance decreases.

For resistances in parallel the total resistance is less than the smallest resistance of any branch. The greater the resistance, the smaller the current.

Example:

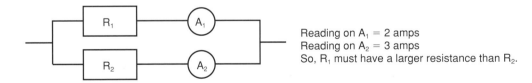

Reading on A_1 = 2 amps
Reading on A_2 = 3 amps
So, R_1 must have a larger resistance than R_2.

Circuits must not be overloaded, a temptation with adaptor plugs. The connecting wires carry too large a current, get very hot and can cause a fire.

Figure 13.25
Overloaded circuits cause fires

The connecting wire has a very small resistance. If it is put in parallel across a device, the wire takes most of its current in a 'short circuit' (Figure 13.26).

Figure 13.26
Short circuit

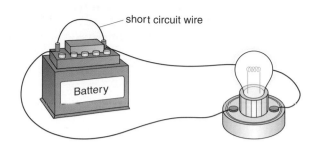
short circuit wire
Battery

Wires of different materials have different resistances. Poor conductors have very high resistances so almost no current flows. But even materials that conduct do not all conduct to the same extent. Copper is a very good conductor – a copper wire will let a large current flow, it has very low resistance. But a steel wire of the same thickness and length, and with the same voltage across it, will not let such a large current flow. It is not such a good conductor – it has a higher resistance.

Figure 13.27
Graph showing differences in resistance for two wires

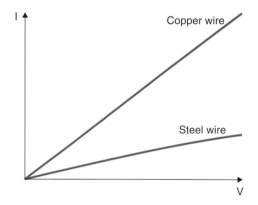

Did you know?

For solids, resistivity gives the difficulty with which electricity is conducted. For liquids, scientists usually use the conductivity. The better a substance conducts electricity, the lower its resistivity, and the higher its conductivity. Because conductance is the opposite of resistance, the unit for measuring conductance used to be the MHO, which is OHM spelt backwards.

Conducting devices

1 Resistors

Resistors pass a certain amount of current for a given voltage. High value resistors let a small current pass, and low value resistors allow a large current to pass. Standard resistors have their resistance clearly marked and some value showing the maximum current allowed without overheating.

For a resistor at constant temperature the current – potential difference graph is a straight line through the origin. This means the resistance remains constant (Figure 13.29).

Reversing the resistor, so the current flows in the opposite direction, does not change its resistance.

A variable resistor has a resistance that is changed from 0 ohms to its maximum value by sliding the connector across the wire coils. Sometimes the wire coils are in a small closed box and the slider returns.

Figure 13.28
A variable resistor

Figure 13.29
Current and potential difference for a resistor

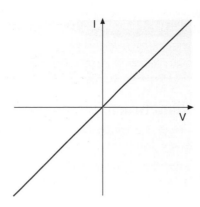

2 Diodes

Diodes are made of a non-metallic material called a semiconductor. A diode only conducts a current in one direction. If the direction of the current is reversed the diode does not let a current pass. The resistance of a diode depends on which way round it is connected. It has a very large resistance when connected in the reverse direction.

Figure 13.30
Current and potential difference in a diode

Figure 13.31
Lamp P lights but lamp Q will not as its diode is connected in the reverse direction to the current

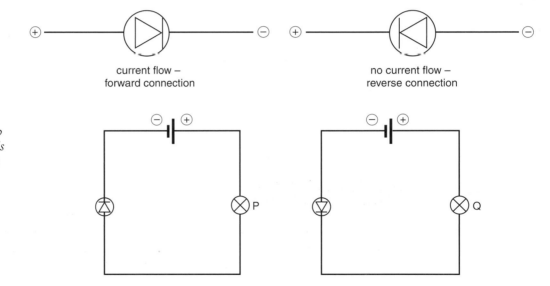

current flow –
forward connection

no current flow –
reverse connection

The graph linking current and voltage for a diode would look like Figure 13.32.

No current flows in the reverse connection as resistance is infinitely large, for normal potential differences.

Figure 13.32 ▶

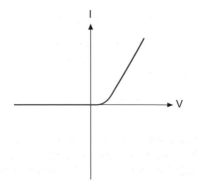

Figure 13.33 ▲
A light emitting diode

3 Filament lamp

A filament lamp uses a very hot wire to give light. As the voltage across the lamp increases, the light gets brighter. This is because the filament gets hotter. As the filament heats up its resistance increases. This means the current flow is less than expected.

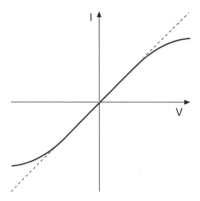

Figure 13.34
Resistance for a filament lamp

The lamp has the same resistance increase when it is connected in the opposite way.

4 Light Dependent Resistors (LDR)

LDRs are also made of semiconductor material. They have a greatly reduced resistance when light falls on them.

The LDR is used in light operated circuits, such as security lighting.

Light level	Resistance, in ohms
dark	1 000 000
bright	500

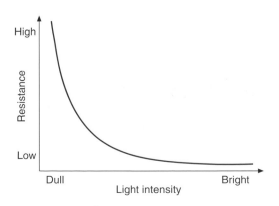

Figure 13.35
Variation of resistance with light intensity for an LDR

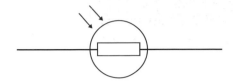

Figure 13.36
Symbol for an LDR

5 Thermistors

The **thermistor** is another semiconducting device. Its resistance decreases as the temperature increases. The drop in resistance causes a larger current to flow. The thermistor can be used in temperature sensing circuits.

Figure 13.37
Symbol for a thermistor

Summary

- For components connected in **series**:
  - the total **resistance** is the sum of their resistances
  - the same amount of **current** flows through each component
  - the total voltage (potential difference) is shared between the components.

- For components connected in parallel:
  - the total current in the circuit is the sum of the currents in each component
  - the current through each component is dependent on its resistance
  - there is the same voltage (potential difference) across each parallel branch.

- Current, voltage (potential difference) and resistance are related by the equation:

  potential difference = current × resistance
      (volt, V)   (ampere, A)   (ohm, Ω)

- The current in a resistor (at constant temperature) is proportional to the voltage across the resistor.

- For a filament lamp the resistance increases as the temperature of the filament increases.

- Diodes only allow a current to flow in one direction. For a diode the resistance is very high in the reverse direction.

- For a **light dependent resistor** (LDR) the resistance decreases as the amount of light shining on it increases.

- For a **thermistor** the resistance decreases as the temperature increases.

Topic questions

1 Match the circuit symbols with each component: ammeter, cell, fuse, lamp, variable resistor, switch.

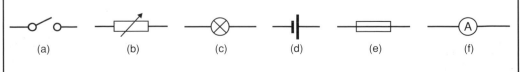

    (a)   (b)   (c)   (d)   (e)   (f)

2 Which switches have to be closed to make these lamps light?

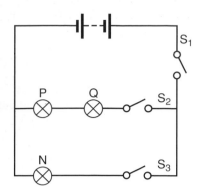   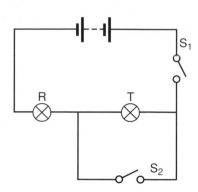

a) P and Q only
b) R and T
c) R only

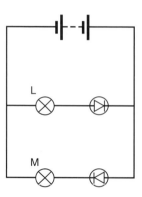   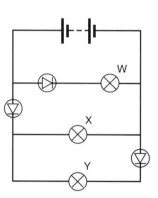

3 Which lamps are lit up when these circuits are switched on?

4 Complete the missing words
The unit of resistance is the _____. A resistor has a _____ resistance provided the _____ stays the same. A filament lamp has an increase in its _____ as it gets _____ and brighter. The resistance of a _____ decreases as its temperature increases. A light dependent resistor or _____ has its highest resistance in the _____ .

5 a) Write down the equation linking current, potential difference and resistance.
   b) What is the potential difference across a 20 ohm resistor if a current of 0.5 A flows?
   c) What current flows through a 4 ohm resistor if the potential difference across it is 12 V?
   d) What is the resistance of a heater if the current flowing is 10 A and the mains voltage is 230 V?

6 a) What happens to the resistance of a filament lamp as it gets brighter?
   b) Why does a filament lamp usually 'blow' at the moment you switch it on?

7 a) When is the resistance of an LDR greatest?
   b) When is the resistance of a thermistor greatest?

Co-ordinated	Modular
DA 12.2	DA 10
SA n/a	SA n/a

Energy and power in a circuit

Electrical energy is used to work many machines; it is also used for heating lighting, televisions and communication systems. This energy is provided by a source of electricity, such as those shown in Figure 13.38.

Figure 13.38
Sources of electricity
a) Batteries
b) Power Station

A **cell** transfers its chemical energy to electrical energy when its terminals are joined by wires in a circuit. By joining a conductor from one end to the other, the negative charges can move to the positive terminal.

As the charge moves, an electric current flows through the circuit. The current stops flowing when the chemical reaction stops. This is when the cell is 'flat'. Some cells' chemical reactions can be renewed when they are recharged.

Figure 13.39 ▼
The current flows and the lamp is lit

Sometimes one cell is wrongly called a **battery**. A battery is really a set of cells in series. A typical car battery has six cells each 2V, with a total 12V. The battery recharges when the car engine turns.

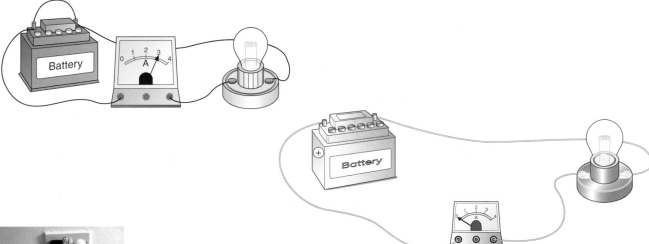

Figure 13.40
When the cell is 'flat', no current flows and the lamp stays off

As charge moves through the lamp, electrical energy is transferred to heat and light energy by the lamp.

The total amount of energy transferred depends on the total amount of charge moved and the potential difference of the source of electricity. The amount of energy is measured in a unit called a **joule** (J).

Figure 13.41
A battery with charger

267

Heating effect

Heating elements are usually made of special wire. They can transfer the electrical energy to heat energy efficiently without damaging the wire. Nichrome wire is frequently used in heating elements.

Figure 13.42
Some appliances designed to deliver heat energy

cooker

iron

Measuring electric charge

The amount of charge transferred in a circuit depends on the size of the current and the time for which it flows. The **coulomb**, C, is the unit of **electric charge**.

$$\begin{array}{ccc}
\text{charge} & = & \text{current} & \times & \text{time} \\
\text{(coulomb, C)} & & \text{(ampere, A)} & & \text{(seconds, s)} \\
Q & = & I & \times & t
\end{array}$$

Energy transferred by a charge

The more charge that is moved, and the bigger the potential difference across the terminals of the electrical source, then the larger the electrical energy delivered to the circuit components.

$$\begin{array}{ccc}
\text{energy transferred} & = & \text{potential difference} & \times & \text{charge} \\
\text{(joule, J)} & & \text{(volt, V)} & & \text{(coulomb, C)} \\
E & = & V & \times & Q
\end{array}$$

Potential difference, or voltage, gives the amount of energy transferred by one unit of charge. A voltmeter records 1 volt when 1 joule of energy is transferred by 1 coulomb of charge.

Example: A liquidiser is used from a mains supply at 230 volts. How much electrical energy is transferred to the liquidiser by a current of 1 A in 3 minutes?

$$\begin{array}{lcl}
\text{charge} & = & \text{current} \times \text{time} \\
& = & 1 \times 3 \times 60 \\
& = & 180 \text{ C} \\
\text{energy transferred} & = & \text{potential difference} \times \text{charge} \\
& = & 230 \times 180 \\
& = & 41\,400 \text{ J}
\end{array}$$

Electric power

How quickly energy is transferred is as important as how much energy there is in the transfer. Walking slowly up a hill is difficult, but running up it is much harder. The same amount of energy is used each time but running uses the energy more quickly. It requires more **power**. Electrical power measures how quickly electrical energy is transferred by various devices to heat or light or movement, for example.

Figure 13.43
Electrical devices

Power, whether muscle power or electrical power, is measured in a unit called the **watt** (W). A power of 1 watt is the transfer of 1 joule of energy in 1 second. Electrical power is related to the current and potential difference as:

power (watts, W)	=	potential difference (volt, V)	×	current (ampere, A)
P	=	V	×	I

The kettle shown in Figure 13.43 has a high power rating, 2000 W. It transfers electrical energy to heat energy quickly and so heats the water quickly. A **kilowatt** (kW) is 1000 W, so 2000 W can be written as 2 kW.

Example: A TV and video draws a current of 2 A from a 230 V power supply. What is the power rating of the TV and video?

Power	=	potential difference	×	current
	=	230	×	2
	=	460 W		

Summary

◆ In many electrical appliances electrical energy is transferred as heat.

◆ **Power** is the rate at which energy is transferred.

◆
power (watt, W)	=	potential difference (voltage, V)	×	current (ampere, A)

◆ For a given charge, the higher the voltage the more energy is transferred, so

transferred energy (joule, J)	=	potential difference (voltage, V)	×	charge (coulomb, C)

◆ Charge, current and time are related by the equation:

charge (coulomb, C)	=	current (ampere, A)	×	time (seconds, s)

Topic questions

1   Complete the table for the quantity and its unit

Quantity	Unit
Current	
	Watt
	Joule
Resistance	
Potential difference	

2   What is the potential difference across these?
a) torch lamp, R = 5 ohm, I = 0.25 A
b) resistor, R = 75 ohm, I = 3 A
c) heater, R = 45 ohm, I = 5 A

3   What is the power rating in watts for:
a) a motor on a 125 V supply, taking a 2 A current?
b) a kettle on a 230 V supply, taking a 10 A current?
c) an electric oven on a 230 V supply, taking a 15 A current?

4 a) In the diagram what is the potential
difference across:
  i) the resistor
  ii) the lamp
b) What is the resistance of the lamp?

5 A 24 W set of Christmas lights has 20
lamps in series worked from a 120 V
supply. What is the current through the
lamps? What is the potential difference
across one lamp? What is the power
transferred to each lamp?

6 A 12 V battery drives a motor taking a
current of 2 A for 20 seconds. What is the
electric energy transferred to the motor?

7 A 60 W lamp used on the mains supply
of 230 V is switched on for 2 hours. How
much electric charge was transferred in
this time?

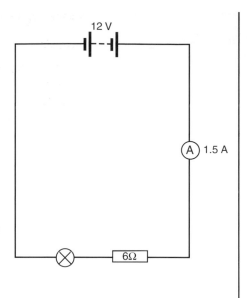

⌶∃.Ь Mains electricity

Co-ordinated	Modular
DA 12.3	DA 10
SA 12.3	SA 17

Alternating current and direct current

A **direct current** (d.c.) always flows in the same direction, from a fixed positive
terminal to the fixed negative terminal of a supply. A cell or battery gives a constant
direct current (Figure 13.44). A cathode ray oscilloscope (CRO) shows how the
potential difference (voltage) changes with time.

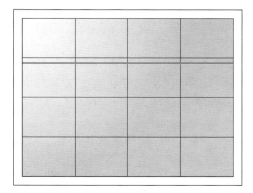

Figure 13.44
*The wave trace on a CRO for a d.c. supply
from a cell. The zero line is shown in red*

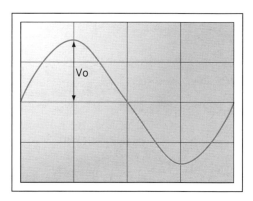

Figure 13.45
*Alternating current trace from an a.c. as
seen on a CRO. The zero line is shown in
red. The height of the wave gives the peak
voltage, V_O*

An **alternating current** (a.c.) in a circuit changes direction as the terminals of the
supply change from positive to negative.

When the CRO trace shows a potential difference below the zero line, the current is
in the opposite direction to that above the zero line (Figure 13.45). You can use a
CRO to measure the frequency with which the current changes direction. Frequency
is measured in hertz (Hz).

Taking measurements from a CRO trace for an a.c. supply

For the CRO trace in Figure 13.46, each vertical division represents 2 volts of potential difference. Each horizontal division represents 5 milliseconds of time. Therefore, this trace shows a maximum or peak potential difference of 6 volt and the time for one cycle of 40 milliseconds, giving the frequency as 25 hertz.

Figure 13.46

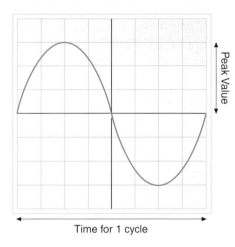

Time for 1 cycle

Peak value	=	3 divisions	×	2 volts per division
	=	6 V		
Time for one cycle	=	8 divisions	×	5 millliseconds per division
	=	40 milliseconds		
	=	0.04 seconds		
So, frequency	=	$\frac{1}{0.04}$		
	=	25 hertz		

Comparing CRO traces for two different a.c. supplies

Traces from two a.c. supplies were seen on a CRO, with the same settings for the vertical, potential difference reading and the horizontal, time reading.

Figure 13.47
a) 1 cycle takes 8 divisions
b) 2 cycles take 8 divisions

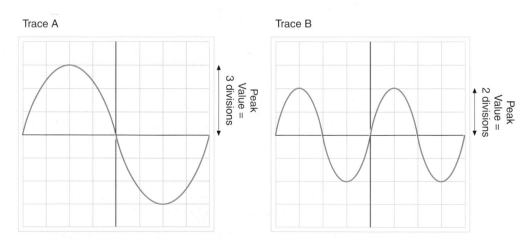

Trace A

Trace B

If the supply shown in trace A has a peak value of 6 V, then supply B will have a peak value of 4 V. The frequency of supply B is 2 times that of supply A. So if A has a frequency of 50 Hz, then B has a frequency of 100 Hz.

The mains electricity supply

In the United Kingdom, the mains supply delivers electric power at an average potential difference of 230 volts. The supply provides an alternating current with a frequency of 50 hertz. This means the current flows one way, then the other, 50 times each second.

Safety devices and electrical appliances

Potential differences greater than 50 V and currents as small as 50 milliamps can be fatal to humans. Safety measures use insulation and a way of stopping or diverting the current.

The three pin plug

These days new appliances have plug already attached. On older appliances, plugs need to be fitted. Care must be taken to connect the wires from the appliances to the correct pins in the plug. The colour coded wires help in this. The green/yellow striped earth wire protects the user should the live wire loosen and touch the appliance. By earthing the appliance with a wire leading directly to the ground, the current from the loose live wire flows to earth along the earth wire and not through the person touching the appliance!

Figure 13.48

Correct wiring in a three-pin plug

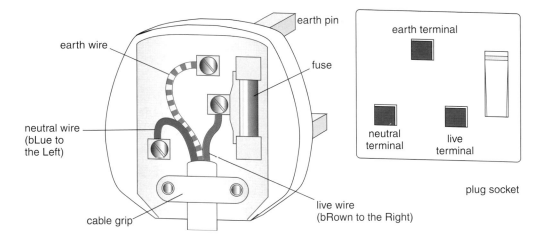

- The outside case of the plug is made of plastic or rubber, because these are good insulators.

- Pushing the plug into a socket connects the pins of the plug to the terminals of the socket.

- The live terminal of the mains supply alternates between a positive and negative voltage when compared with the neutral terminal. The neutral terminal stays at a voltage close to zero.

Fuses

A **fuse** is connected to the live wire before it joins the appliance. It is there to shield the appliance from too large a current. If a fault causes a larger current to flow through the appliance than that marked on the fuse, then the fuse melts and so breaks the circuit.

Each appliance has a power rating such as

Figure 13.49
Fuses

> Power 460 W
> Supply 220-240 V
> $\sim$ 50Hz
>
> ($\sim$ indicates an a.c. supply)

The working current for the appliance is found using

$$
\begin{aligned}
P &= V \times I \\
460 &= 230 \times I \\
I &= 460/230 \\
&= 2 \text{ A}
\end{aligned}
$$

Figure 13.50
The power rating label on the side of an electric jig-saw

This is the normal current for the device and any larger current can destroy it. A 3 A fuse would allow a normal working current to flow and protect the appliance from larger currents. A 13 A fuse would allow a dangerously high current to flow and still not 'blow'. So it is important to use the correct size of fuse. A fuse does not protect the person using the appliance. It can take 1 to 2 seconds for a fuse to melt – enough time for the user to receive a fatal electric shock.

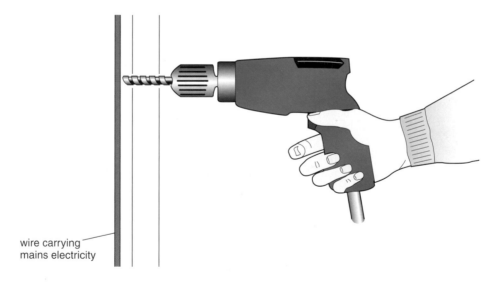

Figure 13.51 ▶
The current will flow through the person to the ground if there is no earth wire connected to the casing of the drill

wire carrying
mains electricity

Double insulation

Insulation is the first protective barrier for the user. Wires have an outer plastic or rubber coat. Many appliances have a plastic casing and do not have the earth wire. A loose live wire does not conduct through the plastic casing. The insulation of both the casing and the wires is shown by a double insulation label $\boxed{\boxdot}$.

Circuit breakers

A residual current device, RCD, is a fast acting circuit breaker. It stops current flowing in less than 0.05 seconds. If the RCD detects any difference between the current in the live and neutral wires it will break the circuit.

Modern domestic wiring methods use a set of miniature circuit breakers (mcb) instead of a fuse box. Each mcb protects one circuit from too large a current overheating the wires and starting a fire. Once the fault is corrected, the mcb is reset (see section 13.6).

Summary

◆ **Alternating current** (a.c.) changes direction but **direct current** (d.c) flows in one direction only.

◆ In the UK the mains supply is at 230 V a.c. with a frequency of 50 hertz.

◆ Traces on a CRO can be used to compare the voltages for a.c. and d.c. supplies.

◆ All electrical circuits should include safety devices because even a very low current could be fatal.

◆ The three-pin plug correctly wired contains a fuse and an earth wire.

◆ The earth wire is connected to the metal casing of many appliances. Other appliances use double insulation.

◆ Electrical circuits can be protected by circuit breakers.

Topic questions

1 Give two features about:
   a) alternating current
   b) direct current.

2 Complete the gaps in the table to provide information about the wiring of a three-pin plug.

wire	Colour
	Yellow/green
neutral	
	brown

3 a) To which part of an electric fire would an earth wire be fitted?
   b) How does an earth wire stop someone getting an electric shock?

4 Explain two ways a circuit breaker is more efficient as a safety device than a fuse?

5 Fuses for household appliances can be bought in these values 2 A, 3 A, 5 A, 10 A and 13 A.

   Domestic electricity is supplied at 230 V. Which fuse is correct for each of the following appliances?

Appliance	Power rating
Lamp	150
Television	200
Iron	800
Kettle	2400

13.5   Paying for electricity

Co-ordinated	Modular
DA 12.4	DA 9
SA 12.4	SA 17

There are many examples of electrical appliances that transfer electrical energy into other forms of energy, such as:

● heat (thermal energy)
● light
● sound
● movement (kinetic energy).

The cost of using an electrical appliance depends upon the particular energy transfer.

Heating appliances are expensive to run. The cost also depends on how quickly the energy transfer takes place. This is the electrical power delivered.

Figure 13.52
Some appliances designed to transfer energy from one form to another

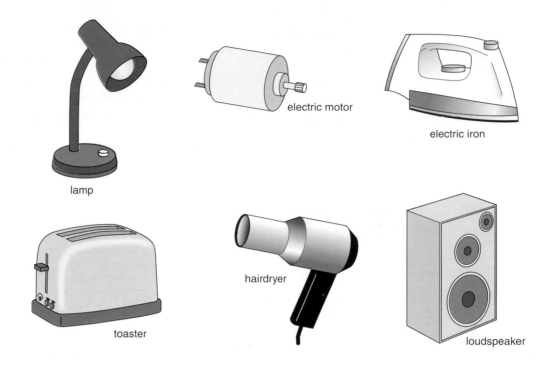

lamp

electric motor

electric iron

toaster

hairdryer

loudspeaker

Figure 13.53

Appliance	Useful energy transfer
Toaster	Electricity to heat
Hairdryer	Electricity to heat and movement
Electric motor	Electricity to movement
Lamp	Electricity to light
Loudspeaker	Electricity to movement and sound
Electric iron	Electricity to heat

Did you know?

1 kW is equivalent to 3.6 million joules of energy.

The power rating of an appliance is given in watts or **kilowatts**. This gives the amount of electrical energy transferred in 1 second by a working appliance (1 kW = 1000 W).

Electrical companies bill customers for electrical energy in special units of **kilowatt hours** (kWh). One kilowatt hour is the energy transferred when 1000 W is delivered for 1 hour.

Figure 13.54
Electricity meter showing two readings after an interval of time

kWh
5 5 6 5 2 1
February

kWh
5 7 1 3 9 6
May

The readings on the meters are used to calculate the number of Units transferred in a given period of time. These meters show that 57139−55652=1487 Units or kilowatt hours of electrical energy was transferred.

$$1 \text{ Unit} = 1 \text{ kilowatt hour} = 1000 \text{ watt for 1 hour}$$

electrical energy transferred = power × time
(kilowatt hour, kWh)   (kilowatt, kW)   (hour, h)

Example: How many units of electrical energy were supplied to a 100 W lamp left on at night for 8 hours?

energy transferred = power × time
= 0.1 × 8
= 0.8 kWh

The cost of 1 unit (1 kWh) varies. If the price is 7p per Unit:

total cost = number of Units × cost per Unit

So for lamp left on all night:

cost = 0.8 × 7
= 5.6p

Heating appliances are the greatest energy consumers. An immersion heater, rated at 3000 W and used for 8 hours at 7p per Unit:

cost = energy transferred × price
= 3 × 8 × 7
= 168p

The total amount of electrical energy transferred, measured in joules, is calculated using the equation (see section 17.4):

energy transferred = power × time
(joule, J)   (watt, W)   (seconds, s)

Example: A kettle with a power rating of 2500 W is left on for 5 minutes. How much electrical energy has been transferred in that time?

energy transferred = power × time
= 2500 × (5 × 60)
= 750 000 J
= 750 kJ

Summary

◆ The power of an appliance is measured in watts (W) or **kilowatts** (kW).
 1000 W = 1 kW

◆ The amount of energy transferred from the mains is measured in kilowatt hours or Units.
 1 Unit = 1000 W delivered for 1 hour − **1 kilowatt hour** (kWh)

◆ The amount of energy transferred can be calculated using:
 energy transferred = power × time
 (kilowatt hour, kWh)   (watt, W)   (hour, h)

◆ The readings on an electricity meter can be used to calculate the number of Units used.

◆ The cost of the energy used can be calculated using:
 total cost = number of Units × cost per Unit

◆ The amount of energy transferred by an appliance can be calculated using:
 energy transferred = power × time
 (joule, J)   (watt, W)   (second, s)

Topic questions

1 How many units of electrical energy in kWh are used for a
   a) 100 W lamp for 10 hours?
   b) 200 W TV for 4 hours?
   c) 2000 W kettle for 5 minutes?

2 If the price for 1 unit is 7p. What is the running cost of each item in question 2?

13.6

Co-ordinated	Modular
DA 12.20	DA 9
SA n/a	SA n/a

Electromagnetic forces

Magnetic repulsion and attraction

As long ago as 1000 BC it was known that certain rocks attracted iron objects. The mineral that did this was named magnetic after the region where it was first discovered. Pieces of this mineral would point to the north and could be used for navigation. These days **magnets** are manufactured. A freely-suspended magnet will line itself up along the north–south line of the Earth.

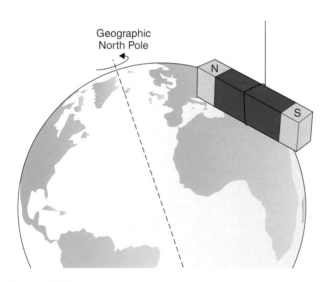

Figure 13.55
A bar magnet suspended in the Earth's magnetic field

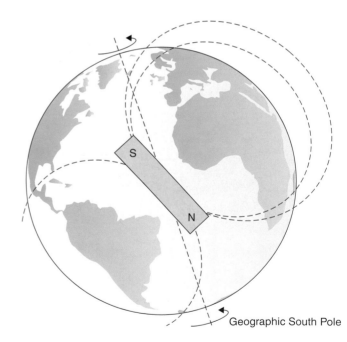

Figure 13.56
The theoretical bar magnet in the Earth that creates the Earth's magnetic field

Navigational compasses like this have been used since the eleventh century. Compasses work because the Earth behaves like a giant magnet with its north pole a little eastwards of the geographic south pole. The direction and strength of this magnet gradually changes over very long periods of time.

Experiments with magnetic forces show that:

- like magnetic poles repel
- unlike magnetic poles attract.

This is like the result for electrostatic charges. In magnetic substances, the atoms line up in groups to form tiny magnets. A larger magnetic force is produced when all the tiny magnets point the same way.

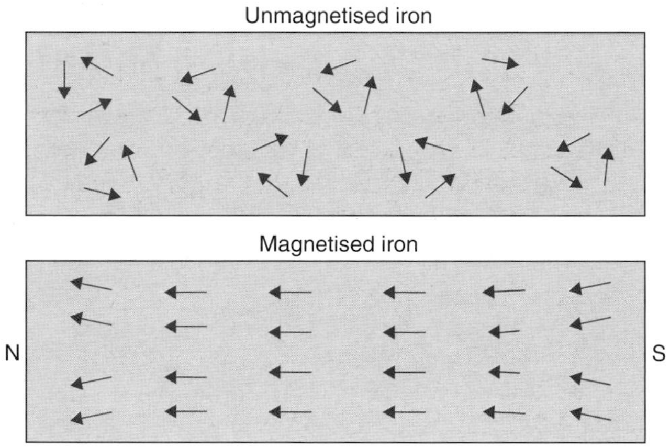

Figure 13.57
Arrangement of 'atomic magnets' in unmagnetised and magnetised iron

Unmagnetised iron

Magnetised iron

N   S

Magnetic fields

The metals iron, steel, nickel and cobalt are strongly attracted to magnets. Some magnetic substances can be made into permanent magnets. Temporary magnets can be made by passing a current along a wire.

The magnetic force is increased if the wire is coiled. Putting an iron bar in the coil strengthens the magnet even more. The coil of wire is called a **solenoid**. This is a simple **electromagnet**.

current-carrying wire

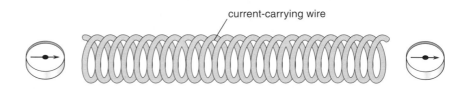

Figure 13.58
A simple electromagnet

Magnets attract small iron filings and change the direction in which a compass needle points. The space around the magnets where the force acts is called the **magnetic field**. The shape of a magnetic field is shown by the iron filings. The magnetic force is strongest at the poles. A compass needle shows the direction of the magnetic lines of force.

Figure 13.59
The magnetic field around a) a straight wire, b) a solenoid

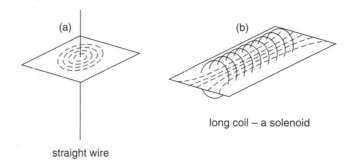

(a)

straight wire

(b)

long coil – a solenoid

The direction of the magnetic field is shown by the pointing of the north pole of a plotting compass.

Increasing the number of turns (coils) of wire around an iron **core** increases the strength of an electromagnet. If the current in the wire is increased, an even stronger force is produced. Changing the direction of current reverses the magnetic poles. When the current stops, the magnetic field disappears. Any iron or steel objects held by the magnet fall off, as the iron core does not stay a magnet. (Steel and some other magnetic materials can be made into permanent magnets by putting them in an electric coil.)

Figure 13.60
The direction of the magnetic field around a) a straight wire, b) a solenoid

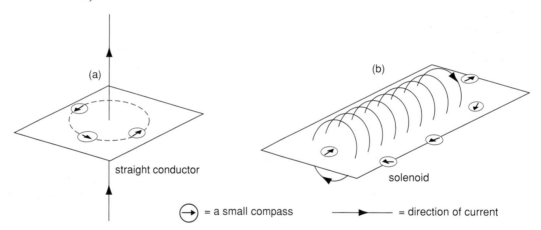

(a)

straight conductor

(b)

solenoid

$\bigoplus$ = a small compass $\longrightarrow$ = direction of current

Using electromagnets

Large electromagnets are used to pick up steel and iron objects in a scrap yard or a factory. Small electromagnets are used in buzzers, bells, relay switches and miniature circuit breakers.

How a simple circuit breaker works

Figure 13.61 shows a simple **circuit breaker**. When the current through the circuit is 15 A or less, the force of attraction from the electromagnet is not sufficient to compress spring A. The iron bolt remains in place.

However, if there is a fault in the rest of the circuit then a current greater than 15 A will flow. The force of the electromagnet is now greater and sufficient to overcome the force of spring A, and attract the iron bolt. The iron bolt moves towards the electromagnet. Spring B can now push the plunger away from the contacts in the circuit.

Electricity and magnetism

Because the metal ends of the plunger are no longer touching the contacts the circuit has been broken so everything connected in this circuit, including the electromagnet is immediately switched off. Later, when the fault has been corrected, pushing the reset button, which is fixed to the plunger, resets the circuit breaker. The metal end closes the contacts, a current flows and the iron bolt clicks back into the slot in the plunger.

Many older houses have individual fuses for each wiring circuit, held in the fuse box. These are slower in their action and take longer to repair.

Figure 13.61
A circuit breaker

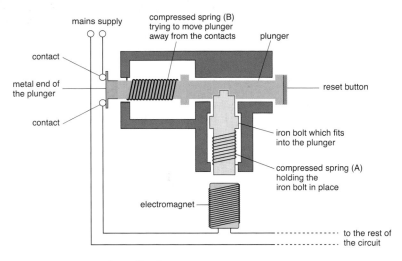

The motor principle

When a wire with a current flowing through it is placed between the poles of a magnet, on one side the magnetic field of the wire strengthens the magnetic field of the magnet, on the other side it weakens it. So one side of the wire has a stronger magnetic field than the other. The wire is pushed towards the weaker side of the field.

Figure 13.62
The motor principle

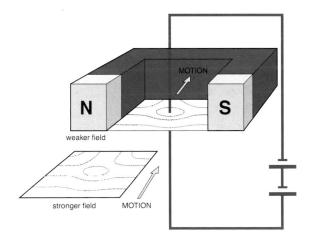

The force causing motion is greatest when the wire is at right angles to the field. No motion occurs if the wire is parallel to the field. Reversing the direction of the current, or the magnetic field, reverses the direction of motion. A stronger force is produced if the magnetic field strength or the current is increased.

If a coil of wire with a current flowing through it is placed in a magnetic field, the coil tends to turn. As one side of the coil moves up, the other goes down. If the coil is vertical, when the current direction changes, the coil will continue to turn. This is how a direct current electric motor works. A split ring ensures that the current flow changes direction at the right time.

Figure 13.63
A direct current motor

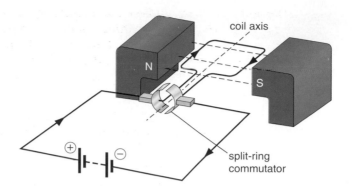

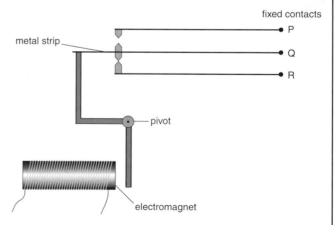

Summary

◆ A magnetic effect is produced by a current in a wire.

◆ An electromagnet is used in circuit breakers.

◆ A wire carrying current in a magnetic field experiences a force.

◆ This effect is made use of in d.c. motors.

Topic questions

1 Draw the magnetic field pattern for a:
   a) straight wire conductor
   b) solenoid.

2 Fill in the missing words.

   In magnets, like poles _____ and _____ poles attract. A temporary magnet is made by _____ wire around an _____ bar.

3 The diagram shows an electromagnet in a relay switch when no current flows.
   a) Which of the contacts are joined together?

   The electromagnet is then turned on

   b) In which direction does the metal strip move?
   c) Which contacts are now joined together?

4 A student makes a simple motor with 10 coils of wire around a square plastic frame. The frame is on a spindle so it can turn between opposite poles of two magnets. The wire ends are either side of the spindle, touching metal contact strips connected to one cell.
   a) What happens if the magnetic poles are the same?
   b) What are the three different ways of making the coil turn faster?

281

13.7	

Electromagnetic induction

Just as moving charges cause a magnetic field, so a changing magnetic field about a conductor produces a current. This effect is easily shown by moving a wire through the magnetic field between the poles of a strong magnet. A potential difference is induced across the ends of the wire and a current is made to flow. This is **electromagnetic induction**.

Electromagnetic induction transfers kinetic (movement) energy to electrical energy. A current flows only when the wire moves in the magnetic field.

Figure 13.64
Electromagnetic induction

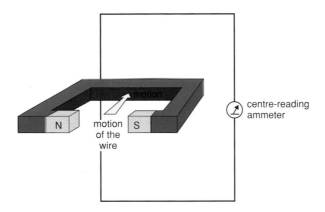

The dynamo principle

A coil of wire has a current (potential difference) induced when a magnet is moved into or out of the coil. No potential difference or current is induced when the magnet is still. The current flows the opposite way when the magnet is moved in the opposite direction.

If the pole entering the coil is changed, the direction of current is reversed. The induced potential difference or current is increased by moving the magnet faster or using a stronger magnet or by having more turns on the coil cycle.

Figure 13.65
Electromagnetic induction in a coil

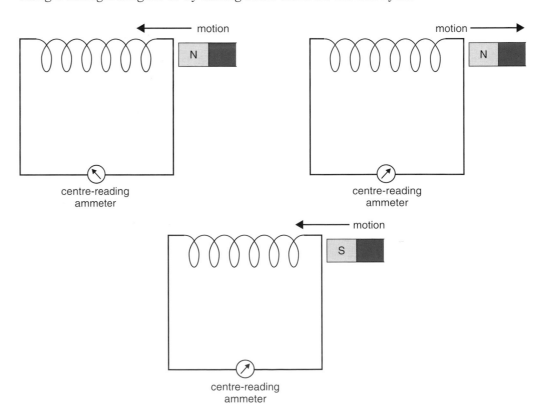

Figure 13.66
The effect of altering the pole on the direction of current

This idea is used in making a **dynamo** to produce a current. It is not necessary to move a magnet in a coil to get a current induced. Rotating a coil of wire between the poles of a magnet will also induce a current to flow in the wire.

The induced potential difference or current is increased when:

- the coil rotates faster (or if the coil is stationary the magnet rotates faster).
- the area of the coil is increased.
- there are more turns on the coil.
- the strength of the magnetic field is increased.

Figure 13.67
A bicycle dynamo

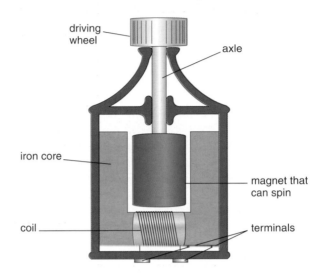

Figure 13.68
The dynamo principle

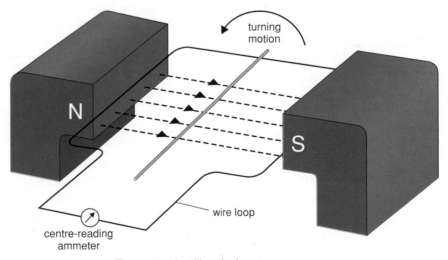

The magnet is still as the loop turns.
Each side cuts across the magnetic field.
An a.c. current is induced in the loop.

Simple a.c. generator

Figure 13.69
A simple a.c. generator

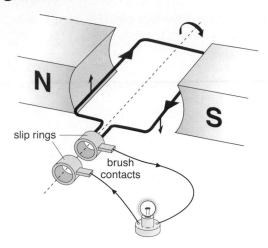

slip rings

brush
contacts

The a.c. generator has a fixed magnet and a coil that rotates. As the coil rotates a voltage is induced in the wire if the coil. Each end of the coil is connected to a conducting ring (slip ring) that also turns with the coil. The slip rings come into contact with two fixed carbon brushes. As the coil turns, the induced voltage changes direction for each half turn of the coil. The alternating current so produced passes via the slip rings and the brushes to the rest of the circuit.

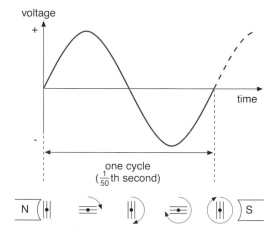

voltage

time

one cycle
($\frac{1}{50}$th second)

Figure 13.70
As the coil turns an a.c. wave is produced

Figure 13.71
A simplified mains a.c. generator

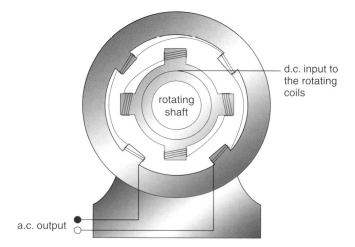

d.c. input to
the rotating
coils

rotating
shaft

a.c. output

Other types of electricity generator use large stationary coils. The movement is provided by an electromagnet rotating inside the stationary coils. The rotating electromagnet is usually driven by steam turbines using coal, oil, gas or nuclear energy as the energy source. The current produced in the coils is an alternating current (a.c.).

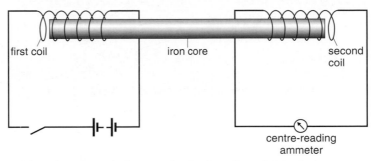

The transformer principle

A changing magnetic field in a fixed coil will induce a current in a second fixed coil, if there is a magnetic link between the two coils. An iron core will provide such a link.

Figure 13.72
Obtaining an induced current

There is a current in the second coil only as the switch is closing or opening in the first coil. The magnetic field in both coils then changes.

Transformers and electromagnetic induction

Transformers use magnetic linking between two coils to step-up or step-down alternating voltages. They can only work using an alternating current. The changing magnetic field created by the alternating current in the **primary coil** causes an alternating current in the **secondary coil**. One of the reasons why mains electricity generators give an a.c. supply is to allow the use of transformers.

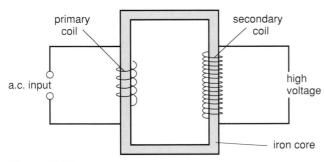

Figure 13.73
A step-up transformer

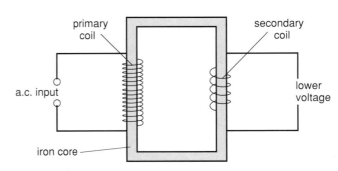

Figure 13.74
A step-down transformer

If the number of turns on the secondary coil of a transformer is more than on the primary coil, the output voltage is stepped up (increased). If the number of turns on the secondary coil is less than on the primary coil, the output voltage is stepped down. The output current then increases.

This equation can be used to calculate the output voltage:

$$\frac{\text{voltage across primary coil(volt, V)}}{\text{voltage across secondary coil(volt, V)}} = \frac{\text{number of turns on primary}}{\text{number of turns on secondary}}$$

$$\frac{V_p}{V_s} = \frac{N_p}{N_s}$$

Example: A transformer is designed to step-down 230 V to 11.5 V. There are 1000 turns of wire on the primary coil. Calculate the number of turns of wire on the secondary coil.

Figure 13.75

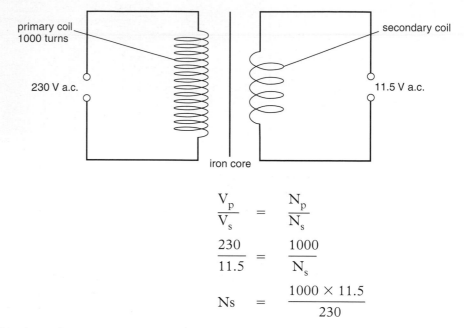

$$\frac{V_p}{V_s} = \frac{N_p}{N_s}$$

$$\frac{230}{11.5} = \frac{1000}{N_s}$$

$$Ns = \frac{1000 \times 11.5}{230}$$

Number of turns in secondary = 50 turns

Did you know?

Transformers can be constructed with very small energy losses. The large currents flowing in the coils produce unwanted heat energy. This is minimised by using low resistance windings in the coils. Large transformers have these windings specially cooled.

The changing magnetic fields linking the coils can produce **eddy currents** in the core material. These small electric currents would heat the core, so wasting energy. By building up the core with a large number of thin insulating sheets between the many magnetic layers, the eddy currents are greatly reduced.

Some transformers are almost 100% efficient, and many others approach this. The power transferred from the primary coils to the secondary coil stays almost constant.

Transmission of electricity

Power station generate electricity at voltages of 25 000 V (25 kV). The power is transferred by cables in the transmission of lines of the National Grid to all parts of the country.

A typical power station produces 10 MW (1 megawatt = 1 000 000 W). If the power is transmitted at a low voltage then the current is high as P = VI. Much of the electrical energy is wasted as heat in the cables. The current is much reduced when the voltage is stepped up by a transformer to 400 000 V or 275 000 V and power losses in the cables are greatly reduced.

Figure 13.76
Simplified diagram of power transmission

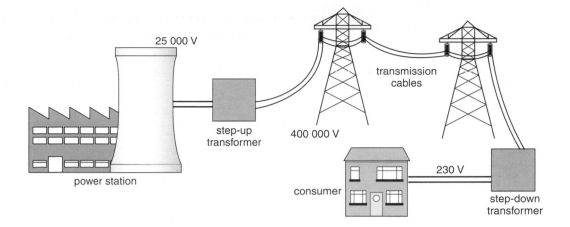

Did you know?

Aluminium transmission cables have a steel core. This is because aluminium is not strong enough to support the weight of the cable by itself.

At the consumer end, the voltage is stepped down to 230 V for houses. Every few streets in a town have an electricity substation, which is a step-down transformer serving the houses. The input voltage at these substations is very high and very dangerous. Most substations are open to the air to help them cool down.

The overhead transmission cables are usually made of pure aluminium because it has a low resistance. To reduce power losses in heating, the cables are thick. (Resistance is smaller for a thicker wire.) To reduce the weight and increase the rate of cooling, the cables are not covered by an insulation layer.

Figure 13.77
An electricity substation

They are held far above the ground by ceramic insulating supports on pylons. Underground cables need costly insulation and maintenance work is also more difficult.

Summary

◆ If a magnet is moved into or out from a coil of wire, or if a coil moves in a magnetic field a current is induced in the wire. This effect is called **electromagnetic induction**.

◆ The direction of the induced current depends on the direction of movement of the magnet or the coil.

◆ This effect is made use of in the a.c. generator.

◆ **Transformers** step-up or step-down voltages. They are used in the National Grid.

◆ If electrical energy is transmitted at very high voltages, thinner cables can be used because the current will be reduced.

◆ Transformers consist of two coils wound on a single iron **core**.

◆ An alternating voltage in one coil induces an alternating voltage in the second coil.

◆ The voltages and the number of turns are related by the equation:

$$\frac{\text{voltage across primary}}{\text{voltage across secondary}} = \frac{\text{number of turns on primary}}{\text{number of turns on secondary}}$$

Topic questions

1 What would you notice if a wire, connected to a sensitive ammeter, is moved through a magnetic field?

2 A coil of wire with 20 turns spinning 60 times per minute between the poles of a magnet induces a current to flow in the coil. What three changes can be made to increase the size of the induced current?

3 The diagram shows a coil connected to a centre reading ammeter and a magnet about to be pushed into the coil.
   What will happen to the meter needle when the magnet:
   a) moves into the coil?
   b) stays still in the coil?
   c) moves out of the coil?

4 The diagram shows the percentage of original energy transferred in the production and distribution of electricity

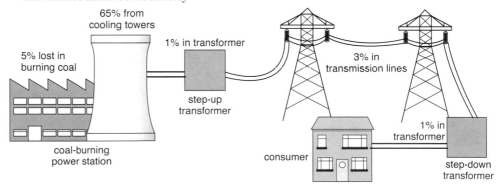

What percentage of the energy from the coal is
   a) useful energy to the consumer?
   b) lost in the transmission lines and transformers?
   c) lost during production in the power station?

5 A simple transformer has 20 turns on the primary coil and 120 turns on the secondary coil. An alternating voltage of 12 V is supplied to the primary coil. What is the voltage across the secondary coil?

Examination questions

1 A student did an experiment with two strips of polythene. She held the strips together at one end. She rubbed down one strip with a dry cloth. Then she rubbed down the other strip with the dry cloth. Still holding the top ends together, she held up the strips.

a) i)   What movement would you expect to see?
   (1 mark)
   ii) Why do the strips move in this way?
   (2 marks)
b) Complete the **four** spaces in the passage. Each strip has a negative charge. The cloth is left with a _____ charge. This is because particles called _____ have been transferred from the _____ to the _____ .
   (4 marks)

2 a) The diagram shows a 13 amp plug.

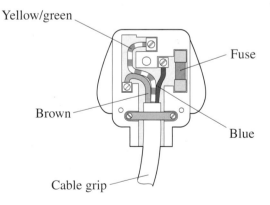

   i)   What is wrong with the way this plug has been wired?   *(1 mark)*
   ii) Why do plugs have a fuse?   *(1 mark)*

b) The diagram shows an immersion heater which can be used to boil water in a mug.

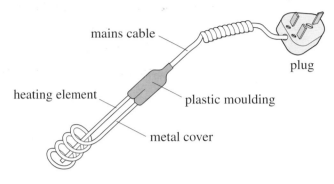

   i)   Which part of the immersion heater should be connected to the earth pin of the plug?
   (1 mark)
   ii) Complete the sentence by choosing the correct words from the box. Each word may be used once or not at all.

chemical	electrical	heat	light

   When the immersion heater is switched on _____ energy is transferred to _____ energy.   *(2 marks)*

3 a) Look at this table of results.

VOLTAGE (V)	0.0	3.0	5.0	7.0	9.0	11.0
CURRENT (A)	0.0	1.0	1.4	1.7	1.9	2.1

   i)   Plot a graph of current agains voltage. Place current, in amps, on the vertical axis and voltage, in volts, on the horizontal axis.
   (3 marks)
   ii) Use your graph to find the current when the voltage is 10 V.   *(1 mark)*
   iii) Use your answer to (ii) to calculate the resistance of the lamp when the voltage is 10 V.   *(2 marks)*
b) i)   What happens to the resistance of the lamp as the current through it increases?
   ii) Explain you answer.   *(2 marks)*

4 The drawing shows an experiment using a low voltage supply, a joulemeter, a small immersion heater and a container filled with water.

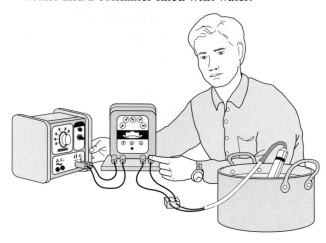

The potential difference was set at 6 V d.c. The reading on the joulemeter at the start of the experiment was 78 882 and 5 minutes later it was 80 142.

a) Use the equation:

$$\text{potential difference} = \frac{\text{energy transferred}}{\text{charge}}$$

to work out the total charge which flowed through the immersion heater in five minutes. Clearly show how you get to your answer and give the unit. *(3 marks)*

b) Calculate the current through the immersion heater during the 5 minutes. Write the equation you are going to use, show clearly how you get to your answer and give the unit.

(3 marks)

5 The diagram shows a simple electricity generator. Rotating the loop of wire causes a current which lights the lamp.

State **three** ways to increase the current produced by the generator. *(3 marks)*

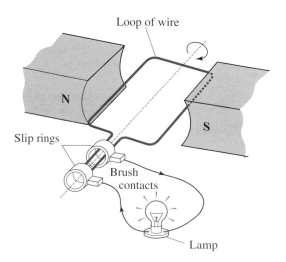

6 A fault in an electrical circuit can cause too great a current to flow. Some circuits are switched off by a circuit breaker.

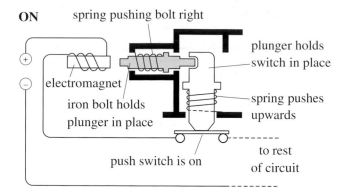

One type of circuit breaker is shown above. A normal current is flowing.
Explain, in full detail, what happens when a current which is bigger than normal flows.

(4 marks)

7 a) The diagram represents a simple transformer used to light a 12 V lamp. When the power supply is switched on the lamp is very dim.

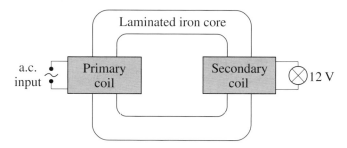

Give **one** way to increase the voltage at the lamp with without changing the power supply.

(1 mark)

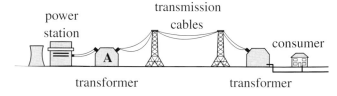

b) Electrical energy is distributed around the country by a network of high voltage cables.
   i) For the system to work the power is generated and distributed using alternating current rather than direct current. Why?

(1 mark)

   ii) Transformers are an essential part of the distribution system. Explain why. *(2 marks)*

   iii) The transmission cables are suspended high above the ground. Why? *(1 mark)*

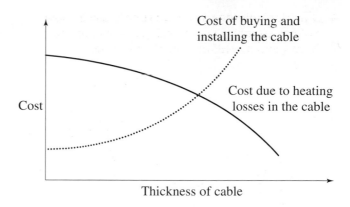

c) The power station generates 100 MW of power at a voltage of 25 kV. Transformer **A**, which links the power station to the transmission cables, has 44 000 turns in its 275 kV secondary coil.

   i) Write down the equation which links the number of turns in each transformer coil to the voltage across each transformer coil.

   (1 mark)

   ii) Calculate the number of turns in the primary coil of transformer **A**. Show clearly how you work out your answer.

   (2 marks)

d) The diagram shows how the cost of transmitting the electricity along the cables depends upon the thickness of the cable.

   Why does the cost due to the heating losses go down as the cable is made thicker?   *(1 mark)*

Cost of buying and installing the cable

Cost due to heating losses in the cable

Cost

Thickness of cable

Chapter 14
Forces and motion

14.1

Co-ordinated	Modular
DA 12.6	DA 11
SA n/a	SA n/a

Speed, velocity and acceleration

If you travel at a fast **speed**, it takes less time to finish a journey than when you travel at a slow speed. Some passenger trains are so fast they can travel 280 kilometres in just one hour. In other words, the train has a speed of 280 kilometres per hour (280 km/h). This is the average speed of the train. During a one hour journey the train will sometimes go faster and sometimes slower.

Average speed can be worked out using this formula:

$$\text{average speed (in metres per second)} = \frac{\text{distance travelled (in metres)}}{\text{time taken (in seconds)}}$$

Example: In a race, a horse travels 1280 metres in 80 seconds. What is the average speed of the horse in metres per second?

distance travelled   =   1280 metres (m)

time taken   =   80 seconds (s)

$$\text{average speed} = \frac{\text{distance travelled}}{\text{time taken}} = \frac{1280\text{m}}{80\text{ s}} = 16 \text{ m/s}$$

Figure 14.1
Race horses in action

On a journey, it's not just speed that is important, direction also counts. Figure 14.2 shows two routes which can be taken by a pupil going to school.

Figure 14.2

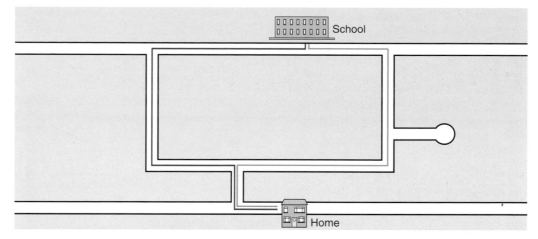

Both routes are the same distance and take the same time, so the speed is the same even though the directions are different. But each time the pupil changes direction, the **velocity** changes. This is because velocity is the speed of an object in a particular direction. If direction changes, velocity changes, even though the speed may stay the same. So to give a value for velocity, both speed and direction must be given.

Distance–time graphs

A distance–time graph can be used to work out the speed of an object.

Figure 14.3 shows that it takes 5 seconds to travel 30 metres. So the average speed can be calculated using:

$$\text{average speed} = \frac{\text{distance travelled}}{\text{time taken}} = \frac{30 \text{ m}}{5 \text{ s}} = 6 \text{ m/s}$$

This value is equal to the slope (or gradient) of the line of the graph. So, the slope of a distance–time graph equals the speed.

A distance–time graph can also be used to describe the movement of an object.

In Figure 14.4 the distance travelled is staying the same. This object is not moving – it is stationary. So, when an object is stationary, the distance–time graph is flat.

In Figure 14.5 the object is travelling the same distance each second. This object is described as moving at a constant speed. So, the steeper the slope of the straight line, the faster the speed.

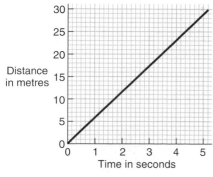

Figure 14.3

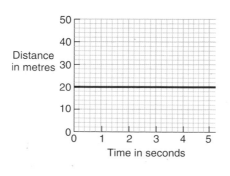

Figure 14.4

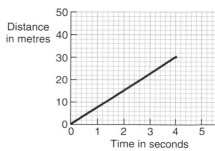

Figure 14.5

293

Figure 14.6

In Figure 14.6 the distance the object is travelling each second is increasing. This shows that the speed of the object is increasing. The object is said to be accelerating.

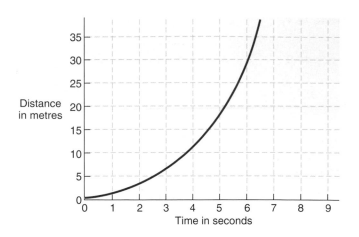

Using a distance-time graph

Figure 14.7 shows a distance–time graph for a cyclist. The slope of the line can be used to calculate the cyclist's speed.

Figure 14.7

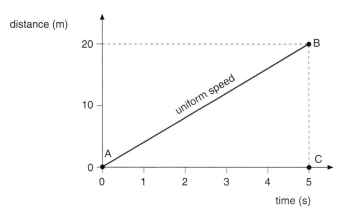

The slope of the line AB in Figure 14.7 $= \dfrac{B - C}{C - A}$

B – C = distance travelled (m)

C – A = time taken (s)

So, $\dfrac{B - C}{C - A} = \dfrac{\text{distance travelled}}{\text{time taken}} = $ speed (m/s)

In this example the slope of AB $= \dfrac{20}{5} = 4$, so the cyclist was moving at a steady speed of 4 m/s.

Velocity–time graphs

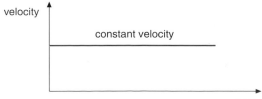

Figure 14.8 shows a velocity–time graph for a car travelling at a constant speed along a straight road. This means that the car is travelling with a constant velocity.

Figure 14.8

Figure 14.9

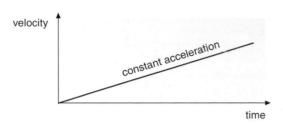

Figure 14.9 shows the velocity–time graph for a car accelerating along a straight road. The steeper the slope the greater the rate of acceleration. A straight line for the slope means that the acceleration was constant.

Figure 14.10

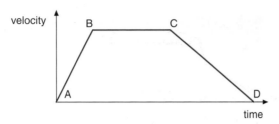

Figure 14.10 shows the velocity–time graph for a train moving from station A to station B along a straight track.

- Between A and B the train has a constant acceleration.
- Between B and C the train is travelling with constant velocity.
- Between C and D the train has a constant negative acceleration (**deceleration**).

Using a velocity–time graph

Figure 14.11 shows a velocity–time graph for a cyclist travelling in a straight line. The cyclist travels down a hill and then along a flat road.

Figure 14.11

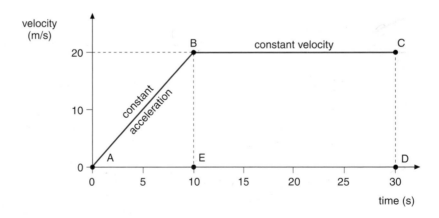

The slope of the line in a velocity–time graph can be used to calculate the acceleration of a moving object.

In Figure 14.11 the slope of line AB $= \frac{B-E}{E-A}$

B − E = change in velocity (m/s)

E − A = time taken (s)

So, $\frac{B-E}{E-A} = \frac{\text{change in velocity}}{\text{time taken}} =$ acceleration (m/s^2).

In this example:

slope AB $= \dfrac{B - E}{E - A} = \dfrac{20}{10} = 2$, so the cyclist is accelerating uniformly at 2 m/s^2

slope BC = 0, the cyclist is not accelerating but moving at a constant velocity.

The area under a line in a velocity–time graph can be used to calculate the distance travelled by the moving object. In Figure 14.11, the rectangular area under the line BC = C – B (or D – E) × C – D (or B – E)

C – B (or D – E) = time taken (s)
C – D (or B – E) = velocity (m/s) $= \dfrac{\text{distance (m)}}{\text{time (s)}}$

So $(C - B) \times (C - D) = \text{~~time (s)~~} \times \dfrac{\text{distance (m)}}{\text{~~time (s)~~}} = \text{distance (m)}$

In this example the distance travelled by the cyclist while accelerating between A and B:

the area under line AB $= \frac{1}{2} \times (E - A) \times (B - E)$
$= 0.5 \times 10 \times 20$
$= 100$
So the cyclist has travelled 100 m.

The distance travelled by the cyclist while going at constant velocity between B and C:

the area under line BC $= (C - B) \times (C - D)$
$= 15 \times 20$
$= 300$
So the cyclist has travelled 300 m.

The total distance travelled by the cyclist is therefore 400 m.

Acceleration

When the velocity of an object changes, the object is accelerating. The faster the speed changes, the larger the **acceleration**. An object which is slowing down has a negative acceleration.

Acceleration can be worked out using this equation:

$$\begin{array}{c}\text{acceleration} \\ \text{(metres per second per second, m/s}^2\text{)}\end{array} = \dfrac{\text{change in velocity (metres per second, m/s)}}{\text{time taken (seconds, s)}}$$

This can also be written as: $a = \dfrac{v - u}{t}$

$$\text{acceleration} = \dfrac{\text{final velocity} - \text{starting velocity}}{\text{time taken (in seconds)}}$$

The unit of acceleration, metres per second per second, or metre/second squared, is usually written as m/s^2.

Example: At the start of a 100 metre race, an Olympic runner can accelerate to 12 metres per second in 2 seconds. What is the acceleration of the runner in m/s^2?

Figure 14.12 ▲
Olympic runners accelerating off the blocks

$$\text{starting velocity (u)} = 0 \text{ m/s}$$
$$\text{final velocity (v)} = 12 \text{ m/s}$$
$$\text{time taken (t)} = 2 \text{ s}$$

$$\text{acceleration} = \frac{\text{change in velocity}}{\text{time taken}} = \frac{12 - 0}{2} \text{ m/s}^2 = \frac{12}{2} \text{ m/s}^2 = 6 \text{ m/s}^2$$

Remember that velocity also involves direction. When the direction of an object changes, the velocity of the object changes. So whenever an object changes direction it has accelerated, even if the speed stays the same.

Summary

♦ Distance–time graphs can be used to show when an object is stationary or moving with a constant speed.

♦ The gradient of a distance–time graph represents the speed of a moving object.

♦ The velocity of an object is its speed in a particular direction.

♦ Velocity–time graphs can be used to show when an object is moving with constant velocity or with constant acceleration.

♦ The slope of a velocity–time graph can be used to calculate the acceleration of a moving object.

♦ The area under the graph can be used to calculate the distance travelled.

♦ **Acceleration** is the rate at which the velocity of an object changes.

♦ Acceleration (m/s²) $= \dfrac{\text{change in velocity (m/s)}}{\text{time taken (s)}}$

Topic questions

1 Complete the following sentences.
   a) To work out speed you need to know the _____ travelled and the _____ taken.
   b) Velocity is the speed of an object in a particular _____ .

2 a) Complete the following sentence by crossing out the two lines in the box which are wrong.

This graph shows an object which is moving
| zero |
| a constant |
| an increasing |
at speed.

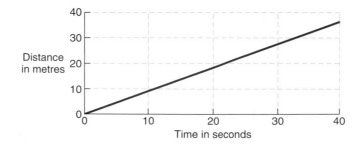

b) Calculate the average speed of this object.

3 For each of the animals, work out the missing quantity.

   a) Cheetah: distance travelled = 75 m
                       time taken = 2.5 s
                           speed = ?
   b) Antelope: distance travelled = 100 m
                       time taken = 4 s
                           speed = ?
   c) Snail: time taken = 4000 s
                  average speed = 0.0005 m/s
               distance travelled = ?
   d) Swift: distance travelled = 200 m
                  average speed = 50 m/s
                    time taken = ?

4 A pupil cycles to school. The journey is shown on the distance–time graph below.

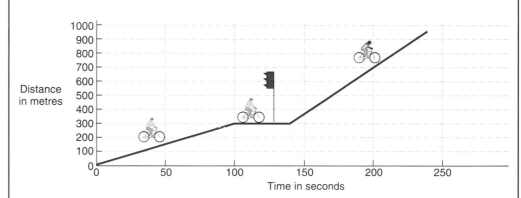

   a) How far does the pupil live from school?
   b) How long does it take for the pupil to cycle to school?
   c) Work out the average speed of the pupil for the whole journey.
   d) How do you know that the pupil had to stop at the traffic lights?
   e) During which part of the journey was the pupil cycling the fastest?

5 A train starts from rest and accelerates uniformly along a straight track to reach a velocity of 30 m/s after 120 seconds. Draw a velocity–time graph for the train and use it to calculate:

   a) the acceleration of the train
   b) the distance travelled by the train.

Co-ordinated	Modular
DA 12.7	DA 11
SA n/a	SA n/a

Force and acceleration

A force has both size and direction. The size of a force is measured in **newtons** (N). In diagrams, a force is represented by an arrow. The longer the arrow, the larger the force.

If the forces are equal in size and opposite in direction, then the forces are balanced. Balanced forces do not change the velocity of an object. If an object is stationary, it will remain stationary. If an object is moving, it will continue to move at a constant speed in a straight line.

Figure 14.13 shows the forces acting on a flying aircraft.

Figure 14.13

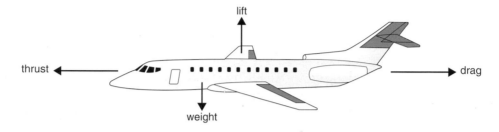

When the aircraft is flying at constant height and speed, lift is equal to **weight** and thrust is equal to **drag**. The vertical forces balance and the horizontal forces balance.

In a tug of war, two teams pull against each other. When both teams pull equally hard, the forces are balanced and the rope does not move. But when one team starts to pull with a larger force the rope moves. At this point the two forces are no longer balanced.

Figure 14.14

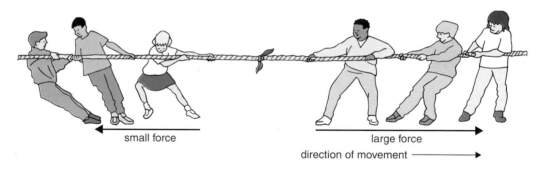

Unbalanced forces will change the velocity of an object. Since velocity involves both speed and direction, unbalanced forces can make an object speed up, slow down or change direction.

Unbalanced forces applied to the handlebars will make the cyclist change direction. This means the velocity of the cyclist will change even though the speed may stay the same.

Figure 14.15

Forces and motion

An object will only accelerate when an unbalanced force acts on it. It then accelerates in the direction of the unbalanced force.

A car with a flat battery can usually be push started. With only one person pushing, the acceleration of the car is small, but the more people that push, the larger the acceleration. So, the larger the force the larger the acceleration.

Figure 14.16

Even four people would find it difficult to push start a van. This is because the **mass** of a van is far larger than the mass of a car. The larger the mass, the smaller the acceleration.

Figure 14.17

The force needed to accelerate a mass can be worked out using this equation:

$$\text{force} = \text{mass} \times \text{acceleration}$$
(newton, N)   (kilograms, kg)   (metre second squared m/s^2)

$$F = m \times a$$

Example: Calculate the force needed to give a train of mass 500 000 kg an acceleration of 0.5 m/s.

$$F = m \times a$$
$$\text{So, } F = 500\ 000 \times 0.5$$
$$F = 250\ 000 \text{ N}$$

This equation also shows that 1 newton is the force needed to give a mass of 1 kg an acceleration of 1 m/s^2.

Summary

◆ Balanced forces have no effect on the movement of an object. If it is stationary it will remain stationary, if it is moving it will carry on moving at the same speed and in the same direction.

◆ Unbalanced forces will affect the movement of an object.

◆ An unbalanced force on an object causes its velocity to change – it accelerates. The greater the force the greater the acceleration.

◆ The greater the mass of an object the greater the force needed to make it accelerate.

◆ One newton is the force needed to give a mass of one kilogram an acceleration of one metre per second squared.

◆ force (N) = mass (kg) × acceleration (m/s²)

Topic questions

1 Explain why a canoeist slows down when they stop paddling.

2 Why does a lorry need more powerful brakes than a car?

3 The diagram shows the four forces acting on a flying aircraft.
   a) When the plane is flying at constant speed, which two forces must be equal?
   b) When the plane is flying at a constant height above the ground, which two forces must be equal?

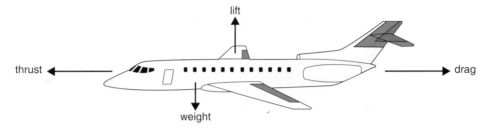

lift

thrust ◄————————————————►————► drag

weight

4 The diagram shows a ball which is about to be kicked

   Describe two different effects that the force of the kick will have on the ball.

5 The part of the space shuttle which returns to Earth has a mass of 78 000 kg and lands at a speed of 100 m/s. After touchdown it takes 50 s to decelerate and come to a halt.
   a) Calculate the deceleration of the orbiter.
   b) Calculate the force needed to bring the orbiter to a halt.

6 A cyclist travelling along a flat road stops pedalling. The speed drops from 8 m/s to 5 m/s in 6 s.
   a) Calculate the deceleration of the cyclist.
   b) The mass of the cyclist and cycle is 90 kg. Calculate the resistance force which slows the cyclist down.

14.3 Frictional forces and non-uniform motion

Co-ordinated	Modular
DA 12.8	DA 11
SA n/a	SA n/a

An engine produces the force needed to keep a car moving forwards. If the car is not accelerating, the force from the engine must be balanced by an equal force backwards. This force, called **air resistance** or drag, is a force of **friction**.

Figure 14.18

force backwards

force forwards

If the car engine stops, the car will slow down (decelerate). This happens because the force of friction opposes the motion of the car.

Friction also acts when one surface rubs against another. Without friction, the tyres of a car would not grip the road. The car would not be able to move forwards, backwards or turn.

Car brakes rely on friction. Pushing the brake pedal causes friction between the brakes and the wheels. The friction slows the wheels and stops the car. The total stopping distance of a car is made up of two parts: the **thinking distance** and the **braking distance**. The thinking distance is how far the car travels while the driver reacts to an emergency and applies the brakes. The braking distance is how far the car travels once the brakes have been applied. The force applied by the brakes will affect the braking distance. The larger the braking force, the shorter the braking distance.

Road conditions will also affect the braking distance. In icy or wet conditions the friction between the car tyres and the road is reduced. This reduces the grip and increases the braking distance. A rough surface will increase the friction and so reduce the braking distance.

Drivers' reactions are much slower if they have been drinking alcohol or if they are tired. The slower a driver's reaction time, the greater the thinking distance.

Speed affects both the thinking distance and the braking distance.

The faster the car, the greater the total stopping distance.

Figure 14.19

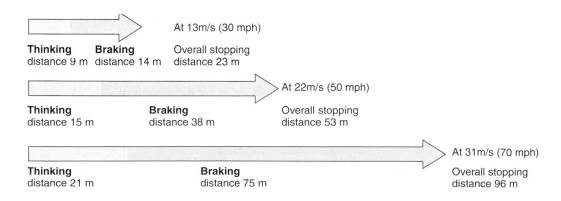

At 13m/s (30 mph)

Thinking distance 9 m **Braking** distance 14 m Overall stopping distance 23 m

At 22m/s (50 mph)

Thinking distance 15 m **Braking** distance 38 m Overall stopping distance 53 m

At 31m/s (70 mph)

Thinking distance 21 m **Braking** distance 75 m Overall stopping distance 96 m

Force and non-uniform motion

If you want to know your weight, you will probably stand on some bathroom scales.

If the bathroom scales measure in kilograms, it does not show your weight, it shows your mass. Mass and weight are not the same thing. Mass is the amount of matter that makes up an object while weight is the force which **gravity** exerts on a mass. Mass is measured in kilograms (kg) while weight, like all forces, is measured in newtons (N). But more mass does mean more weight, because there is more mass for the force of gravity to pull on.

Figure 14.20

On Earth, gravity pulls on every one kilogram of mass with a force of 10 newtons. This is called the gravitational field strength (g).

$$g = 10 \text{ N/kg}$$

So someone with a mass of 50 kg will weigh

$$50 \times 10 = 500 \text{ N}$$

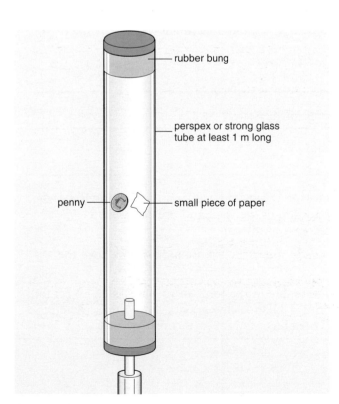

— rubber bung

— perspex or strong glass tube at least 1 m long

penny —

— small piece of paper

Gravity is a force of attraction which acts between objects. A ball thrown into the air will be attracted back towards the ground by gravity. As the ball falls downwards, gravity will cause it to accelerate. If there is no air resistance, the ball will accelerate at 10 m/s^2.

The acceleration due to gravity (g) = 10 m/s^2

This means that if there is no air resistance, the speed of a falling object will increase by 10 m/s every second. In a vacuum, where there is no air resistance, all falling objects accelerate at the same rate.

When the tube in Figure 14.21 is evacuated and then turned upside-down, the penny and piece of paper fall together. This is because the penny has more mass than the piece of paper. But this means gravity will exert a larger force on the penny, giving both objects the same acceleration.

Figure 14.21

303

Forces and motion

Usually air resistance does act on a falling object. Air resistance can only be ignored if the force it exerts is very small.

When sky-divers jump from a plane, the forces on them are unbalanced so the sky-divers accelerate.

Figure 14.22
Sky-diver at terminal velocity

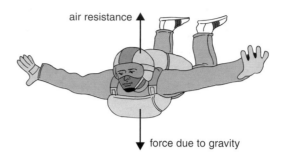

air resistance

force due to gravity

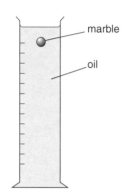

marble

oil

Figure 14.23

The faster the sky-diver falls, the larger the air resistance, so the smaller the acceleration. Eventually the downward force due to gravity and the upward force due to air resistance will be balanced. The sky-diver will stop accelerating and continue to fall at a constant speed. We say that the sky-diver has reached their **terminal velocity**.

Opening a parachute increases the air resistance. Since the force due to gravity stays the same, it will take less time before the two forces are balanced. This gives the sky-diver a lower terminal velocity.

Terminal velocity depends on the size of the air resistance force.

Air resistance is affected by the shape of an object. If a sky-diver curls up into a ball, air resistance will be reduced. It will take longer before the upward and downward forces balance, so the sky-diver reaches a higher terminal velocity. Terminal velocity therefore depends on the shape of an object.

Objects falling through liquids experience much larger resistance or drag forces.

A marble falling through oil will have a much lower terminal velocity than a marble falling through the air.

Did you know?

Vesna Vulovic, an air hostess, fell over 10 kilometres without a parachute and survived.

Summary

◆ The stopping distance of a moving vehicle depends on the braking force, the driver's reaction time, the condition of the road and the car, weather conditions, the speed and the mass of the car.

◆ For a vehicle travelling at a steady speed the frictional forces balance the driving force.

◆ Falling objects initially accelerate due to the force of gravity. As an object falls the air resistance increases until it balances the gravitational forces. When the resultant forces are balanced the object falls at its terminal velocity.

Topic questions

1. Explain the effect on the stopping distance of a car when there is ice on the road.

2. Why does a parachute slow down a falling parachutist?

3. Complete the following sentence by crossing out in each box the two lines that are wrong.

   When an astronaut goes into space the mass

   will | increase
   stay the same | and the
   decrease

   weight will | increase
   stay the same
   decrease.

4. An astronaut is standing on the Moon and is about to let go of a hammer and a feather at the same instant.

   What will happen, and why?

5. What is wrong with the following statement 'my weight is 80 kg'?

Examination questions

1. a) Two sky-divers jump from a plane. Each holds a different position in the air.

   Complete the following sentence.
   Sky-diver _____ will fall faster because
   _____
   (*2 marks*)

A

B

b) The diagram shows the direction of the forces acting on one of the sky-divers.

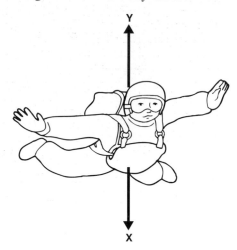

In the following sentences, cross out in each box the two lines that are wrong.

i) Force X is caused by

| air resistance |
| friction |
| gravity |

(1 mark)

ii) Force Y is caused by

| air resistance |
| gravity |
| weight |

(1 mark)

iii) When force X is bigger than force Y, the speed of the sky-diver

will
| go up |
| stay the same |
| go down |

(1 mark)

iv) After the parachute opens, force X

| goes up |
| stays the same |
| go down |

(1 mark)

c) How does the area of an opened parachute affect the size of force Y? *(1 mark)*

2 Two students Anna and Graham took part in a sponsored run. The distance–time graph for Graham's run is shown. Four points have been labelled A, B, C and D.
  a) Between which pair of points was Graham running the slowest? *(1 mark)*
  b) Anna did not start the run until 10 minutes after Graham. She completed the whole run at a constant speed of 4 m/s.

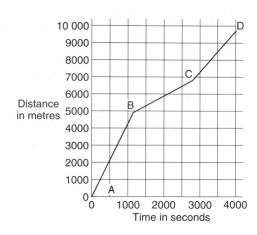

i) Write down the equation that links distance, speed and time. *(1 mark)*
ii) Calculate, in seconds, how long it took Anna to complete the run. Show clearly how you work out your answer. *(2 marks)*
iii) Draw a line on the graph to show Anna's run. *(2 marks)*
iv) How far had Graham run when he was overtaken by Anna? *(1 mark)*

3 Five forces, **A**, **B**, **C**, **D** and **E** act on the van.

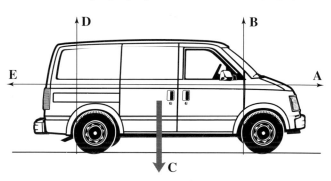

a) Complete the following sentences by choosing the correct forces from **A** to **E**.
  Force _____ is the forward force from the engine.
  Force _____ is the force resisting the van's motion. *(1 mark)*
b) The size of forces **A** and **E** can change. Complete the table to show how big force **A** is compared to force **E** for each motion of the van. Do this by placing a tick in the correct box. The first one has been done for you.

Motion of van	Force A smaller than force E	Force A equal to force E	Force A bigger than force E
Not moving		✓	
Speeding up			
Constant speed			
Slowing down			

(3 marks)

c) When is force **E** zero? *(1 mark)*

d) The van has a fault and leaks one drop of oil every second.

The diagram below shows the oil drops left on the road as the van moves from **W** to **Z**.

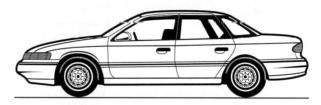

Describe the motion of the van as it moves from **W** to **X**, **X** to **Y** and **Y** to **Z**.

(3 marks)

e) The driver and passengers wear seatbelts. Seatbelts reduce the risk of injury if the van stops suddenly.

**backwards   downwards   force
forwards   mass   weight**

Complete the following sentences, using words from the list above, to explain why the risk of injury is reduced if the van stops suddenly.

A large _____ is needed to stop the van suddenly.
The driver and passengers would continue to move _____ .
The seatbelts supply a _____ force to keep the driver and passengers in their seats.

(3 marks)

f) The van was travelling at 30 m/s. It slowed to a stop in 12 seconds. Calculate the van's acceleration. *(3 marks)*

4 The graph shows three stages of a van's journey.
a) During which stage of the journey **A–B**, **B–C** or **C–D**:
i)   is the van stationary? *(1 mark)*
ii)   is the van moving at a constant speed? *(1 mark)*

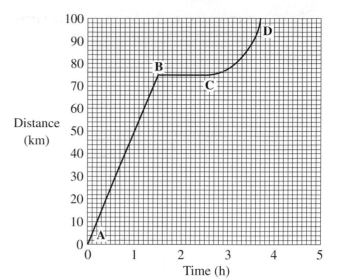

b) Calculate the gradient of the graph from **A** to **B**. *(2 marks)*
c) What does this gradient measure? *(1 mark)*

5 The diagram shows a parked car.

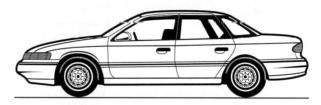

When the car is driven away, its engine gives a constant forward force.
The speed increases quickly at first, then more slowly. After a time the car reaches a constant speed.
Explain why the motion of the car changes in this way. *(3 marks)*

Chapter 15
Waves

15.1 Characteristics of waves

15.1	
Co-ordinated	Modular
DA 12.9	DA 12
SA 12.5	SA 18

If a rope is shaken up and down a wave travels along it. A person holding the other end of the rope will feel the pulse when it arrives. The pulse has carried energy but when the pulse has passed, the rope remains exactly as it was before. None of the material of the rope has moved permanently. So **waves** carry energy from one place to another without transferring matter.

Figure 15.1
A wave moving along a rope

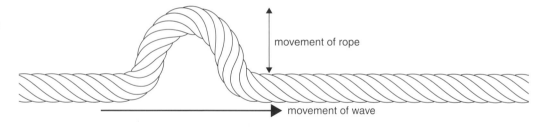

As the pulse passes, the rope moves up and down. The movement of the rope is at right angles to the movement of the pulse. This is an example of a **transverse wave**.

Energy can also be transmitted as a pulse in a spring by shaking it backwards and forwards

When the spring is shaken backwards and forwards, each coil moves backwards and forwards rather than up and down. The movements are in the same direction that the energy travels. In some places the coils bunch together (areas of **compression**) and in other places they are further apart (areas of **rarefaction**). This type of wave is called a **longitudinal wave**.

Figure 15.2
A wave moving along a spring

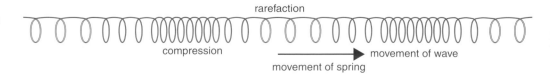

Light and sound travel as waves. Energy is transmitted but there is no movement of mass. Sound waves travel through solids, liquids and gases as longitudinal waves. Most waves, for example water waves, waves in a rope and light waves, are transverse waves. Light waves can travel through a vacuum.

For any wave the number of waves made each second is called the **frequency**. Frequency is measured in units called **hertz** (Hz). One hertz is one **vibration** or **cycle** per second. So 2000 vibrations in a second would be described as 2000 Hz or 2 kHz (kilohertz).

For any wave system there are two factors needed to describe the wave. For a transverse wave these are the distance between the **crests** or the distance between the **troughs** of a wave – which is called the **wavelength** – and the height of the wave crest, which is called the **amplitude**.

Figure 15.3
A transverse wave in water showing the wavelength and amplitude

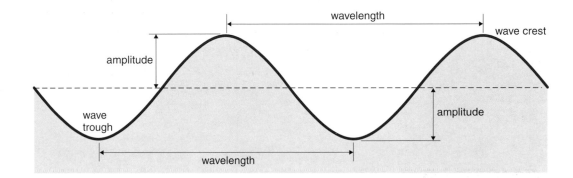

The amplitude depends upon the energy in the wave. The greater the amplitude the greater the energy of the wave.

The behaviour of waves

Waves and refraction

Light can pass through transparent materials like air, water, glass and 'perspex'. When light crosses the boundary between two transparent materials at an angle greater than 0° to the normal it changes direction – this is called **refraction**.

When light goes from one material to another it changes speed. This change of speed causes the light to change direction. Light travels slower in glass than it does in air.

The diagram shows that when light travels from air to glass it bends towards the **normal** on its way into the glass because it slows down. When light travels from glass to air it bends away from the normal on its way out of the glass because it speeds up.

Figure 15.4
Refraction of light through a glass block

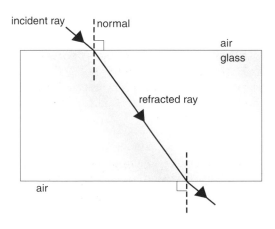

Waves

When light changes speed as it moves from one transparent substance to another, the frequency does not change but the wavelength does.

Water waves also show refraction effects. Figure 15.5 shows water waves being slowed down as they pass over an area of shallow water. They are being refracted but they are not changing direction, because they are meeting the shallow water straight on – along the normal. When the waves leave the shallow water they speed up again.

Figure 15.6 shows the water waves hitting the shallow water at an angle. This time the waves are not only slowed down but they have changed direction.

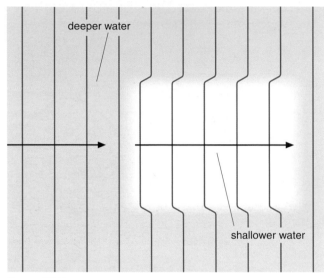

Figure 15.5

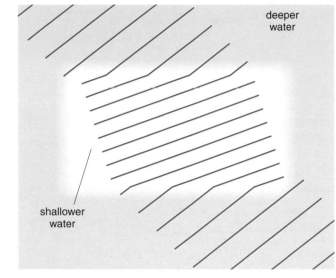

Figure 15.6

Waves and diffraction

Figure 15.7 shows water waves passing through gaps of different sizes. When they pass through the gap that is about the same width as their wavelength they spread out as they emerge on the other side. If the gap is very wide compared to the wavelength then very little spreading out occurs. Waves will also spread out when they pass the edge of an object. This spreading out of a wave as it passes the edge of an object or as it moves through a gap is called **diffraction**.

Figure 15.7
Diffraction of waves at a narrow and at a wide gap

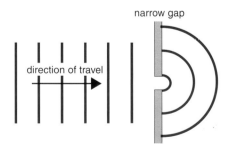

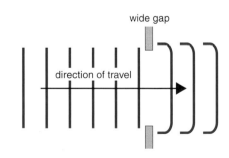

Therefore the degree of diffraction depends on the wavelength of the wave and the width of the gap it passes through. Because of this, some diffraction effects are not easy to see. Visible light with its very short wavelengths (0.000005 m) would need a gap with a width to match the wavelength in order for diffraction effects to be seen.

VHF radio waves have wavelengths of a few metres, whereas long wave radio waves have wavelengths of thousands of metres. Long wave radio signals are readily diffracted by the large gaps between buildings and as they pass over hills. This means that long wave radio signals can be received in the shadow of a hill. VHF waves are less likely to show diffraction effects and spread out, so the receiving of these signals is difficult in hilly areas.

Figure 15.8
The diffraction of
a) VHF waves
b) long wave radio
waves

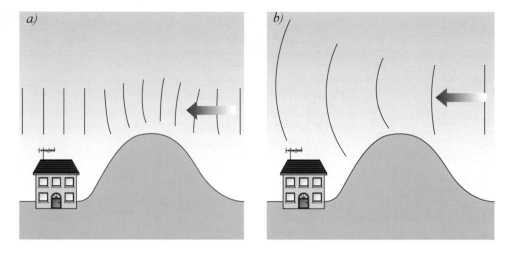

Sound waves have wavelengths of about one metre they can readily spread out as they pass through an open doorway or as they hit an obstacle such as the corner of a building. So someone outside a room may hear a conversation although they cannot see the speakers.

Figure 15.9
Diffraction of sound
waves through a
doorway

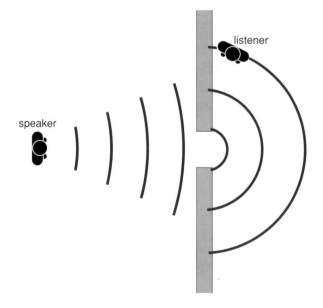

Total internal reflection

When a ray of light emerges from glass, 'Perspex' or water into air it bends away from the normal. As the internal angle gets bigger so does the external angle. When the internal ray reaches about 42°, the external angle reaches 90°. When this happens the interior angle is called the **critical angle**. So for glass the critical angle is 42°.

If a ray of light hits the inside of a glass block at an angle above the critical angle it reflects inside the block. Because ALL the light reflects inside the block, this is called **total internal reflection**.

Figure 15.10
Demonstrating total internal reflection

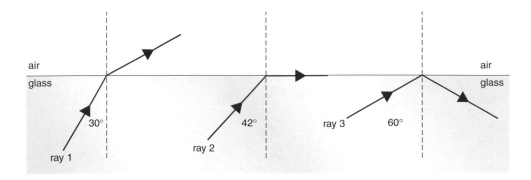

Light entering the prism hits the first angled face at 45°. This angle is greater than the critical angle so the light ray is totally internally reflected. The ray then hits the second face at 45°. Once again the ray is totally internally reflected. It comes out of the prism parallel to the direction it went in.

Figure 15.11
Total internal reflection inside a 45°/45°/90° glass prism

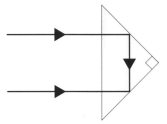

Prisms are used in 'prismatic' binoculars because the quality of the image is much better with a prism than a mirror. The reflectors on cars and bicycles and in some 'cat's eyes' in the road use total internal reflection in a 45°/45°/90° prism.

A more recent application of total internal reflection is in fibre optics.

Fibre optics allow light rays entering one end of the fibre to be totally internally reflected numerous times until they leave at the other end. No matter how much the fibre optic gets twisted or bent. the light will always pass all the way along the fibre.

Figure 15.12
Passage of light through fibre optic

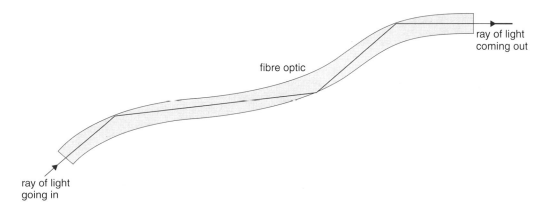

Total internal reflection in fibre optics has two very important practical uses:

* *Medical uses.* An endoscope is a bundle of thin optical fibres which can be inserted into the body. Light is sent down some fibres and reflected back so that the medical observer can view the inside of the patient. This has enabled 'key-hole' surgery to be developed where very small incisions can be made rather than using more invasive surgery.

Figure 15.13
An endoscope in use

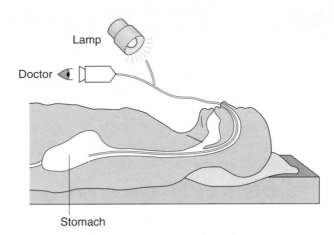

Lamp

Doctor

Stomach

- *Communications* e.g. cable telephones and cable television. The use of fibre optic cable means that as many as 500 000 telephone conversations can be carried at the same time. This is hundreds of times more than can be carried by a copper cable of similar size. The fibre optic cable is also cheaper.

Summary

- Waves transfer energy not matter.

- The **amplitude** of a transverse wave is the maximum distance of the wave above or below the middle position.

- The **wavelength** of a transverse wave is the distance between the tops of successive crests (or the bottoms of successive troughs).

- The **frequency** is the number of waves produced in 1 second. It is measured in **hertz** (Hz).

- In a **transverse** wave the disturbance causing the wave is at right angles to the direction in which the energy travels.

- In a **longitudinal** wave the disturbance causing the wave is in the same direction in which the energy travels.

- **Refraction** is caused by changes in the speed of a wave as it passes through materials with different densities.

- When waves move through gaps or pass the edges of objects they can spread out. This effect is called **diffraction**.

- Under certain conditions light can be totally internally reflected.

Topic questions

1 Draw a water wave and mark on it the wavelength and the amplitude.

2 a) Which of the following is the unit of frequency?
   amp   hertz   metre
   b) What is meant by the frequency of a wave?

3 a) What are longitudinal waves?
   b) Give an example of a longitudinal wave.

4 a) What are transverse waves?
   b) Give two examples of transverse waves.

5 a) What is value of the critical angle for a ray of light in a glass block?
   b) When will a ray of light show total internal reflection?

6 a) What is diffraction?
   b) What happens to the speed of a water wave as it passes over shallow water?
   c) Why is it difficult to show that light waves can be diffracted?
   d) Why can sounds easily spread around corners?

15.2 The wave equation

Co-ordinated	Modular
DA 12.9	DA 12
SA n/a	SA n/a

Imagine a wave with a wavelength of λ metres travelling with a **wave speed** of v metres per second.

In 1 second the wave travels v metres.

Each wave has a length of λ metres so in v metres there are $\frac{v}{\lambda}$ waves.

So in 1 second there are $\frac{v}{\lambda}$ waves.

This means the frequency of the wave (f) is $\frac{v}{\lambda}$ Hz so $f = \frac{v}{\lambda}$

This equation can be rewritten as:

$$\begin{array}{ccc} \text{wave speed} & = & \text{frequency} \times \text{wavelength} \\ \text{(metre/second, m/s)} & & \text{(hertz, Hz)} \quad \text{(metre, m)} \end{array}$$

$$v = f \times \lambda$$

Example: What is the speed of waves which have a frequency of 200 Hz and a wavelength of 0.5 m?

$$\begin{array}{rl} \text{frequency} & = 200 \text{ Hz} \\ \text{wavelength} & = 0.5 \text{ m} \\ \text{So, wave speed} & = 200 \times 0.5 \text{ m/s} = 100 \text{ m/s} \end{array}$$

Summary

◆ **Wave speed** = frequency × wavelength
   (m/s)   (Hz)   (m)

Topic questions

1   Draw two waves, one of which has double the wavelength and half the amplitude of the other.

2   What is the speed of a sound wave having a frequency of 512 Hz and wavelength 0.6 m?

3   Radio 4 has a wave speed of 300 000 000 m/s and a wavelength of 1500 m. What is its frequency?

15.3 The electromagnetic spectrum

Co-ordinated	Modular
DA 12.10	DA 12
SA 12.6	SA 18

The **electromagnetic spectrum** contains a range of transverse waves all of which can travel through a vacuum at the same speed (300 000 000 m/s, or 3×10^8 m/s). These waves are classified into several groups according to their wavelength and frequency and how they behave. Light is one of these groups.

Because each part of the electromagnetic spectrum has a different wavelength and frequency, each part will be reflected, refracted, absorbed or transmitted differently.

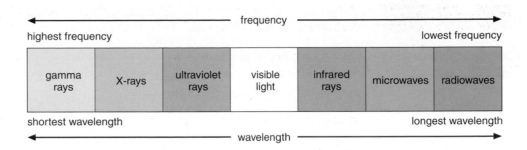

Figure 15.14
Wavelengths and frequencies of the electromagnetic spectrum

When electromagnetic radiation is absorbed, the energy it carries is likely to make the absorbing object hotter, it may even produce an alternating current with the same frequency as the radiation.

Uses and dangers of different parts of the electromagnetic spectrum

Radio waves

Radio waves have wavelengths between about 1500 metres (long-wave radio signals) and a few metres (very high frequency radio signals – VHF signals). Television transmissions use even shorter wavelengths (about 0.1 metres). Radio waves are used for communications.

The longer wavelength radio waves are reflected from an electrically charged layer high up in the Earth's atmosphere. This enables them to be sent between distant points despite the curvature of the Earth's surface.

Microwaves

Microwaves are also used for communication purposes. They have shorter wavelengths than radio waves and therefore have a greater frequency. This enables them to carry more information. Microwaves are used for mobile phones and for carrying television channels between city centres via very tall telecommunication towers.

Microwaves can penetrate the upper atmosphere so can be sent to satellites and back to Earth. Many people receive some of their television programmes via radio waves which have travelled from the Earth to a satellite and then back to a dish aerial on the side of their house.

Figure 15.15
Sending of a signal via a satellite

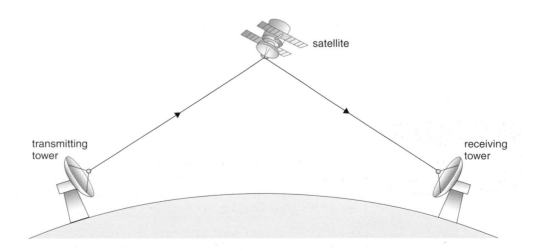

Waves

Microwaves can also be used to cook food. The microwave's energy is absorbed easily by water molecules in the food. This makes the water molecules vibrate more, so increasing their kinetic energy. In this way the temperature is raised and the food cooks very quickly. In a microwave oven liquids heat up more quickly than solids. This is why a microwaved doughnut may only be warm on the outside but have scalding hot jam inside. Microwaves can seriously damage living tissue. Microwave ovens have to be very carefully screened to prevent radiation escaping and the oven will turn off immediately if the door is opened.

Infrared radiation

Infrared radiation is used in the remote controls for televisions and stereo systems.

It is the heat radiated from grills, toasters and radiant heaters.

Infrared waves are absorbed by skin giving a sense of warmth but excessive exposure will lead to skin burns.

Did you know?

Infrared cameras can detect objects at different temperatures and are used by the army and police to spot people in the dark. Rescue teams use infrared cameras to detect living people in rubble after explosions.

Ultraviolet radiation

Ultraviolet rays are responsible for sun tans. The ozone in the upper atmosphere shields us from an excess of ultraviolet, but there are concerns that 'holes' are developing in the upper ozone layer due to the effects of CFCs.

Ultraviolet radiation is dangerous because the energy can penetrate living tissue and may cause skin cancer. So it is important that appropriate creams are used to protect the skin from this radiation.

Ultraviolet rays create visible light in fluorescent lamps. In these lamps the ultraviolet rays transfer their energy into light by hitting fluorescent coatings on the inside of the glass.

Ultraviolet radiation is also used in security coding. Numbers or names are written on to valuable objects using special ink that is visible only in ultraviolet light.

X-rays

X-rays affect a photographic plate causing it to be exposed. Medical uses of X-rays rely on the fact that living tissue is almost transparent to X-rays but bones absorb most of the rays. It is thus possible to study broken bones with X-rays.

X-rays are also useful for security scanning at airports because metal objects stop the X-rays.

Gamma rays

Gamma rays (see section 18.1) are used to kill harmful micro-organisms in food, sterilise surgical instruments and kill cancer cells.

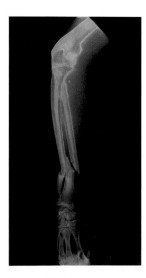

Figure 15.16
An X-ray of a broken arm

Dangers of exposure to electromagnetic radiation

It is important to realise that although each of the waves of the electromagnetic spectrum can be used for beneficial purposes there are many dangers that could arise from their misuse. Many of these dangers affect the cells of living organisms, including humans. It is necessary therefore that measures be taken to reduce the chances of exposure to many types of radiation. Figure 15.17 outlines some of the dangers caused by over exposure to the different types of radiation and some of the measures taken to reduce such exposure.

Figure 15.17
The effects of different types of radiation on living cells

Radiation	Effect
microwave	These are absorbed by water in the cells, so cells can be damaged or killed by the heat produced.
infrared	These are absorbed by the skin and felt as heat which can damage or kill cells.
ultraviolet	These can pass through the skin and reach the deeper tissues. This radiation causes the skin to darken and can cause skin cancers. The darker the skin the more ultraviolet radiation it absorbs and less reaches the deeper tissues. Over-exposure to ultraviolet radiation should be avoided and the use of various sun-creams that have been designed to protect the skin from this radiation is recommended.
X-rays and gamma radiation	Although most of these pass through soft tissue some of their energy is absorbed causing cells to become cancerous. Workers who are likely to be operating machinery that produce X-rays and gamma radiation wear protective clothing that includes lead shielding to protect their reproductive organs from excessive exposure. Workers who are likely to come in contact with gamma radiation and other forms of radiation from radioactive sources wear special badges that contain photographic film (see section 11.5). This film is regularly developed to check whether any over exposure to radiation has occurred.

Figure 15.18▼
Using X-rays to treat a cancerous tumour

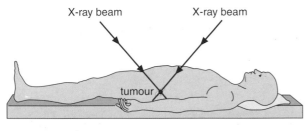

Radiation and cancers

Although low level doses of ultraviolet radiation, X-rays and gamma radiation can cause normal cells to become cancerous, high doses of these types of radiation can be used to kill cancerous cells.

The cancer is exposed to X-rays projected from several directions. This makes sure that the beams are concentrated in the area of the cancer and that the surrounding cells receive only a low dose of the radiation.

With gamma ray treatment the source moves around the cancer. Again, this makes sure that the radiation is concentrated in the area of the cancer and that the surrounding cells receive only a low dose.

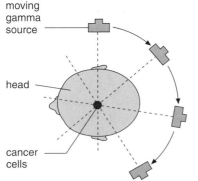

Figure 15.19▲
Using gamma rays to treat a cancerous tumour

Sending messages

The telegraph was the first invention to use an electric current as a message carrier. Morse code was invented for this first message carrier. Morse code is made up from short and long pulses. These pulses were the first form of **digital signals**. Digital signals have only two states, 'on' or 'off'.

Figure 15.20
Morse code as a series of digital signals

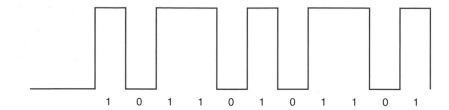

1 0 1 1 0 1 0 1 1 0 1

The telephone, invented in 1876, consists of a microphone and a loudspeaker. The microphone changes sound waves into electrical signals that can be displayed as transverse waves on a CRO (see section 15.4). These signals are called **analogue signals**. Analogue signals are continuous waves that vary in amplitude and frequency.

Figure 15.21
Analogue signals produced by a microphone

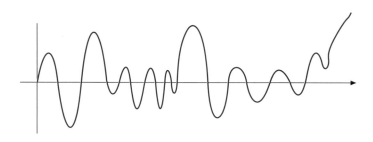

The radio was invented in 1901 and for many years radio signals were transmitted as analogue signals. One form of these signals consisted of a carrier wave sent at a fixed frequency. The signal being carried was sent by changing the carrier wave's amplitude to match the amplitude of the message being transmitted.

Figure 15.22
Carrier wave and signal wave

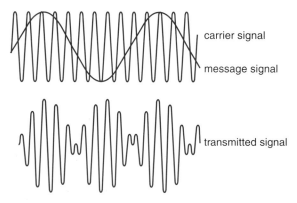

carrier signal

message signal

transmitted signal

Today the air is crowded with radio signals and the telephone wires can hardly cope with the number of conversations going on. Optical fibres (see section 15.1) are being used to replace the main telephone cables. These optical fibres will carry coded pulses of light instead of electrical signals. A standard telephone copper cable can carry up to about 10 000 coded conversations at any one time. An optical fibre can carry at least 110 times as many conversations coded as pulses of light.

The digital radio signals or the digital pulses of light do have certain advantages over analogue signals.

- The information carried as a digital signal is unlikely to get changed – the signal is either 'on' or 'off'. This means the signals received are of a high quality and sound like the original.

- More information in a given time can be transmitted because digital signals take up less wave space than analogue signals.

As analogue signals travel from the transmitter circuit to the receiver circuit they gradually weaken and often collect unwanted signals called 'noise'. Different frequencies of an analogue signal weaken by different amounts, so during any amplification process to make the weak signal stronger, the original signal gets more and more distorted.

Digital signals will also weaken as they travel from the transmitter to the receiver. However they are still recognisable as 'on' or 'off' states. Any unwanted 'noise' usually has a very low amplitude so is interpreted by the receiver circuit as 'off'.

Summary

◆ The **electromagnetic spectrum** is a range of transverse waves that travel at 300 000 000 m/s in a vacuum.

◆ Electrical messages can be sent either as **digital signals** or as **analogue signals**.

Topic questions

1 Use the types of electromagnetic radiation in the box to answer the following questions. Some of the questions have more than one correct answer.

| radio   microwaves   infrared   light   ultraviolet   X-rays   gamma rays |

Which type(s) of radiation can be used a) for communications b) for cooking c) to give a sun tan d) cause cancer e) to sterilise equipment?

2 Why are microwaves so dangerous to living tissue?

3 a) What are analogue signals? b) What are digital signals?

4 Why are copper telephone cables being replaced by optic fibre cables?

15.4   Sound and ultrasound

Co-ordinated	Modular
DA 12.11	DA 12
SA 12.7	SA 18

Sound is not an electromagnetic wave. Sound waves differ from electromagnetic waves in many ways.

Sound is produced whenever an object vibrates at a frequency which the ear can detect. The vibrating object creates very small pressure waves in the air. These pressure waves are longitudinal waves so the air is alternately compressed and rarefied. Each air particle vibrates but does not move permanently along the sound wave. The energy of a sound wave is passed from one particle to another by collision.

Sound waves cannot pass through a vacuum as there are no particles to carry the vibrations.

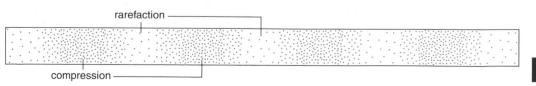

Figure 15.23
Variation in air pressure in a sound wave

319

Waves

Sound can travel through solids and liquids as well as through gases like air. The greater the density of the medium, the faster the sound travels. Thus sound travels faster through water than it does through air. This is because in water the particles are closer together so can pass on their vibrations quicker. Sound travels even faster through solids such as wood because the particles are even closer together.

Looking at sound waves

A microphone transfers sound energy into electrical energy. If the microphone is connected to a cathode ray oscilloscope (CRO) the sound can be displayed on a screen.

The shape of the sound wave is similar to that of Figure 15.3. Like all waves, sound waves have amplitude and frequency.

Figure 15.24
Displaying sound waves on an oscilloscope screen

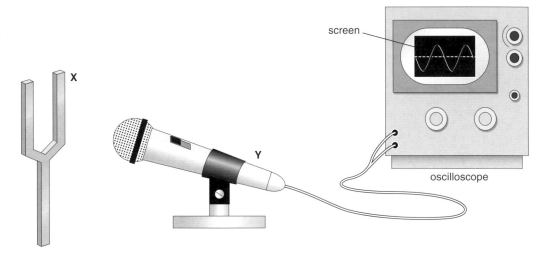

The amplitude of a sound wave is a measure of its loudness. The louder the sound, the bigger the difference in pressure between the areas of compression and the areas of rarefaction in the sound wave. This means that loud sounds carry more energy. On the CRO screen, the louder the sound the higher the crests and deeper the troughs of the wave on the screen.

The frequency of a sound wave is a measure of its pitch. The higher the frequency, the higher the pitch. A higher frequency means the wavelength is shorter. So on a CRO screen, a higher pitch sound means the waves are closer together and you will see more waves on the screen.

The traces shown in Figure 15.25 are for 'pure' or regular sounds. A tuning fork produces a reasonably pure sound.

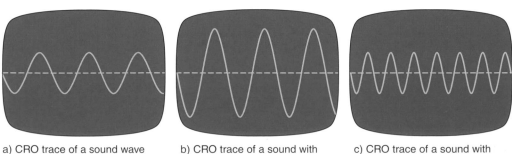

Figure 15.25
Sound waves on oscilloscope screens

a) CRO trace of a sound wave

b) CRO trace of a sound with the same pitch as a) but much louder

c) CRO trace of a sound with the same loudness as a) but a higher pitch

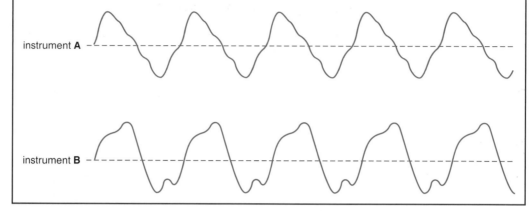

Two different musical instruments playing the same note do not sound exactly alike. This is because the sound produced is not a pure sound. The diagram shows the trace of the sound of two different instruments playing the same note. The frequencies and amplitudes are the same but the shape of the note is different. This is what gives a musical instrument its particular sound (timbre).

instrument **A**

instrument **B**

Ultrasound

Ultrasound is any sound which has a frequency or pitch which is too high for humans to hear, although other animals may be able to. Because of its short wavelength, the ultrasound produced from a small source is unlikely to spread out by diffraction.

Ultrasound can have a frequency as high as 15 000 000 Hz. At this high frequency the waves can travel through most substances as a narrow beam. Some of the ultrasound is always reflected back when the wave hits the boundary between two different types of materials. A detector placed near the source of the waves measures how long it takes for the waves to be reflected back by various boundaries. The time taken for the reflections to occur is used to generate a visual display.

Ultrasound can be used to produce the image of a **fetus**. The ultrasound passes directly through the skin of the mother into the fetus. The reflected sounds are converted into pictures and enable medical staff and prospective parents to get an early image of the developing baby. This is called **fetal imaging**. The advantage of ultrasound is that it does not harm living cells.

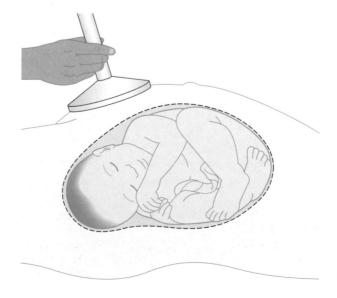

Figure 15.26
Using ultrasound in fetal imaging

Waves

Ultrasonic waves can be used for cleaning delicate equipment without the need to take it apart. The ultrasonic waves cause the dirt particles to vibrate rapidly and fall off. Ultrasound can also be used to check for cracks or defects in metal castings. The ultrasonic waves are distorted by imperfections in the structure of the metal.

Summary

◆ **Ultrasound** (ultrasonic waves) has a frequency above the upper limit of the human hearing range.

Topic questions

1 Ultrasonic sounders do not use a continuous signal. The ultrasound is sent out in a series of pulses with silence in between. Why is this necessary?

2 A CRO trace of a sound wave looks like this.
   a) The beam takes 0.005 s to travel across the screen. What is the frequency of the sound wave?

b) How would the trace differ if the sound
   i) had twice the frequency
   ii) was much quieter.

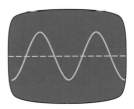

15.5   Seismic waves

Co-ordinated	Modular
DA 12.12	DA 12
SA n/a	SA n/a

The Earth is a layered structure with a **core**, a **mantle** surrounding the core and a thin solid **crust**.

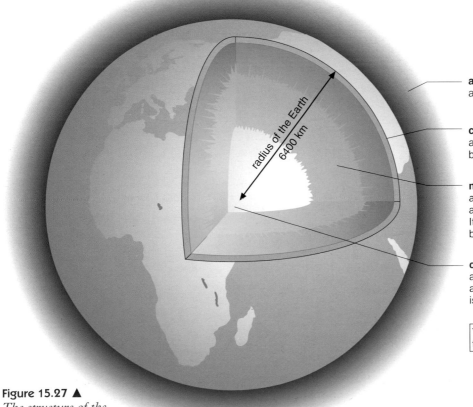

radius of the Earth 6400 km

atmosphere:
a capsule of gases between -50 and 50°C

crust:
a thin layer of less dense solid rock between -50 and 1500°C

mantle:
a thick layer of moderately dense solid and molten rock between 1500 and 4000°C. It has all the properties of a solid except that because it is hot it can flow slowly

core:
a central ball of very dense, hot molten nickel and iron at 4000°C. The outer part of the core is liquid, the inner part is solid

The **lithosphere** is made up of the crust and the solid upper layer of the mantle

Figure 15.27 ▲
The structure of the Earth

Evidence for this layered structure comes from studying the density of the Earth and analysing the movement of **seismic waves**.

The density of the Earth

Because the overall density of the Earth is much greater than that of the materials making up the crust, the interior of the Earth must be made up of different materials to the crust. These materials must be much denser than the materials in the crust. The high density of the interior can be explained if the **core** is considered to be made mainly of metals.

Seismic waves and the Earth

Earthquakes are caused by shock waves created when two parts of the Earth's crust move relative to each other. The shock waves are called seismic waves. The waves are detected using **seismographs**.

Figure 15.28
Quakes recorded on a seismograph

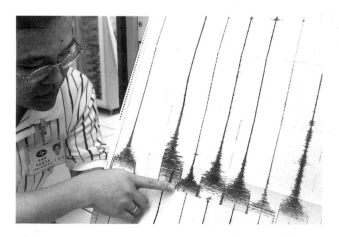

Two types of seismic waves are P waves and S waves.

- **P (primary) waves.** These are longitudinal and can pass through solids and liquids. They are called primary waves because they travel faster and are detected first.

- **S (secondary) waves.** These are transverse waves which can only travel through solids. S waves travel more slowly than P waves.

It is easy to remember which waves are longitudinal and which are transverse. P waves are Pushing and Pulling waves (longitudinal) and S waves are Shaking Sideways waves (transverse).

Although both waves travel at different speeds their speed increases with depth. Because the density of each layer of the Earth changes gradually the speed of each wave changes gradually and so the waves follow curved paths. As the waves pass from one layer to another the waves are refracted (bent).

The outer core is liquid so only P waves can pass through it. (Even though the inner core is solid, only P waves pass through it because it is completely surrounded by liquid.) The mantle is solid and allows both waves to pass through it.

Waves

Figure 15.29
A simplified diagram to show the way that P waves and S waves travel through the Earth

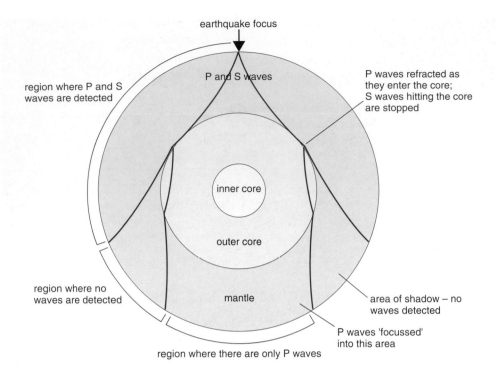

earthquake focus

P and S waves

region where P and S waves are detected

P waves refracted as they enter the core; S waves hitting the core are stopped

inner core

outer core

region where no waves are detected

mantle

area of shadow – no waves detected

P waves 'focussed' into this area

region where there are only P waves

Because the P waves are refracted as they enter the outer core there is a 'shadow' zone where no seismic activity can be detected. It was this fact that enabled scientists to work out how big the Earth's core was and that the Earth had a layered structure.

Summary

◆ The Earth has a layered structure made up of the **crust, mantle** and **core**.

◆ The **lithosphere** is made up of the crust and the top solid layer of the mantle.

◆ Seismic waves can be longitudinal (**P waves**) or transverse (**S waves**).

Topic questions

1 Earthquakes cause two types of waves.
a) What are both types of waves called?
b) What instrument is used to detect them?
2 One type of wave is called a P wave. The other is an S wave.
a) Which type is transverse?
b) Which can travel through solids and liquids?
c) Which can only travel through solids?
d) Which travels faster?

3 Why do S waves not travel through the core?

15.6   **Tectonics**

Co-ordinated	Modular
DA 12.13	DA 12
SA n/a	SA n/a

At one time it was believed that the continents were formed as the once-molten Earth's crust cooled and shrank. For many years most scientists and philosophers also believed that these continents, once formed, were too vast to move so have always been in the same position on the Earth's surface.

However, as early as 1620 an English philosopher, Francis Bacon, realised that there was a great similarity in the shapes of the east coast of South America and the west coast of Africa.

Figure 15.30

Two maps made in 1858 showing how the continents of South America and Africa may once have fitted together

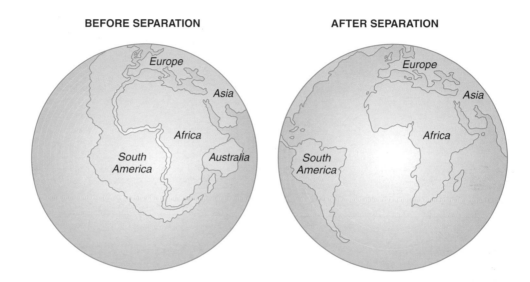

In the early 1900s a German scientist, Alfred Wegener, became intrigued about this remarkable fit of the South American and African continents. He was also interested in the similarities in the types of rocks and of plant and animal fossils that were to be found in the coastline areas of continents that were many thousands of miles apart.

In 1912 Wegener developed the idea that continents were moving. He believed that originally there was just one supercontinent, which he called 'Pangaea', and that about 2000 million years ago 'Pangaea' began to break apart (Figure 15.31).

Evidence to support Wegener's ideas have been collected so that today scientists agree with his theory. It is now accepted that:

- The Earth's outermost layer (the **lithosphere**) is made up of a number of large blocks (**tectonic plates**) that are moving relative to each other at speeds of a few centimetres per year. The lithosphere is the more or less rigid outer shell made up from the crust and the uppermost layer of the mantle.

- The plates are moving because of powerful convection currents created within the mantle. These currents are caused by the tremendous heat released by natural radioactive processes taking place in the core of the Earth.

It took some time for scientists to accept Wegeners theory. This is explained later in the chapter.

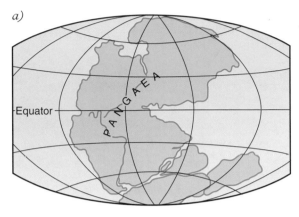

a)

225 million years ago

b)

200 million years ago

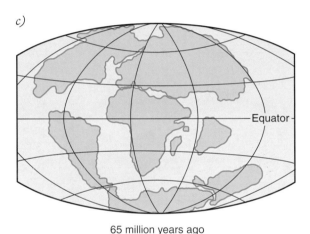

c)

65 million years ago

Figure 15.31
a) The supercontinent of 'Pangaea' 225 million years ago
b) 'Pangaea' breaking up into Laurasia – in the northern hemisphere and Gondwanaland in the southern hemisphere 200 million years ago
c) The land masses 65 million years ago

Earthquakes, volcanoes and the plates

Most earthquakes are caused by stresses set up by movements of the tectonic plates. For this reason most of them occur at the boundaries of the plates in areas where one plate slides past another or where one plate slides under the other.

Magma and gases try to reach the surface of the Earth through weak areas in the lithosphere. Where magma and gases do reach the surface they form volcanoes. Most of the weak areas are found along the boundaries of the tectonic plates.

Figure 15.32
*The Earth, showing
plate boundaries and
direction of movement*

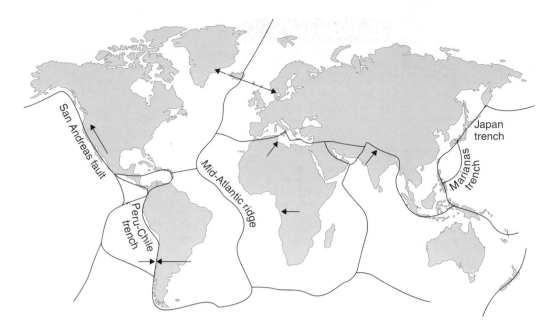

Figure 15.33
Earthquake zones

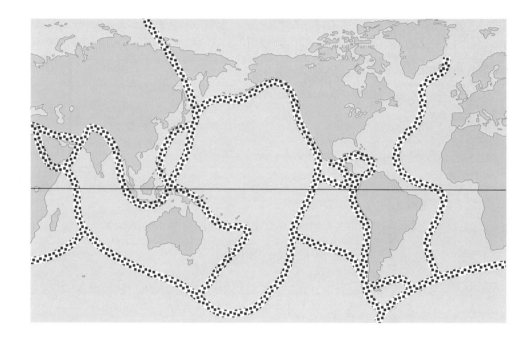

Why the accurate prediction of earthquakes and volcanic activity is difficult

Predicting earthquakes

Although most earthquakes occur in predictable areas scientists cannot give adequate warning of an impending disaster.

Much research has been undertaken to improve the seismic recording equipment and much time has been spent monitoring the many thousands of earthquakes that occur each year. Because earthquakes are caused by stresses building up in rocks below the Earth's surface it is not possible to do much more than monitor the past and present earthquake history of an area and make an intelligent guess as to when the next major earthquake may occur.

Predicting volcanic eruptions

Unlike earthquakes, a volcano will often give a variety of warning signs many years before it erupts. An increase in seismic activity being only one of the signs. Other signs include:

● a change in ground level – as pressure builds up within the magma below the volcano

● an increase in the amount of sulphur dioxide in the area – this indicates that the magma level is rising

● an increase in temperature

● an increase in the flow of magma at the surface.

Even though there are several signs that can be monitored to indicate an increase in volcanic activity, just as with earthquakes a prediction is no more than an intelligent guess.

Why Wegener's theory of Continental Drift was only gradually accepted

Wegener was convinced that all the continents were once joined together as one big 'supercontinent'. He felt that this explained how the continents 'fit' together like a jigsaw, as well as why similar fossils are found in different continents. Another important piece of evidence was that the types of fossils found on one continent suggested that it was once located in a completely different climatic zone. For example, coal deposits (the fossils of tropical plants) in Antarctica led him to believe that this land, now frozen, must once have been much closer to the equator.

However, in 1912 when Wegener published his theory – called Continental Drift – it was not well received even though it seemed to be supported by the scientific data then available. Too many scientists still believed that the continents had been, and still were, permanent features. Part of the reason was because the mechanism of plate tectonics was unknown. Wegener spent the rest of his life trying to find more evidence to support his theory. Wegener died in 1930 and it was not until the 1950s that the exploration of the ocean floor provided evidence which aroused renewed interest in Wegener's theory.

In the 1950s surveys of the ocean floor showed that it contained an enormous mountain range, called the mid-ocean ridge. This range is more than 50 000 km long, sometimes more than 800 km across and in places rises to 4500 m high.

At the same time, other scientists who were using magnetic instruments based on those used to detect submarines, found that there were strange magnetic variations in the rocks on the ocean floor.

Scientists knew that grains of an iron-rich mineral, called magnetite, are found in volcanic rock and that when the magma from a volcano cools, the grains of magnetite get locked into the rock crystals and line up in the direction of the Earth's magnetic field (see section 13.6). What was surprising about the results was that they showed not only that there were stripes of differently magnetised rocks on either side of the mid-ocean ridge but that the stripes were arranged in the same pattern on either side of the ridge.

Figure 15.34
Magnetic striping on either side of the mid-ocean ridge

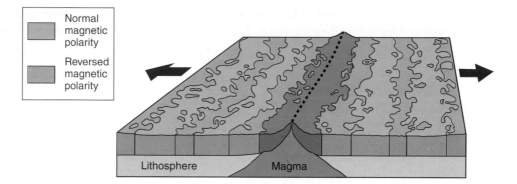

In 1961 scientists began to understand the reasons for the magnetic striping. They considered that the mid-ocean ridge was a weak area in the ocean floor. Along the 50 000 km length of this ridge, magma was erupting and forming new ocean floor. This idea they called sea-floor spreading. They believed that the rocks near the crest of the ridge were the youngest and that these rocks showed the present-day polarity (normal polarity) of the Earth's magnetic field. Stripes of rocks parallel to the crest were older and showed alternate stripes of normal polarity and opposite polarity (reversed polarity) suggesting that the direction of the Earth's magnetic field has been reversed many times during its history.

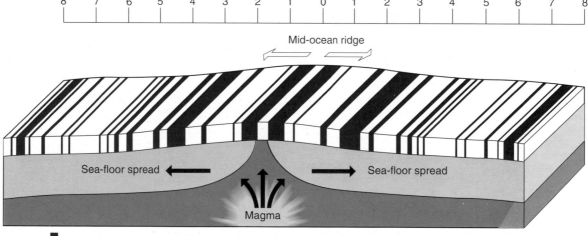

Figure 15.35 ▲
The magnetic striping on the ocean floor

The explanation of the formation of the mid-ocean ridge and the magnetic striping finally convinced scientists that Wegener's ideas had been nearly correct. It is not the continents that move but the plates on which they are fixed. Today the concept of 'Continental Drift' has been replaced with the theory of 'Plate tectonics'.

Movement of the plates

Figure 15.36 shows hot molten magma reaching and bursting through the oceanic crust to form a very long chain of underwater volcanoes – the mid-ocean ridge. The molten magma solidifies to make new crust.

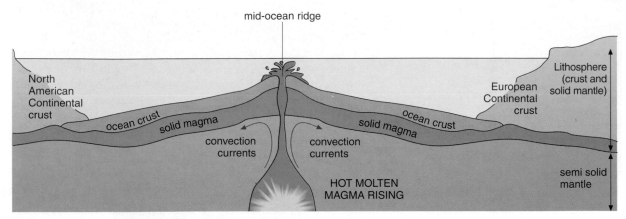

Figure 15.36▲
The Atlantic Ocean

Because of the rapid underwater cooling the material of the ocean crust is mainly basalt. The lithosphere (the very top of the mantle and the crust) is a rigid layer around the Earth that is floating on the semi-solid mantle. Convection currents rise through the mantle and reach the crust along the line of the mid-ocean ridge. These convection currents move outwards from the ridge and carry with them the newly formed ocean crust. This effect is called sea-floor spreading.

Because the mid-ocean ridge marks the boundaries between tectonic plates, the formation of new ocean crust and its movement outwards from the ridge means that the continents, such as those of North America and Europe, are moving apart. The Atlantic Ocean is getting wider at a rate of about 4 cm per year.

Figure 15.37 ▼
The Pacific Ocean

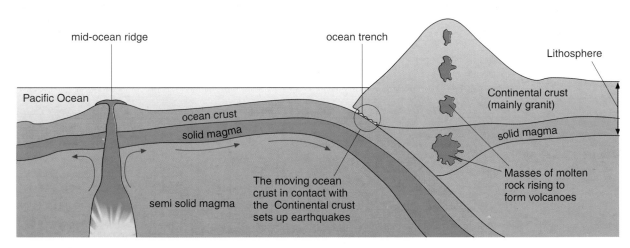

Figure 15.37 shows the ocean crust being carried by convection further and further away from the mid-ocean ridge. As it does so it cools, becoming more dense. Where the ocean crust comes into contact with the continental crust the denser ocean crust gets pushed down by the thicker continental crust. Areas where this happens are called **subduction zones**. The pressures and heat created by the subduction of the ocean crust are sufficient to cause the continental crust to become folded and faulted and the oceanic plate to partially melt. The resulting magma may cause nearby rocks to metamorphose and some of the magma may move up the rock faults to become volcanoes. This is happening along the western side of South America (the Andes).

Did you know?

The Mariana Trench in the Pacific Ocean is caused by a subduction zone. At its deepest it is 11 034 metres below sea level.

In some areas of the world the plates are trying to slide past each other. This is happening in California along the San Andreas Fault.

Figure 15.38
San Andreas Fault

Summary

◆ The lithosphere is cracked into a number of sections called **tectonic plates**.

◆ The tectonic plates move as a result of convection currents in the mantle.

◆ Earthquakes and volcanic activity occur at the plate boundaries.

◆ Wegener proposed a theory to explain the movements of the continents that was not accepted until surveys of the seabed proved that the lithosphere was made up of slowly moving plates .

Topic questions

1 The diagram represents a simplified section through the Earth. Label the parts marked by the lines.

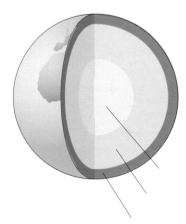

2 What is the lithosphere?

3 Explain what P waves are and how they behave.

4 Explain what S waves are and how they behave.

5 Which travels faster: P waves or S waves?

6 Why do waves from earthquakes passing through one of the layers in the Earth follow a curved path?

7 Why do waves from earthquakes change direction as they pass from one layer to another?

8 What was Pangaea?

9 What are tectonic plates?

10 What causes tectonic plates to move?

11 What causes a volcano?

12 What causes an earthquake?

13 How do we know that the Earth's magnetic field has varied many times during the Earth's history?

Examination questions

1 The boxes on the left show some types of electomagnetic radiation. The boxes on the right show some uses of electromagnetic radiation.

Draw a straight line from each type of radiation to its use. The first one has been done for you.

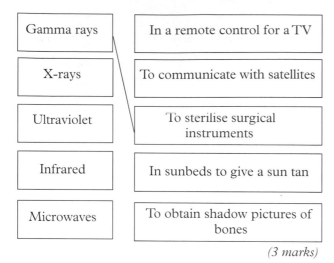

Gamma rays	In a remote control for a TV
X-rays	To communicate with satellites
Ultraviolet	To sterilise surgical instruments
Infrared	In sunbeds to give a sun tan
Microwaves	To obtain shadow pictures of bones

(3 marks)

2 a) The diagram represents the electromagnetic spectrum. Four of the waves have not been named. Copy the diagram and draw lines to join each of the waves to its correct position in the electromagnetic spectrum. One has been done for you.

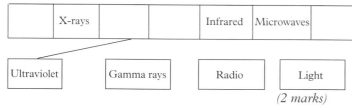

| | X-rays | | | Infrared | Microwaves | |

| Ultraviolet | Gamma rays | Radio | Light |

(2 marks)

b) Copy and complete the following sentence by choosing the correct answer from the three lines in the box.

The speed of radio waves through a vacuum is

| faster than |
| the same as | the speed of light through a vacuum.
| slower than |

(1 mark)

c) i) Before sunbathing it's a good idea to apply a sun cream to your exposed skin. Why?

(1 mark)

ii) From which type of electromagnetic wave is sun cream designed to protect the skin?

(1 mark)

d) The diagram shows an X-ray photograph of a broken leg.
Bones show up white on the photographic film. Explain why. *(2 marks)*

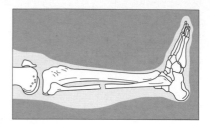

3 a) The diagram shows an electric bell inside a glass jar. The bell can be heard ringing.

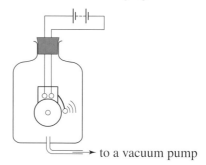

to a vacuum pump

Copy and complete the following sentences, by choosing the correct line in each box.

When all the air has been taken out of the glass jar,

the ringing sound will
| stop. |
| get louder. |
| get quieter. |

This is because sound
| travels faster |
| travels slower | through
| cannot travel |
a vacuum. *(2 marks)*

b) The microphone and cathode ray oscilloscope are used to show the sound wave pattern of a musical instrument.

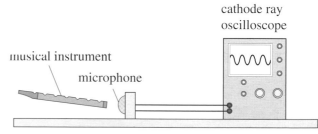

cathode ray oscilloscope

musical instrument

microphone

One of the following statements describes what a microphone does. Identify the correct statement.

(1 mark)

● A microphone transfers sound energy to light energy.

● A microphone transfers sound energy to electrical energy.

● A microphone transfers electrical energy to sound energy.

c) Four different sound wave patterns are shown. They are all drawn to the same scale.

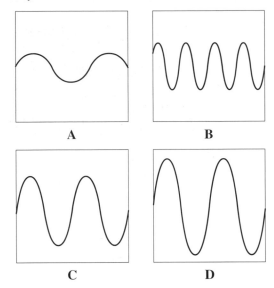

i) Which sound wave pattern has the highest pitch? Give a reason for your answer. *(2 marks)*

ii) Which sound wave pattern is the loudest? Give a reason for your answer. *(2 marks)*

d) i) The frequency of some sounds is too high for humans to hear. Underline the word which describes this sound.

microwave   ultrasound   ultraviolet

(1 mark)

ii) Give **one** use for this type of sound wave. *(1 mark)*

4 a) The student is using a microphone connected to a cathode ray oscilloscope (CRO).

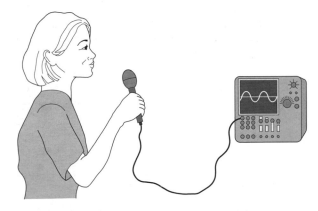

The CRO displays the sound waves as waves on its screen. What does the microphone do? *(2 marks)*

b) The amplitude, the frequency and the wavelength of a sound wave can each be either increased or decreased.

i) What change, or changes, would make the sound quieter? *(1 mark)*

ii) What change, or changes, would make the sound higher in pitch? *(1 mark)*

c) People can generally hear sounds in the frequency range 20 Hz to 20 000 Hz.

i) What are very high frequency, and inaudible, sounds with frequencies **greater** than 20 000 Hz called? *(1 mark)*

ii) Give **two** uses for very high frequency sounds. *(2 marks)*

d) The diagram shows sound waves approaching a gap in a wall.

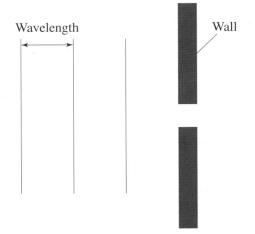

i) Copy and complete the diagram to show what will happen to the sound waves on the other side of the wall. *(2 marks)*

ii) What is the name of this effect? *(1 mark)*

iii) What would the width of the gap need to be for this effect to be most pronounced? *(1 mark)*

5 The diagram represents the structure of the Earth.

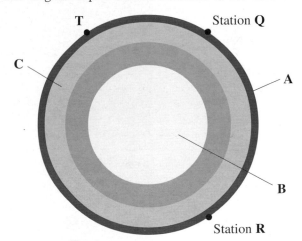

a) On the diagram, name the parts **A**, **B** and **C**.

b) An earthquake occurs at the point **T** on the Earth's surface. Two types of shock wave are produced by the earthquake, P waves and S waves.

Describe **two** similarities and **two** differences between P waves and S waves as they travel through the Earth. *(4 marks)*

c) State whether P waves or S waves or both will reach:

i) Station Q *(1 mark)*

ii) Station R. *(1 mark)*

Chapter 16
The Earth and beyond

16.1

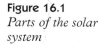

Co-ordinated	Modular
DA 12.14	DA 11
SA 12.8	SA 18

The solar system

Within our **solar system** there is one **star** (the **Sun**) and nine **planets**. Each of the nine planets follows a regular path, called an **orbit**, around the Sun. Mercury and Venus are the only planets which do not have a **moon**.

Also in orbit around the Sun are **comets**.

Figure 16.1
Parts of the solar system

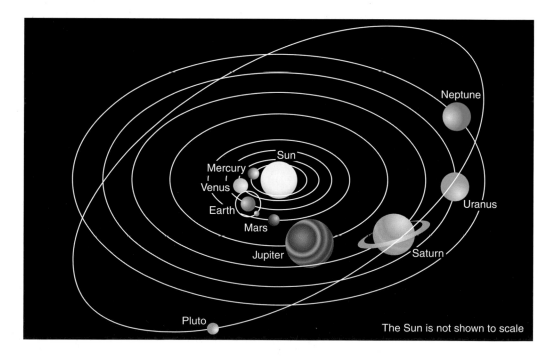

The planets

Each planet travels on a path or orbit around the Sun. Some planets have an orbit which is almost a circle, with the Sun at the centre of the circle.

Other planets have orbits like squashed circles, these are elliptical orbits. The orbit of Pluto is so elliptical that at times it is closer to the Sun than Neptune.

Figure 16.2

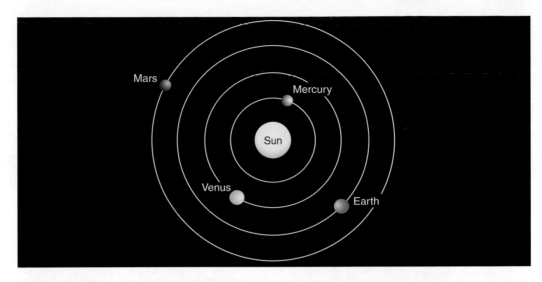

Except for Pluto, all the planets orbit the Sun in the same plane.

Figure 16.3

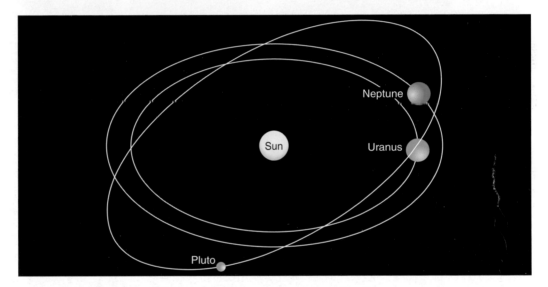

The Moon

The Moon is the Earth's natural **satellite**. Like the Earth orbits the Sun, so the Moon orbits the Earth. One orbit of the Moon takes 27 days or 1 lunar month. Like the planets, the Moon is non-luminous. We see the Moon when it reflects light from the Sun towards us.

Figure 16.14

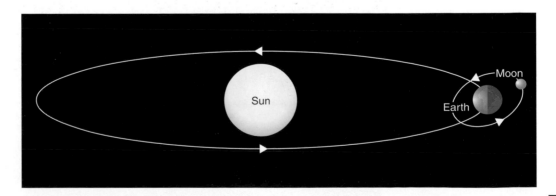

Many of the other planets are orbited by one or more moons. Saturn has 20 moons.

Comets

Comets are thought to be chunks of frozen rock covered by huge amounts of frozen water and gases. Comets also orbit the Sun, however compared to the planets, their paths are more elongated and in different planes. Comets can only be seen for the short amount of time that their orbit passes close to the Sun. At this time energy from the Sun causes some of the frozen gases and water to vaporise, giving the comet its spectacular tail.

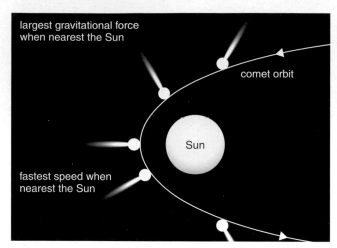

Figure 16.5
The path of a comet

Figure 16.6
The comet Hyakutake showing its glowing tail

Figure 16.7 ▶
Small masses, small force of attraction

Gravitational forces in the solar system

Gravity is a force which tries to pull objects together. It is a force of attraction. Gravity acts between all objects, no matter how big or how small. The bigger the mass of the objects, the bigger the gravity force. The force is only big enough to feel if one of the objects has a very large mass, such as one of the planets or a star.

The gravitational pull of the Sun affects all objects in the solar system. It is the gravitational force between the Sun and a planet which keeps the planet in its orbit around the Sun. If the Sun's gravitational pull suddenly stopped pulling, the Earth and all the other planets would shoot off into space.

Figure 16.8
Large masses, large force of attraction

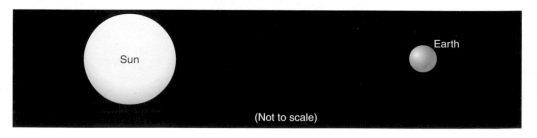

(Not to scale)

The orbit of a moon around a planet is due to the gravitational pull between the moon and the planet.

As the distance between objects in the solar system increases, so the gravitational force between them decreases. The force of gravity between planets is weak because of the great distances between the planets.

Figure 16.9

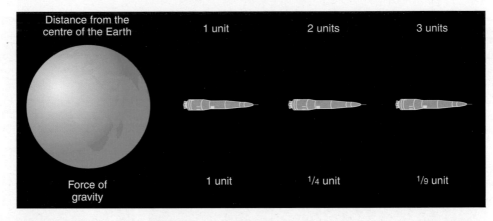

Figure 16.9 shows how the force of gravity between two masses changes with distance. If the distance between the two masses doubles, the force of gravity goes down by a factor of four. If the distance between the two masses trebles then the force of gravity goes down by a factor of nine. This is called the inverse square relationship.

Gravity can be used to change the direction of a spacecraft. Figure 16.10 shows the path taken by the Cassini spacecraft. Each time the spacecraft approaches a planet, the gravitational force will swing and accelerate it for the next part of its journey.

> **Did you know?**
>
> The nuclear powered spacecraft Cassini, launched in 1997, will reach Saturn in the year 2004. A probe will then explore the atmosphere and surface of Titan, Saturn's largest moon.

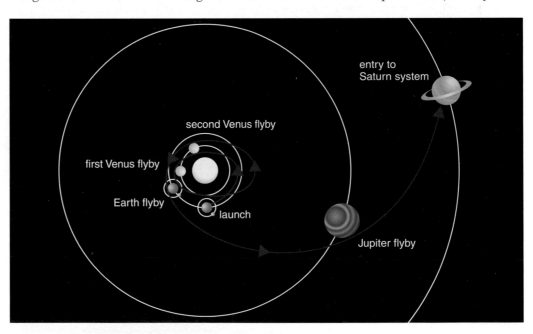

Figure 16.10

Artificial satellites

These are held in orbit by the gravitational pull of the Earth and the speed at which they are moving. The first **artificial satellite**, Sputnik 1, was put into orbit in 1957. Since then, hundreds of satellites have been launched into space. Artificial satellites have many different uses.

1 Observation of the Earth

Satellites used to observe the Earth are usually put into a low polar orbit. Passing over the Earth they scan the surface sending back detailed pictures. Such things as volcanic activity, the position of an oil slick or the path of a hurricane can all be watched and monitored. Because the Earth spins, different parts of the Earth are seen on each rotation.

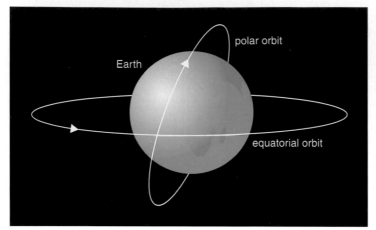

Figure 16.11

Figure 16.12
A satellite picture of the River thames in London

2 Weather monitoring

Satellites, such as the European Space Agency's Meteostat, are put into high equatorial orbits. They monitor the changing weather patterns over one part of the Earth's surface. With this information, weather forecasts can be made more accurate.

Figure 16.14
The weather pattern seen from a satellite

Figure 16.13
Meteostat in orbit

3 Exploration of the solar system

Satellites can produce sharp images of the solar system. This is because they do not have any atmospheric interference. Space probes can also travel to other planets, take photographs and send the images back to Earth.

Figure 16.15
A satellite transmission system for television

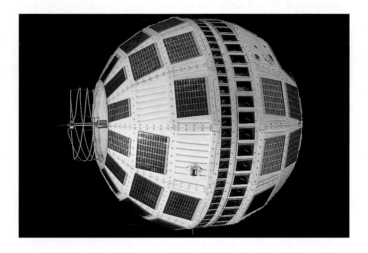

4 Communication systems

Geostationary satellites used in communication systems are put into high equatorial orbits. They travel at a speed which takes them once around the Earth every 24 hours. This means that the satellite moves around the Earth at the same rate as the Earth spins.

In this way the satellite stays above the same point on the Earth's surface. These satellites can be used to send television programmes and telephone messages around the world.

Figure 16.16
Satellite dishes receive the TV signal from the satellite

Television programmes are transmitted to the satellite using microwaves. The satellite then transmits the programmes back to an area of the Earth as large as Europe. A dish aerial fixed in the correct direction can then be used to pick up the programme.

Summary

◆ The **orbits** of the inner planets around the **Sun** are more or less circular; those of the outer planets are elliptical.

◆ **Comets** have very elliptical orbits.

◆ All bodies attract each other with a gravitational force.

◆ Communication and monitoring satellites are put into geostationary orbit.

Topic questions

1 Use words from the box to complete the following sentences. Each word may be used once or not at all.

galaxy	moons	star	universe

The solar system has one _____ called the Sun. There are nine planets in orbit around the Sun. Some planets have one or more _____ in orbit around them.

2 Complete the following sentences.

Planets orbit a _____ .

A moon orbits a _____ .

3 The box contains the names of eight of the nine planets in the solar system.

| Earth | Jupiter | Mars | Mercury | Neptune | Pluto | Saturn | Uranus |

a) Name the planet which has not got its name in the box.
b) Which planet has the shortest orbit?
c) Name the force which keeps a planet in its orbit.

4 The diagram below, which is not drawn to scale, shows a communications satellite in orbit above the Earth.

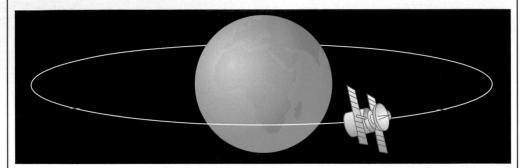

The satellite is in a geostationary orbit. Explain what is special about this sort of orbit.

5 The diagram below, which is not drawn to scale, shows the path of one kind of object in the solar system. The object is not a planet or moon.

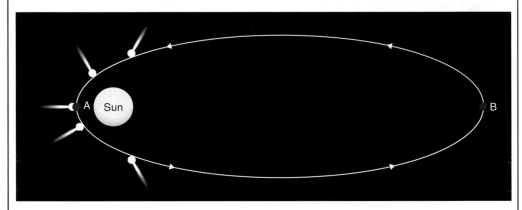

a) What is the name of this kind of object?
b) What is the force which causes the object to change direction?
c) Why is the force greater at point A than at point B?
d) At which point A or B will the object be moving the fastest?

6 Saturn has about 100 times the mass of the Earth, and Saturn is about 10 times further away from the Sun than the Earth. How does the Sun's gravitational pull on Saturn compare to the Sun's gravitational pull on the Earth?

16.2   The wider Universe

Co-ordinated	Modular
DA 12.15	DA 11
SA 12.9	SA 18

A **galaxy** is a vast number of stars held together by the force of gravity. The Sun is just one of the 100 000 million stars which make up the **Milky Way** galaxy. Figure 16.17 shows what the Milky Way would look like if it were being seen from above and from the side. The whole galaxy is so huge that it takes light 100 000 years to travel from one side to the other.

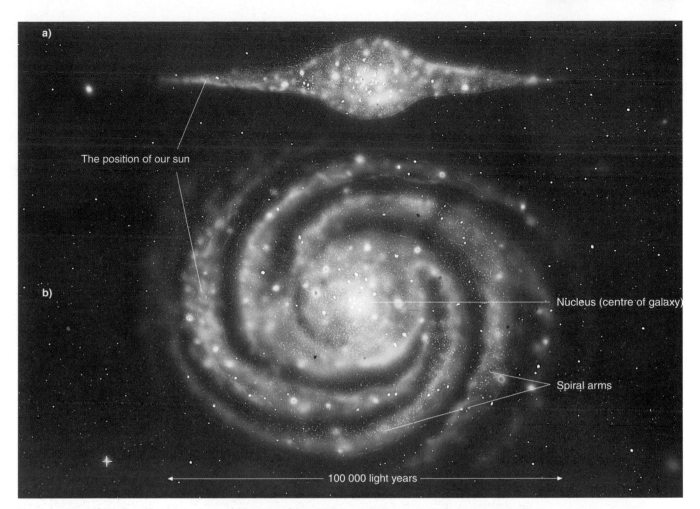

a)

The position of our sun

b)

Nucleus (centre of galaxy)

Spiral arms

← 100 000 light years →

Figure 16.17 ▲
The Milky Way galaxy,
a) viewed from the side,
b) viewed from above

Figure 16.17 ▶
The Andromeda galaxy

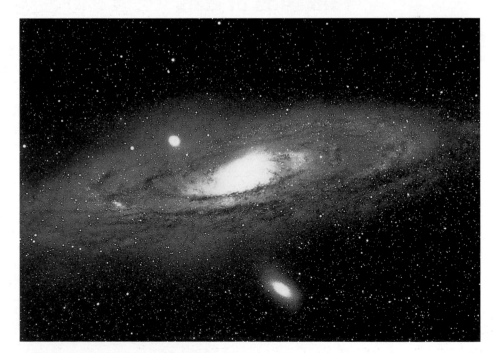

The Milky Way is not the only galaxy. Using large telescopes, millions of similar
galaxies can be seen in all directions in space. The **Universe** may contain billions
(thousands of millions) of galaxies, each having about as many stars as the Milky
Way.

The distance between stars and between galaxies is so big that it is not sensible to measure it in kilometres. The distances are measured by saying how long it takes light from the star or galaxy to reach us. These distances are called **light years**. A light year is the distance that light travels in one year.

Since the speed of light is 300 000 km/s, one light year is roughly 9 500 000 000 000 kilometres.

Evidence for the origin of the Universe

By looking at light sent out by distant galaxies, scientists have come to the conclusion that the Universe is expanding.

In 1929 an American astronomer, Edwin Hubble, examined the spectrum of the light from various galaxies. He noticed that whichever galaxy he looked at, the spectra were composed of light of longer wavelengths, that is, the spectra (spectral lines) were always shifted towards the red end. He called this effect **red shift**. He also noticed that spectra for the closer galaxies showed less red shift than did the spectra for galaxies further away. Scientists have discovered that this shift in the wavelength of light to the red end of the spectrum occurs when a light source is moving away. The faster the light source is moving away, the greater the red shift. Therefore Hubble concluded that all galaxies were moving away from Earth and that the farther the galaxy the faster it was moving away.

In 1948 George Gamow, a Russian physicist, proposed that if all the galaxies are moving away at high speed then there must have been a time when they were all concentrated in a single place. From this comes the idea that the Universe was then created by a massive explosion. This created the dust and gases which went on to form the planets and the stars. He used the term **Big Bang** to describe that moment when all the matter that was concentrated in one place exploded and started to expand outwards. The Universe has been expanding ever since.

Formation of stars

Stars do not last for ever. They go through a life cycle from birth to death. Like all stars our Sun was formed from a huge cloud of mainly hydrogen gas. Over millions of years this cloud was pulled inwards and condensed into a smaller volume by the force of gravity.

Figure 16.19
A cooling mass of gas in the great Nebula in Orion

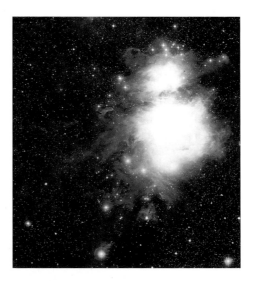

This contraction caused reactions which produced vast amounts of energy. This energy was released as heat and light causing a huge temperature rise. At this stage the star was born and further contraction stopped.

During the main period of its life, a star remains in a stable state. The inward gravitational forces trying to make it collapse are balanced by the enormous outward pressure created by the high temperature at the centre of the star. This period of stability is likely to last for billions of years. Our Sun is in its stable period and about half way through its life.

The reaction which produces the vast amounts of energy is called nuclear **fusion**. In this process hydrogen nuclei fuse together to form helium nuclei. This process continues during the stable period of a star, allowing it to radiate heat and light energy. As the hydrogen is used up, the mass of the star decreases, causing the next phase in the life cycle of the star – the formation of a red giant.

Figure 16.20
A stable star

Eventually the reactions that were responsible for the release of the vast amounts of energy begin to stop and the mass of the star decreases. The inward gravitational forces are no longer able to balance the outward pressure created by the high temperatures. The star gradually expands and cools to become a **red giant**. When our Sun reaches this stage it will be so large that the Earth will end up inside it!

What happens in the final stages of the life cycle of a star depends on its size. Gravitational forces cause the star to contract, causing another rise in temperature and further energy release. The star is now a **white dwarf** and the matter from which it is made may be many millions of times more dense than any matter on Earth.

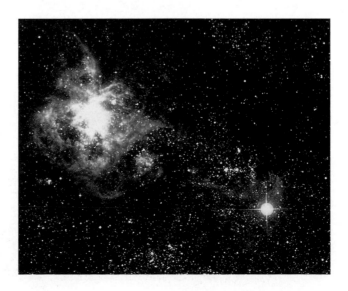

Figure 16.21
A supernova

The Earth and beyond

If a white dwarf is formed from a small star, like the Sun, it will fade and become a black dwarf.

However if a star is massive enough, then the final stage after becoming a very dense white dwarf is for the star to explode as a supernova, scattering gas and dust into space. During a supernova stage the dying star emits more energy into space in a few seconds than is produced by the Sun in millions of years.

The matter left behind from the exploding white dwarf may form a very dense neutron star.

Did you know?

Neutron stars are thought to have very strong magnetic fields which cause splits in the surface of the star allowing it to emit vast amounts of gamma radiation.

Neutron stars are so dense that if the ball on the end of a ball-point pen were made of matter packed as densely, the ball would have a mass of about 91 000 tonnes.

Figure 16.22
The life cycle of a star

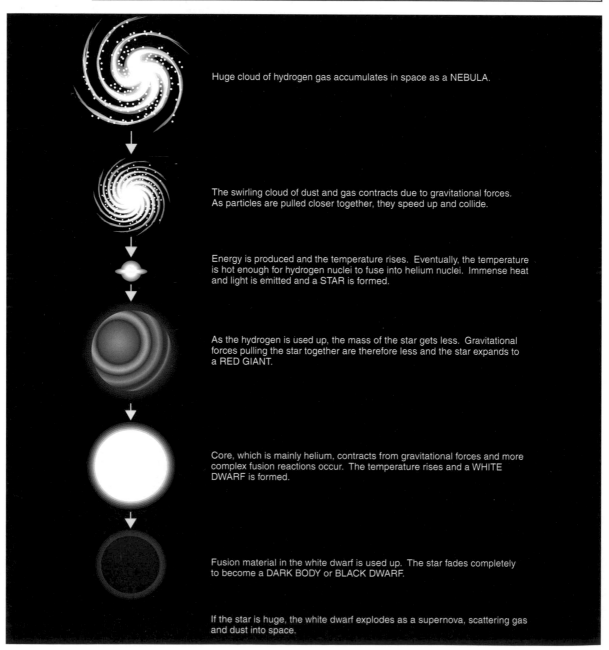

Huge cloud of hydrogen gas accumulates in space as a NEBULA.

The swirling cloud of dust and gas contracts due to gravitational forces. As particles are pulled closer together, they speed up and collide.

Energy is produced and the temperature rises. Eventually, the temperature is hot enough for hydrogen nuclei to fuse into helium nuclei. Immense heat and light is emitted and a STAR is formed.

As the hydrogen is used up, the mass of the star gets less. Gravitational forces pulling the star together are therefore less and the star expands to a RED GIANT.

Core, which is mainly helium, contracts from gravitational forces and more complex fusion reactions occur. The temperature rises and a WHITE DWARF is formed.

Fusion material in the white dwarf is used up. The star fades completely to become a DARK BODY or BLACK DWARF.

If the star is huge, the white dwarf explodes as a supernova, scattering gas and dust into space.

The red giant reaches its maximum size when all the hydrogen has been converted to helium. The temperature at the centre of a red giant causes the helium nuclei to fuse together to form heavier elements. More elements form as the star contracts to become a white dwarf. During a supernova the nuclei of elements heavier than hydrogen are sent into space where they can form part of the contents of future stars. Future stars start life with a richer supply of heavier elements that earlier generations of stars.

Nuclei of the heaviest elements are present in the Sun and atoms of these elements are present in the inner planets of the solar system. This suggests that the solar system was formed from the material produced when earlier stars exploded.

Black holes

Sometimes after a supernova the matter that is left behind and not scattered into space is so dense and the gravitational forces so strong that the matter collapses to form a black hole.

Black holes have been given this name because light can enter but not escape due to the tremendous gravitational forces associated with them. We cannot see black holes but we can observe their effects on their surroundings. For example, we can detect the X-rays emitted when gases from a nearby star spiral into a black hole.

Did you know?

Even the Earth could be made into a black hole if it could be squashed down to the size of tennis ball.

The search for life elsewhere in the galaxy
What should we be looking for?

So far, in the quest for evidence of life beyond our planet the search has been restricted to the sending of manned and unmanned lunar modules to the Moon, unmanned probes to Mars and Venus.

Figure 16.23
An image of the surface of Mars taken by the Mars Pathfinder space probe

Scientists have assumed that if life exists or has existed, any evidence remaining will be based on the life processes associated with living things on Earth. Because they assume that water must be available to support photosynthesis and that oxygen must be available to support respiration, scientists have designed and set up experiments on the surfaces of both the Moon and Mars to test samples of soil and rock.

These experiments were designed:

- to test for the presence of water and oxygen.

- to measure whether there were any changes in the gases that might indicate that photosynthesis and respiration were taking place.

- to examine soil samples to discover if living organisms were present or whether there was fossil evidence that living organisms had once existed.

It is important to realise that our atmosphere would contain little or no oxygen if it were not for the existence of plants.

Much time and money has been spent but so far there has been no success. However, it may be that they have been looking for the wrong evidence, for recently on Earth some organisms have been discovered surviving in conditions that were once thought impossible.

- Microbes have been found in the sulphur-laden atmosphere of volcanoes, hot springs and geysers.

- Deep-diving submarines have found microbes that get their energy from the heat produced by hot rocks.

- Other microbes have been found deep underground that use hydrogen as their energy source.

Did you know?

In 1984 a meteorite was discovered in Antarctica that scientists are convinced came from Mars thousands of years ago. Examination of the meteorite seemed to show the presence of microscopic worm-like structures in the rock.

The Mars meteorite

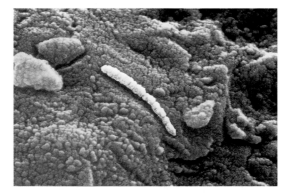

Are these worm-like structures fossilised Martian bacteria?

Should we be looking for intelligent life?

In 1960 an American called Frank Drake was the first scientist to start a careful search for intelligent radio signals from space. He spent six hours each day for about four months using a 25 m radio telescope. He was unsuccessful.

Drake's lack of success has not stopped radio astronomers from trying to search for extraterrestrial life in space (SETI). However, 40 years of SETI have still failed to find anything, even though there have been tremendous advancements in the technology available.

Figure 16.24
Frank Drake and Jill Tarter lead the SETI team. They stand beneath the radio telescope they use to look for radio signals in space

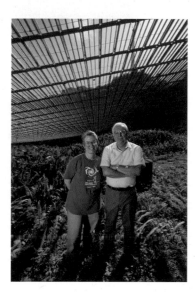

What do we do now?

It is now realised that to send manned craft to other planets and moons is costly and dangerous. For the exploration of deep space a manned craft will take too long. So it seems that any further exploration of space will be undertaken using unmanned craft and robots.

Because it is now known that on Earth there are organisms that can survive in extremely inhospitable conditions, scientists are beginning to reconsider looking for life in what they once considered to be unsuitable areas of the solar system. For example, photographs from the 1979 *Voyager* spacecraft mission to Jupiter and its neighbours showed that the surface of Europa, one of Jupiter's moons, seems to be covered in ice. Further photographic evidence from the recent *Galileo* mission to Jupiter shows that the surface of Europa could be an icy crust between 9–16 km thick. There is some evidence to show that the interior of this moon may just be warm enough to have melted some of the ice and that liquid water may be present below the ice. The Americans are considering sending a lander to Europa with the aim of drilling through the ice and sending a probe called a hydrorobot to investigate the contents of the water.

Did you know?

The planet Pluto is about 5 500 000 000 km from Earth. Even if a spacecraft could travel at half the speed of light, 300 000 km/s, it would take over two years to complete the return trip.

Figure 16.25
The hydrorobot

Summary

◆ The **Universe** is made up of a very large number of **galaxies**.

◆ **Stars** are formed when dust and gas are pulled together.

◆ Stars go through a life cycle from birth to death.

◆ So far, no life seems to have been found anywhere else in the Universe.

◆ The discovery of the concept of the **red shift** led to the idea of the **Big Bang** as being the origin of the Universe.

Topic questions

1 Which one of the following quantities can be measured in light years?

    distance speed time

2 The solar system is part of which galaxy?

3 Rewrite the following in order of size. Start with the smallest.

    galaxy planet solar system star Universe

4 What is a galaxy?

5 The Andromeda Galaxy is about 2.5 million light years away. Why are humans unlikely to ever explore it?

6 Explain how the 'Big Bang' theory accounts for the creation of the Universe.

7 What does the 'Big Bang' theory predict is happening to the size of the Universe?

8 Explain how stars are formed.

9 By what process does a star get its energy?

10 For the millions of years between its 'birth' and its 'death' a star is stable. Describe the two main forces at work in the star during this time.

11 Describe the process by which a red giant changes into a white dwarf.

Examination questions

1 a) Copy and complete each sentence by choosing the correct word or phrase from the box. Each word or phrase should be used once or not at all.

> milky way   moon   planet   solar system
> star   universe

The Sun is the nearest _____ to the Earth.
The Sun is in the galaxy called the _____.
Within the _____ there are millions of galaxies.
Pluto is orbited by one _____.   *(4 marks)*

b) The diagram shows the path taken by the Voyager 2 spacecraft.

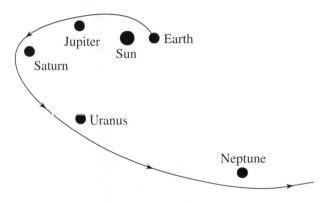

Choosing from the forces in the box, which force caused the spacecraft to change direction each time it got close to a planet?

> air resistance   friction   gravity

(1 mark)

2 a) The table gives some information about four planets.

Planet	Average distance from the Sun in million km	Average time to complete one orbit in Earth years	Average orbital speed in km/sec
Jupiter	800	12	13.0
Saturn	1400	30	9.6
Neptune	4500	165	5.2
Pluto	5900	248	4.7

i) Draw a graph of each planet's average orbital speed against the distance the planet is from the Sun. Plot distance from the Sun on the horizontal axis and orbital speed on the vertical axis.

(3 marks)

ii) How does the average orbital speed of a planet vary with its average distance from the Sun?   *(1 mark)*

iii) The average distance between Uranus and the Sun is 2900 million kilometres. Use the graph to predict the average orbital speed of Uranus.   *(1 mark)*

b) The diagram shows the position of Saturn in July 1984 and July 1986.

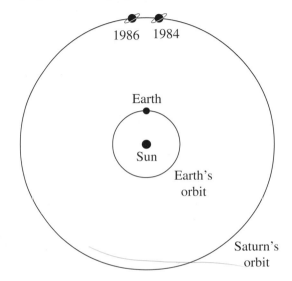

i) Saturn takes 30 Earth years to complete one orbit of the Sun. Copy the diagram and mark the position of Saturn in the year 2000.   *(1 mark)*

ii) Suggest why it was difficult to see Saturn in July 2000.   *(1 mark)*

3 A satellite in a stable Earth orbit moves at constant speed in a circle, because a single force acts on it.

a) i) Name the force acting on the satellite.
(1 mark)
ii) State the direction of this force.   *(1 mark)*

b) Communications satellites and monitoring satellites are placed in different orbits.

i) Describe the orbit of a communications satellite.   *(3 marks)*

ii) Describe the orbit of a weather monitoring satellite.   *(2 marks)*

iii) Explain why the satellites are placed in different types of orbit.   *(3 marks)*

c) Explain, in terms of its orbit, why a comet is rarely seen from Earth.   *(2 marks)*

4 a) The Cassini spacecraft launched in 1997 will take seven years to reach Saturn. The journey will take the spacecraft close to several other planets.

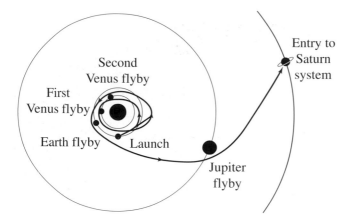

Each time the spacecraft approaches a planet it changes direction and gains kinetic energy. Explain why. *(2 marks)*

b) The Big Bang theory attempts to explain the origin of the Universe.
   i)   What is the Big Bang theory?   *(1 mark)*
   ii) What can be predicted from the Big Bang theory about the size of the Universe?
                                          (1 mark)

c) i)   Explain how stars like the Sun were formed.   *(2 marks)*
   ii) The sun is made mostly of hydrogen. Eventually the hydrogen will be used up and the Sun will 'die'.
        Describe what will happen to the Sun from the time the hydrogen is used up until the Sun 'dies'.   *(3 marks)*

Chapter 17

Using energy and doing work

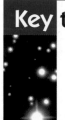

17.1	
Co-ordinated	Modular
DA 12.16	DA 9
SA 12.10	SA 17

Thermal energy transfers

Radiation

Radiation is the energy transfer that takes place without the movement of any particles. Indeed energy transfer by radiation is most efficient when no particles are present. It is the process by which our planet receives heat energy from the Sun through the vacuum of space.

All objects radiate energy. The hotter the object the more heat energy it radiates.

The heat energy radiated is called infrared radiation. Infrared radiation is one of the members of the family of waves called the electromagnetic spectrum (see section 15.3).

Giving out radiation (radiation emission)

- Hot tea in a light-coloured teapot will transfer energy to the air slower than tea in a dark-coloured teapot.

- Hot tea in a highly-polished shiny teapot will transfer energy to the air slower than tea in a non-shiny teapot. (A non-shiny surface is described as having a matt finish.)

So, hot surfaces that are light coloured and shiny will transfer infrared radiation to the air slower than hot, dark-coloured matt surfaces.

Taking in radiation (radiation absorption)

The outside metal of a dark-coloured car warms up quicker in the Sun than does the metal on a light coloured car. A polished shiny car will not get as hot as an unpolished car.

This happens because:

- dark-coloured cold surfaces absorb infrared radiation more quickly than do light-coloured cold surfaces

- shiny cold surfaces reflect more infrared radiation than do matt cold surfaces.

Figure 17.1

Figure 17.2
A summary of radiation

Good absorbers of infrared radiation (if cold these get warm quickly)	Good emitters of infrared radiation (if hot these cool down quickly)	Poor absorbers of infrared radiation (if cold these get warm slowly)	Poor emitters of infrared radiation (if hot these cool down slowly)
dark-coloured surfaces	dark-coloured surfaces	light-coloured surfaces	light-coloured surfaces
non-shiny (matt) surfaces	non-shiny (matt) surfaces	shiny surfaces	shiny surfaces

Infrared radiation and the atmosphere

The carbon dioxide in the atmosphere acts like a blanket around the Earth to keep the Earth warm. Not all of the infrared radiation coming from the Sun and hitting the Earth's surface is absorbed by the Earth's surface. Some of it is reflected back into space.

Figure 17.3
Some of the Sun's radiation is reflected back into space

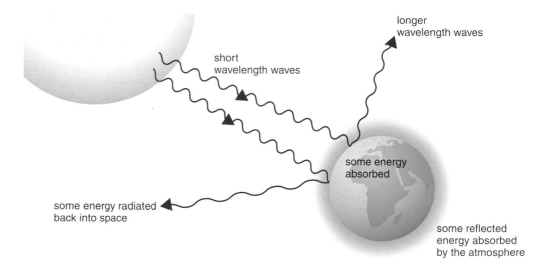

longer wavelength waves

short wavelength waves

some energy absorbed

some energy radiated back into space

some reflected energy absorbed by the atmosphere

The carbon dioxide in the atmosphere absorbs some of this reflected infrared radiation so keeping the Earth reasonably warm. This effect is known as the **greenhouse effect** (see section 6.2). As more and more **fossil fuels** are burned, more carbon dioxide passes into the atmosphere. This has the effect of absorbing more reflected infrared radiation so making the atmosphere even warmer. This warming is called **global warming** (see section 6.2).

Conduction

Conduction is the process by which energy transfers take place in solids. Metals are better conductors than non-metals. If one end of a metal bar is heated in a Bunsen flame, the heat energy from the flame is quickly transferred along the bar from the hot end to the cold end. This happens because metals contain particles that can move, called **free electrons** (see section 7.2), and particles called ions (ions are atoms that have lost an electron) that can only vibrate (Figure 17.6). At the heated end of the bar, the heat energy in the flame:

- increases the rate at which the closely packed ions vibrate. These vibrating ions collide with their neighbours. These collisions give the neighbouring ions more energy so they vibrate more rapidly.

- increases the energy of the free electrons. These free electrons travel long distances very rapidly between collisions and so transfer energy along the bar quickly.

Figure 17.4
Free electrons between the ions in a metal

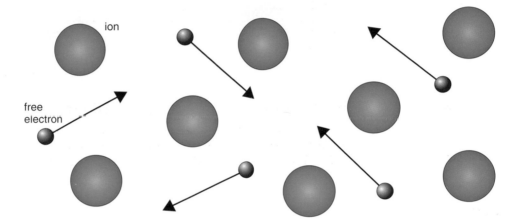

This happens all along the bar and so the heat energy is transferred from the heated end along the bar by a series of collisions between neighbouring ions and by the movement of the free electrons.

Non-metals have no free electrons so the energy transfer process is much slower.

Convection

Convection is a form of energy transfer that takes place in liquids and gases (fluids).

Figure 17.5 shows water being heated.

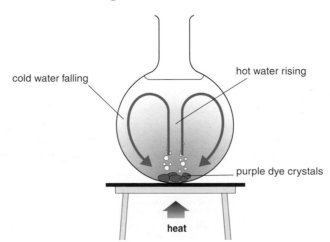

Figure 17.5
Energy transfer in a liquid

As the water at the bottom is warmed up, the particles gain heat energy. This extra energy causes:

- the particles to vibrate with a bigger amplitude
- the particles to take up more space
- the water in the warmer region to expand
- the warm water as it expands to become lighter (less dense) than the cooler water around it
- the warm water to rise.

As the warm water rises, cooler water flows in to replace it. This water gets heated and rises. The movements of hot and cold water due to the changes in density are called convection currents. Convection currents stop when all parts of the water are at the same temperature. Convection currents occur because cold, dense water sinks and warm less dense water rises.

Did you know?

The Atlantic Gulf Stream is caused by convection currents.

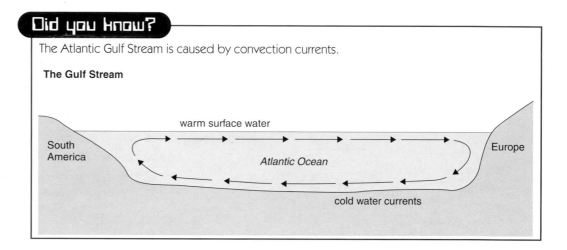

The Gulf Stream

Ways to reduce heat loss

Figure 17.6

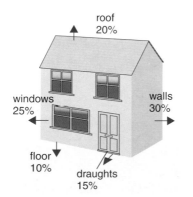

Different amounts of heat are lost through the roof, the walls, the windows, the floor and the doors of a home. These are shown in Figure 17.6.

There are many devices that have been designed to reduce unwanted heat loss from a home. Information about some of these devices is given in Figure 17.7 and the cost of these are given below.

It could take up to 40 years if the money saved in heating bills by having double glazing fitted was used to pay for the fitting.

Ways of keeping the house warm	Pay-back time using money saved on the heating bills
Cavity wall insulation	3 years
Loft insulation	1.5 years
Double glazing	40 years

Although double glazing is very expensive it does act as a sound insulator.

Figure 17.7

Device	How heat energy losses are reduced
Cavity wall insulation convection currents in cavity foam – no convection currents possible	The cavity between two outside walls is filled either with mineral wool or foam. • the foam and mineral wool both trap a large number of tiny air pockets • Because the air is trapped it cannot move, so convection currents cannot occur. • The still air is also a good insulator.
Loft insulation	The fibres in the mineral wool trap a large number of tiny air pockets. • Because the air is trapped it cannot move, so convection currents cannot occur. • The still air is also a good insulator.
Double glazing	• Thick glass is used so reducing heat loss by conduction. • The still air trapped between the two panes reduces heat loss by convection and conduction.
Thick curtains	• These stop cold air blowing into a room. • They trap air between the wall and the curtain so reducing heat loss by conduction.

Summary

◆ **Radiation** is the transfer of energy by infrared radiation.

◆ Radiation can pass through a vacuum and can be reflected.

◆ **Conduction** of heat energy is the result of the movement of free electrons.

◆ Metals are good conductors of heat energy because they have a lot of free electrons.

◆ Non-metallic substances, including trapped air, are poor conductors of heat energy.

◆ **Convection** is the movement of more energetic particles in liquids and gases.

Topic questions

1 Explain why metals are good heat conductors.

2 The elements in an electric kettle are always found near the bottom of the kettle. Explain why.

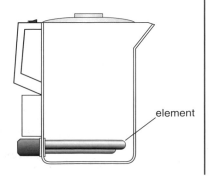

element

3 The freezer compartments of fridges are always found at the top of the fridge. Explain why.

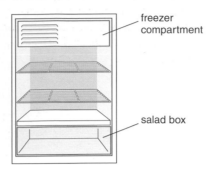

freezer compartment

salad box

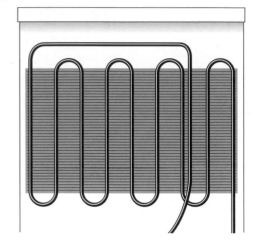

4 As food is cooled in a fridge, heat is transferred from the food to a coolant. This coolant passes through pipes at the back of the fridge. These pipes are painted matt black. Explain why.

5 Large storage containers for gas and liquid fuels are usually painted silver or white. Explain why.

6 Give two reasons, other than they might look good and last a long time, why many people have their windows double-glazed.

7 Cavity walls are often filled with foam to prevent heat transfer through the cavity from a warm room to the air outside. Explain how the foam reduces heat transfer:

  a) by conduction
  b) by convection.

17.2 Efficiency

Co-ordinated	Modular
DA 12.17	DA 9
SA 12.11	SA 17

A light bulb in a torch is designed to transfer **electrical energy** to light energy. However, as with most light bulbs, heat energy is also transferred. This heat is not wanted. In most filament bulbs only 5% of the electrical energy is changed to useful light energy. 95% of the electrical energy is transferred as unwanted heat energy. Such light bulbs are described as being only 5% efficient.

Eventually all energy, both useful and unwanted (wasted) is transferred to the surroundings. This makes the surroundings become warmer. A rise in temperature is not usually noticed because the energy becomes spread out. The more spread out it becomes the more difficult it is for any further useful energy transfers.

Efficiency

Efficiency is a way of calculating how good a device is at transferring the total energy going in compared to useful energy transferred. If the total energy going in is the same as the useful energy transferred the device is 100% efficient. No device can have an efficiency greater than 100%.

Efficiency of a device can be calculated using the equation:

$$\text{efficiency} = \frac{\text{useful energy transferred by device}}{\text{total energy supplied to device}}$$

Efficiency of some electrical devices

Low energy light bulbs v filament light bulbs

A 20 W low-energy light bulb gives out as much light as a 100 W filament lamp. In such bulbs very much more of the electrical energy is transferred as light. Low-energy light bulbs are small fluorescent lamps. They do not have a metal filament that gets hot. Very little electrical energy is transferred into unwanted heat, so these light bulbs are more efficient at transferring electrical energy as light energy.

Figure 17.8
a) filament light bulb and b) a low energy light bulb

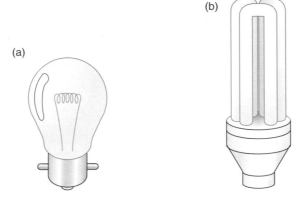

Electric kettles

Electric kettles contain an element that is designed to transfer electrical energy to heat energy. The heat energy makes the water hot enough to boil. The most efficient kettle will be the one in which most heat energy from the element is transferred to the water and not the surroundings.

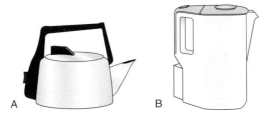

Figure 17.9

Even though each of the kettles shown in Figure 17.9 may have identical elements and hold the same amount of water, they each have some design features that affect the amount of heat lost to the surroundings. These are given in Figure 17.10.

Figure 17.10
Comparing electric kettles

Heat-saving features		Heat-losing features	
Kettle A	**Kettle B**	**Kettle A**	**Kettle B**
Shiny metal surface: this reduces heat loss by radiation	Plastic casing: this reduces heat loss by conduction	Metal casing: this increases heat loss by conduction	Cylindrical shape: this provides a large surface area: volume ratio, so more heat will be transferred to the surroundings through the walls than in the more spherical-shaped kettle

357

Calculating the efficiency of a kettle

An electric kettle, rated at 2500 W, transfers 2500 J of electrical energy to heat energy each second. The kettle takes 150 s to boil some water. In this time 336 000 joules of energy are transferred to the water. Calculate the efficiency of the electric kettle.

The useful energy output is the heat transferred to the water = 336 000 J

The total energy input is the electrical energy supplied to the element in the kettle. So, total energy input = 2500×150

$$= 375\ 000\ J$$

Substituting in efficiency $= \dfrac{\text{useful energy transferred by device}}{\text{total energy supplied to device}}$

$$= \frac{336\ 000\ J}{375\ 000\ J} = 0.89\ J$$

Multiplying this by 100 produces an efficiency for the kettle of 89%.

So:

- 89% of the electrical energy supplied by the element is used to boil the water.

- 11% of the energy supplied by the element is wasted. Most of this will be transferred as unwanted heat to the surrounding air, the material of the kettle and the element. Some, but not much, will be lost as the boiling water evaporates.

Electric motors

Electric motors are designed to transfer electrical energy to kinetic energy. Inside all electric motors are parts that spin very fast. When these parts spin, friction between the moving parts and the rest of the motor often causes heat energy to be transferred to the motor. The production of heat energy reduces the amount of kinetic energy transferred. This reduces the efficiency of the motor. The more friction, the less efficiency.

Summary

◆ The **efficiency** of an energy transfer can be determined by comparing the useful energy output with the total energy input.

◆ efficiency $= \dfrac{\text{useful energy transferred by device}}{\text{useful energy supplied to device}}$

Topic questions

1 An electric oven is described as being 70% efficient. what does this mean?

2 A small electric heating element can be used to boil one cupful of water. If the heating element is rated at 300 W, it transfers 300 J of electrical energy to heat energy each second. It takes 6 minutes for the water in the cup to boil and in this time 84 000 J of heat energy are transferred to the water. Calculate the efficiency of this method.

3 Complete the following sentences using words from the box. Each word can be used once or not at all.

| chemical | electrical | heat | less | more | wasted |

An electric fire is _____ efficient than an electric motor. The fire is designed to transfer _____ energy to heat energy, but the _____ energy produced in a motor is _____ .

17.3	
Co-ordinated	Modular
DA 12.18	DA 9
SA 12.12	SA 17

Energy resources

Producing electricity from renewable and non-renewable energy resources

Some energy resources can be replaced faster than they can be used. These are called **renewable** energy resources. Some energy resources are being used at a faster rate then they are being replaced. These are called **non-renewable** energy resources.

One of the most important uses for renewable and non-renewable energy resources is in the production of electricity. The methods by which some of these resources are used to produce electricity are shown in Figure 17.11.

Figure 17.11

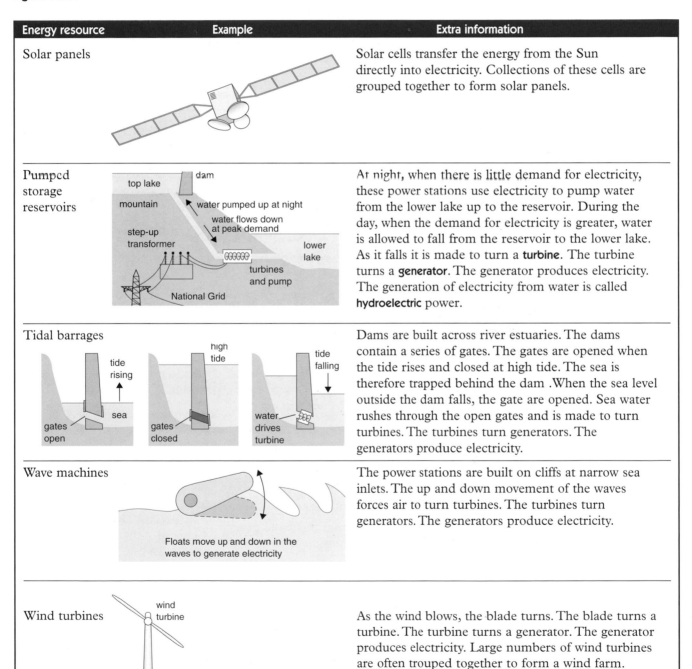

Energy resource	Example	Extra information
Solar panels		Solar cells transfer the energy from the Sun directly into electricity. Collections of these cells are grouped together to form solar panels.
Pumped storage reservoirs		At night, when there is little demand for electricity, these power stations use electricity to pump water from the lower lake up to the reservoir. During the day, when the demand for electricity is greater, water is allowed to fall from the reservoir to the lower lake. As it falls it is made to turn a **turbine**. The turbine turns a **generator**. The generator produces electricity. The generation of electricity from water is called **hydroelectric** power.
Tidal barrages		Dams are built across river estuaries. The dams contain a series of gates. The gates are opened when the tide rises and closed at high tide. The sea is therefore trapped behind the dam .When the sea level outside the dam falls, the gate are opened. Sea water rushes through the open gates and is made to turn turbines. The turbines turn generators. The generators produce electricity.
Wave machines		The power stations are built on cliffs at narrow sea inlets. The up and down movement of the waves forces air to turn turbines. The turbines turn generators. The generators produce electricity.
Wind turbines		As the wind blows, the blade turns. The blade turns a turbine. The turbine turns a generator. The generator produces electricity. Large numbers of wind turbines are often trouped together to form a wind farm.

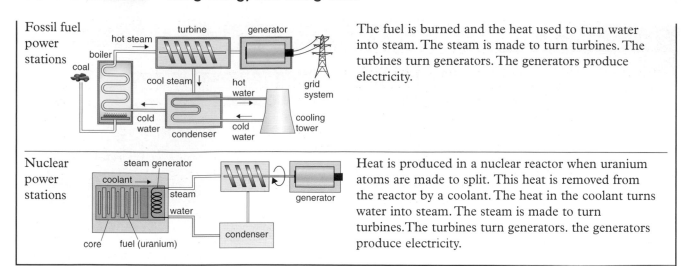

Fossil fuel power stations

The fuel is burned and the heat used to turn water into steam. The steam is made to turn turbines. The turbines turn generators. The generators produce electricity.

Nuclear power stations

Heat is produced in a nuclear reactor when uranium atoms are made to split. This heat is removed from the reactor by a coolant. The heat in the coolant turns water into steam. The steam is made to turn turbines. The turbines turn generators. the generators produce electricity.

Geothermal energy

The heat energy produced by the natural heating processes in the Earth is called **geothermal energy**. The heat energy is produced in the Earth by the decay of radioactive elements such as uranium. This heat energy is transferred to the lithosphere (see section 15.6) by convection currents in the magma. This heat energy is transferred to any water in the underlying rocks. In areas where there are volcanoes, geysers and hot springs this hot water can reach the surface as steam. This steam can be used directly to the drive turbines that turn the generators that generate electricity.

 # The advantages and disadvantages of using different energy sources to generate electricity

Fuel	Advantages	Disadvantages	Financial and economic considerations
Fossil fuels – coal, oil and natural gas	Relatively cheap start up and running costs. There are still large reserves in the world although they are getting more expensive to exploit.	When burnt they release carbon dioxide which is a greenhouse gas so will increase the rate of global warming. There is no way to stop the carbon dioxide entering the atmosphere. Burning coal and oil releases sulphur dioxide which causes acid rain. Sulphur can be removed from the fuels before they are burnt or the sulphur dioxide can be removed from the waste gases before they enter the atmosphere. Non-renewable energy resource.	In generating the same amount of heat, coal produces the most carbon dioxide and natural gas the least. Removing the sulphur or the sulphur dioxide increases the cost of the electricity generated. Coal-fired power stations and most oil-fired power stations need to be kept running all the time because the furnaces are damaged if they are allowed to cool down. Many gas-fired power stations can be switched on and off and so are often used when there is a sudden demand for electricity. About 30% efficient.

Fuel	Advantages	Disadvantages	Financial and economic considerations
Nuclear fuels	Do not produce gases that contribute to global warming or the production of acid rain. When working, normally, little or no radiation or radioactive materials are released into the environment.	Radioactive waste is produced and because much of it remains radioactive for a very long time this waste needs to be stored safely and securely, perhaps for thousands of years. In an accident radiation and radioactive materials may be released over a very wide area. Non-renewable energy resource.	Nuclear fuel is relatively cheap, but the cost of building a nuclear power station is very high, as is the cost of closing it down safely (decommissioning) when it is no longer needed. Must be kept running all the time. About 30% efficient.
Wind turbines/ generators	Use a renewable energy resource. Low running costs.	In large groups (wind farms) on hills or cliffs can be unsightly and can cause noise pollution.	The amount of electricity generated is dependent on the strength of the wind and because this varies they are not as reliable as most other sources of electrical energy. These are about 40% efficient at transferring the kinetic energy in the wind into electrical energy.
Tides	Use a renewable energy resource. Low running costs.	Because tidal barrages need to be built across the mouths of rivers they can cause problems for ships and disturb the flow of water. This may destroy the habitats of organisms such as wading birds and the mud-living organisms on which they feed.	Because the amount of electricity produced depends on the tides, which not only vary during each day but from month to month, they are not a reliable source of electrical energy.
Solar cells and solar panels	Use a renewable energy resource. Low running costs. An ideal energy source for producing electricity in remote areas, e.g. on satellites or where only small amounts of electricity are needed, e.g. calculators.	These have the highest start up costs per unit of electricity produced.	The amount of electricity produced depends on the amount of light falling on them. They are not a reliable source of electrical energy. Only about 15% efficient in transferring light energy into electrical energy.
Hydroelectric schemes	Low running costs. Use a renewable energy resource. They can be made to increase their efficiency by being operated in reverse during the night using surplus electricity from other power stations to pump water from the lower reservoir to the higher one.	Many of these schemes involve flooding river valleys – and that was probably not used for farming or forestry but was the habitat for numerous species.	Generally very reliable. Very short start up time, so are often used when there is a sudden demand for electricity. Suitable only for hilly areas with a reliable rainfall.

361

Summary

◆ The majority of ways in which electricity is generated involve steam/gases/ water turning a **turbine** that turns a **generator**.

◆ The generation of electricity from any energy resource will require choices to be made concerning environmental and economic issues.

Topic questions

1 In what ways is the production of electricity similar in pumped storage power stations, tidal barrages, wave machines and fossil fuel power stations.

2 In what ways is the production of electricity different in a fossil fuel power station and a nuclear power station?

3 Which two gases are produced from a coal or oil-fired power station? What harmful effect does each gas have on the environment?

4 What economic problems need to be considered before building a new nuclear power station?

5 Why does the waste from a nuclear power station need to be stored safely and securely?

6 Why are solar panels a useful energy source for remote areas?

7 How is the efficiency of a hydroelectric power station increased?

8 What environmental problems could be caused by building a new hydroelectric power station?

17.4 Work, power and energy

Co-ordinated	Modular
DA 12.19	DA 9/11
SA n/a	SA n/a

Work

Lifting a book from a table requires a small force to be used. To lift the table requires a larger force. In both cases the forces are making the objects move. Forces that cause movement are said to be doing **work**.

Because energy is transferred whenever work is done, energy and work have the same units.

The joule (J) is the unit of work and the unit of energy.

Energy and work are related by the following equation:

$$\text{work done (J)} = \text{energy transferred (J)}$$

This means that if 100 J of work are to be done, then 100 J of energy need to be transferred.

A lot of work will be done and a lot of energy will be required if a very heavy object is to be moved a very long distance.

The amount of work done in moving any object can be calculated using the equation:

$$\text{work done (J)} = \text{force (N)} \times \text{distance moved in the direction of the force (m)}$$
$$W = F \times d$$

Example: A force of 115 N is used to push a loaded wheelbarrow 100 m. Calculate the amount of work done by the force.

work done (J) = force (N) × distance moved in the direction of the force (m)
work done (J) =   115   ×   100
           = 11 500 J

Two people could be doing exactly the same amount of work by pushing identical wheelbarrows the same distance. The person who does the work the quickest is said to be the most powerful.

Power

Power is a measure of how quickly work is done or how quickly energy is being transferred. The unit of power is the watt (W).

The greater the power in a system, the more energy can be transferred in a given time.

Power can be calculated using the equation:

$$\text{power (watt, W)} = \frac{\text{work done (joule, J)}}{\text{time taken (second, s)}}$$

$$P = \frac{F \times d}{t}$$

Example: An electric motor is used to raise a load of 105 N. The motor takes 3 s to lift the load through a vertical distance of 2 m. Calculate the power of the motor.

$$\text{power (W)} = \frac{\text{work done (J)}}{\text{time taken (s)}}$$

$$\text{power (W)} = \frac{105 \times 2}{3}$$

$$\text{power (W)} = \frac{210}{3}$$

$$\text{power} = 70 \text{ W}$$

Example: An electric motor rated at 40 W lifts a load of 80 N through a vertical distance of 1 m in 4 s. Calculate the efficiency of the electric motor.

The equation needed is

$$\text{efficiency} = \frac{\text{useful energy transferred by device}}{\text{total energy supplied to device}}$$

The useful energy output is the energy transferred from the motor to the load.

This can be calculated using the equation:

work done (J) = force (N) × distance moved in the direction of the force (m)
work done (J) =   80   ×   1
           = 80 J

Using energy and doing work

The total energy input is the electrical energy supplied to the motor. This can be calculated by substituting in the equation:

$$\text{power (W)} = \frac{\text{work done (J)}}{\text{time taken (s)}}$$

$$40 = \frac{\text{work done}}{4}$$

So work done (energy transferred) $= 40 \times 4$
$$= 160\,\text{J}$$

To calculate the efficiency, substitute in:

$$\text{efficiency} = \frac{\text{useful energy transferred by device}}{\text{total energy supplied to device}}$$

$$\text{efficiency} = \frac{80}{160} = 0.5$$

Multiplying this by 100 produces an efficiency for the motor of 50%.

Example: An electric kettle is rated at 2000 W. The kettle takes 3.5 minutes to boil some water. In this time 336 000 J of energy are transferred to the water. Calculate the efficiency of the electric kettle.

The equation needed is

$$\text{efficiency} = \frac{\text{useful energy transferred by device}}{\text{total energy supplied to device}}$$

The useful energy output which is the energy transferred from the heating element to the water $= 336\,000\,\text{J}$.

The total energy input is the electric energy supplied to the kettle.

This can be calculated by substituting in the equation:

$$\text{power (W)} = \frac{\text{work done (J)}}{\text{time taken (s)}}$$

$$2000 = \frac{\text{work done}}{3.5 \times 60}$$

So work done (energy transferred) $= 2000 \times 210$
$$= 420\,000\,\text{J}$$

To calculate the efficiency substitute in:

$$\text{efficiency} = \frac{\text{useful energy transferred by device}}{\text{total energy supplied to device}}$$

$$\text{efficiency} = \frac{336\,000}{420\,000} = 0.8$$

Multiplying this by 100 produces an efficiency for the kettle of 80%.

Kinetic energy (KE)

Kinetic energy is the energy possessed by an object due to its motion. A fast moving car will have more kinetic energy than an identical slow moving car. A lorry with a large mass moving at 20 m/s will have more kinetic energy than a car with a small mass moving at 20 m/s.

In order to make the car or the lorry move, energy needs to be transferred from the fuel. The chemical energy in the fuel is transferred to kinetic energy in the moving vehicles.

The kinetic energy of a moving object depends on:

- its mass
- speed

The kinetic energy of a moving object can be calculated using the equation:

$$\text{kinetic energy} \;=\; \tfrac{1}{2} \;\times\; \text{mass} \;\times\; \text{speed}^2$$
$$\text{(joule, J)} \qquad\qquad \text{(kilogram, kg)} \quad [(\text{metre/second})^2, (\text{m/s})^2]$$
$$\text{k.e.} \;=\; \tfrac{1}{2}\,mv^2$$

Example: A runner of mass 75 kg runs at 10 m/s. What is the kinetic energy of the runner?

Substituting in the equation:

$$\text{kinetic energy} = \tfrac{1}{2}mv^2$$
$$= \tfrac{1}{2} \times 75 \times 10 \times 10$$
$$= 3750 \text{ Joules}$$

Gravitational potential energy (PE)

Gravitational potential energy is the energy stored in an object because of the height to which it is lifted against the force of gravity.

A stone thrown 20 m upwards into the air will return to the ground because of the force of Earth's gravitational field on the stone. To throw the stone 20 m into the air requires energy to be transferred from the muscles to the stone. This energy has to overcome the Earth's gravitational pull on the stone.

When the stone is at the top of the throw, all the energy in the stone is called gravitational potential energy (often called potential energy). The larger the stone or the higher it is thrown, the greater the amount of gravitational potential energy that has been transferred to the stone.

When a weight is lifted, work is done against gravity and the gravitational potential energy of the object increases. Because work is being done against gravity it is the vertical height through which the weight is lifted that is measured. So:

$$\text{change in gravitational potential energy} = \text{weight} \times \text{change in vertical height}$$
$$\text{(joule, J)} \qquad\qquad\quad \text{(newton, N)} \qquad\qquad \text{(metre, m)}$$

But:

$$\text{weight} = \text{mass} \times \text{gravitational field strength}$$
$$\text{(newton, N)} \quad \text{(kilogram, kg)} \quad \text{(newton per kilogram, N/kg)}$$

So,

$$\text{change in gravitational potential energy} = \text{mass} \times \text{gravitational field strength} \times \text{change in vertical height}$$
$$\text{(joule, J)} \qquad \text{(kilogram, kg)} \quad \text{(newton per kilogram, N/kg)} \qquad \text{(metre, m)}$$

This is often written as:

$$\text{g.p.e.} = mg\Delta h$$

On Earth the gravitational field strength is 10 N/kg.

Figure 17.12

Example: The weightlifter raises a bar of mass 70 kg to a height of 2.5 m.

Calculate the gravitational potential energy of the bar.

$$\text{g.p.e.} = mg\Delta h$$
$$\text{g.p.e.} = 70 \times 10 \times 2.5$$
$$= 1750\,\text{J}$$

The link between kinetic energy and gravitational potential energy

Figure 17.13 shows the energy transfers that take place when a pendulum swings. At position A, the pendulum is at the end of its swing and it has potential energy. As it moves to position B, its potential energy is transferred to kinetic energy. At C, all its energy has been transferred to kinetic energy.

The amount of transferred between gravitational potential energy and kinetic energy can be calculated.

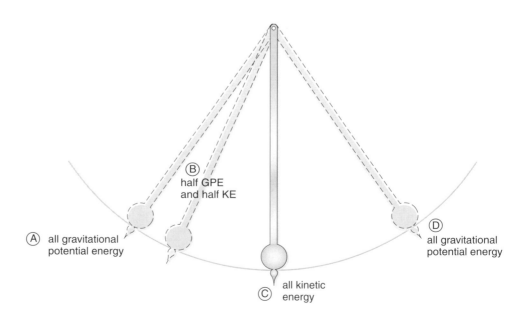

Figure 17.13
Kinetic energy and potential energy for a pendulum

Example: A stone of mass 1 kg is thrown vertically 30 m into the air. Calculate the gravitational potential energy transferred to the stone when it reaches the top of its path.

$$\text{gravitational potential energy} = mg\Delta h$$
$$\text{gravitational potential energy} = 1 \times 10 \times 30$$
$$= 300 \text{ J}$$

It is also possible to calculate the speed with which the stone was thrown into the air. This is because at the moment it was thrown, its speed was greatest. Therefore all its energy was kinetic energy. As it rose through the air it lost speed as all its kinetic energy was gradually transferred to gravitational potential energy. At the top of its flight, all its kinetic energy had been transferred to gravitational potential energy.

$$\text{So kinetic energy at the moment of being thrown} = \text{gravitational potential energy at the top of the throw}$$

$$\text{So,} \tfrac{1}{2}mv^2 = mgh$$

Substituting

$$\tfrac{1}{2} \times 1 \times v^2 = 1 \times 10 \times 30$$
$$0.5 \, v^2 = 300$$
$$v^2 = 600$$
$$v = 24.5 \text{ m/s}$$

Elastic potential energy

Elastic potential energy is the energy stored in an elastic object when work is done to change its shape. In Figure 17.14 energy from muscles is transferred to the stretched elastic cords.

Figure 17.14
Using a chest expander

Summary

◆ Energy is transferred when work is done.

◆ The joule (J) is the unit of work.

◆ Work done (J) = force (N) × distance moved in the direction of the force (m).

◆ Power is the rate of doing work.

◆ The watt (W) is the unit of power.

◆ Power (W) = power = $\dfrac{\text{work done (J)}}{\text{time taken (s)}}$

◆ **Kinetic energy** is the energy of motion.

◆ Kinetic energy = $\frac{1}{2}mv^2$

◆ **Gravitational potential energy** is the energy due to position.

◆ Gravitational potential energy = mgΔh.

◆ **Elastic potential energy** is the energy stored when work has been done to change the shape of an object.

Topic questions

1 What is the equation that links force, distance moved in the direction of the force and work done?

2 a) What are the units of work?
   b) How much work is transferred when a crane lifts a load of 5000 N through a distance of 30 m
   c) When does a force not do any work?

3 a) What is the equation that links power, time taken and work done?
   b) What are the units of power?

4 What power is being used in each of the following?

   a) A crane lifting a load of 5000 N through a distance of 30 m in 10 s.
   b) An electric motor lifting a load of 50 N 2 m in 5 s.
   c) A person weighing 550 N running up stairs in 2 s. The stairs are made of 15 steps each 16 cm high.

5 a) What energy is described as being the energy due to position?

6 A car travelling at a certain speed stops in a distance of 20 m once the brakes have been applied. If the road conditions remain the same, what will be the stopping distance once the brakes are applied if the car travels at:
   a) twice the original speed?
   b) three times the original speed?
   c) half the original speed?

7 Calculate the kinetic energy for a car of mass 1000 kg travelling at a speed of 25 m/s.

8 Calculate the gravitational potential energy of a mass of 3509 kg that has been lifted by a crane a distance of 20 m. (Take the gravitational field strength = 10 N/kg.)

9 A 0.75 kg ball is thrown vertically into the air. It reaches a height of 15 m. At what speed will it hit the ground?

Examination questions

1 a) Use words or phrases from the list to complete the sentences.

    elastic   frictional   gravitational
    less than   more than   the same as

    When a child goes down a slide the
    __friction__ force makes him go faster.
    On a damp day the child takes longer to go
    down the slide. This is because on a damp day
    the force of friction is _____ on a dry
    day.   *(2 marks)*

   b) Use words or phrases from the list to complete the sentence.

    elastic   gravitational potential   sound
    kinetic (movement)   light   thermal (heat)

    When the child goes down the slide, the energy
    transfers are from _____ energy to
    _____ energy and _____ energy.
          (3 marks)

2 a) The list gives energy resources which can be used to produce electricity.

    coal   gas   nuclear fuel   oil
    sunlight   tides   waves   wind   wood

    Circle the **four non-renewable** energy resources.
          (4 marks)

   b) Use words from the list to complete the sentences about generating electricity.

    energy   gas   generator   smoke
    steam   transformer   turbine   water

    In a coal-fired power station, coal is burnt to
    release _____ . This is used to change
    _____ into _____ which drives a
    _____ . Electricity is produced by a
    _____ .   *(5 marks)*

3 a) A weightlifter has lifted a weight of
    2250 newtons above his head. The weight is
    held still.

    i)   In the box are the names of three forms of
        energy.

gravitational potential   kinetic   sound

        Which one of these forms of energy does
        the weight have?   *(1 mark)*

    ii) What force is used by the weightlifter to
        hold the weight still?
        Give a reason for your answer.   *(2 marks)*

   b) To lift the weight, the weightlifter does
    4500 joules of work in 3.0 seconds.
    Use the following equation to calculate the
    power developed by the weightlifter. Show
    clearly how you work out your answer.

$$\text{power} = \frac{\text{work done}}{\text{time taken}}$$

          (2 marks)

4 The diagram shows a high jumper.

In order to jump over the bar, the high jumper
must raise his mass by 1.25m.
The high jumper has a mass of 65kg. The
gravitational field strength is 10N/kg.

a) The high jumper just clears the bar.
    Calculate his gravitational potential energy.
          (4 marks)

b) Calculate the minimum speed the high jumper
    must reach for take-off in order to jump over
    the bar.   *(3 marks)*

5 The drawing shows an investigation using a model steam engine to lift a load.
   In part of the investigation, a metal block with a weight of 4.5 N was lifted from the floor to a height of 90 cm.

   a) i) Calculate the work done in lifting this load. Write the equation you are going to use, show clearly how you get to your answer and give the unit. *(3 marks)*

       ii) How much useful energy is transferred to do the work in part c) i)? *(1 mark)*

   b) In another part of the investigation, 250 J of work is done in one minute.

       Calculate the useful power output. Give the unit. *(2 marks)*

6 State and explain the advantages and disadvantages of using nuclear power stations to produce electricity. *(4 marks)*

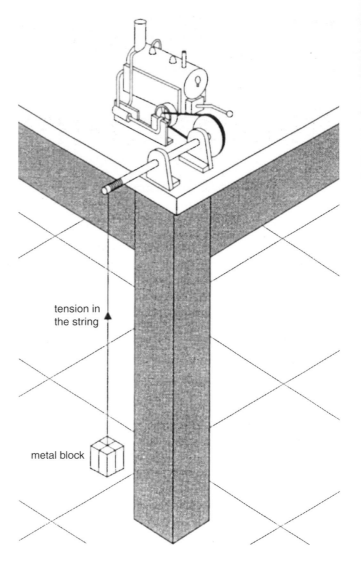

tension in the string

metal block

Chapter 18
Radioactivity

Key terms   activity rate • alpha • atom • atomic number • background radiation • beta • chain reaction • cosmic ray • count rate • decay • electrons • electromagnetic spectrum • element • gamma • Geiger–Müller tube • half-life • ionise • isotope • mass number • neutrons • nuclear fission • nucleon • nucleus • proton • radiation • radioactive dating • radioactive decay • radioactive emissions • radioactive tracer • radioactivity • radioisotope • radionuclides • random

18.1   Types, properties and uses of radioactivity

Co-ordinated	Modular
DA 12.22	DA 12
SA 12.14	SA 18

Radioactivity is the name given to the particles and rays that come from the unstable **nuclei** of certain elements. Radioactivity cannot be seen, heard or felt but it will affect photographic film. This is how it was first discovered in 1896. The rays and particles are emitted at **random** from a radioactive substance. Heating the substance or trying to dissolve it or combining it chemically with another will not affect its radioactivity.

As an unstable nucleus emits a particle or ray, it changes. This process is called **radioactive decay**.

Figure 18.1
Unstable nucleus emitting a particle and a ray

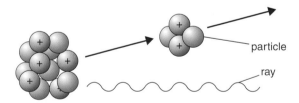

particle

ray

Detecting radioactivity

The average number of emissions in a certain time is called the **activity rate** or **count rate**. This rate depends only on the number of **atoms** of the particular radioactive element present. As the unstable nuclei **decay**, the activity rate decreases because fewer atoms of the original radioactive element are left.

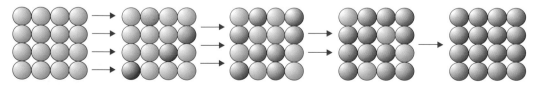

Figure 18.2
As unstable atoms break down, the level of radiation falls

   unstable (radioactive atom)   stable (decay product)   → radiation

Radioactivity

The activity rate can be measured by a **Geiger–Müller tube** and counter.

Figure 18.3
A Geiger–Müller tube counting radiation

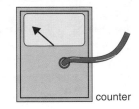

source   absorber   Geiger–Müller tube   counter

Figure 18.3
A Geiger–Müller tube counting radiation

Figure 18.4
A radiation-detecting badge

People who work with radioactive substances wear a special badge containing photographic film. The badge is processed regularly to measure the amount of **radiation** received. This is done to check that the badge wearer has not been exposed to too much radioactivity. The more the exposure to radiation, the darker the film appears after it has been developed.

Did you know?

Henri Becquerel discovered the radioactivity of uranium in 1896. He was awarded the Nobel Prize for Physics in 1903 jointly with Pierre and Marie Curie. The Curies isolated the radioisotopes radium and polonium from uranium ore. They worked on a small budget in dilapidated surroundings. Marie Sklodowska Curie continued this work after her husband's death. She received the Nobel Prize for Chemistry in 1911.

Henri Becquerel   *Pierre and Marie Curie*

Sources of radiation

Radioactive emissions are classed as either natural or man-made. There is a lot of radioactivity from natural sources such as the rocks in the Earth's crust and from deeper inside the Earth. One of the radioactive elements released from rocks is radon gas. This can accumulate to dangerous levels in confined spaces. Certain surface rocks, such as the Cornish and Aberdeen granites, show more activity than others.

Cosmic rays, which constantly bombard the Earth from space, are a major source of the radioactivity in the air.

Man-made sources are the results of the processing and use of radioactive materials in industry, medicine, nuclear reactors and some weapons. Highly radioactive materials are strictly contained and their disposal is controlled.

Radioactive waste material from nuclear reactors, factories and hospitals is either specially stored or, if it has little radioactivity, it is diluted and discharged into the environment. These man-made sources make only a small contribution to the natural sources of radioactivity.

Figure 18.5
Different sources of radiation: a) medical (mainly X-rays), b) air flights increase exposure to cosmic rays

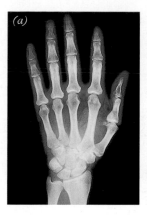

So the air we breathe, the food we eat, the rocks around us, even our bones, become sources of radioactivity. These radiations which are around us all the time are called **background radiation**. The level of background radiation changes from place to place but is always quite low. Occasionally people receive more radiation as a result of their work or from medical treatments. Nuclear accidents are rare but do raise radiation levels at those sites.

Figure 18.6 ▲
The storage of radioactive material

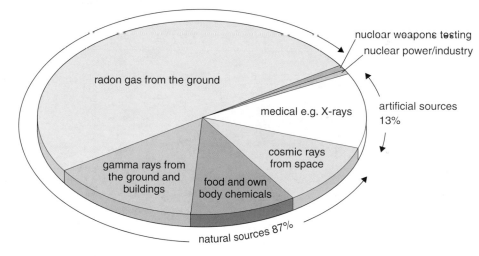

Figure 18.7 ▲
Typical radiations for the UK (NRPB)

Did you know?

The Chernobyl nuclear reactor became supercritical when its reaction went out of control in April 1986. Eventually it caught fire. The remnants are still very radioactive and will be for a long time yet. The shield of thick concrete extends under the reactor. Despite the hazards, courageous engineers continue to monitor the levels of radioactivity and check the safety of this concrete 'tomb'.

Types of emission

When radioactivity was first discovered, the identity of the different particles and rays was not known. They were called **alpha** (α), **beta** (β), and **gamma** (γ). (These are the first three letters of the Greek alphabet.) What they are and how they behave is now better understood.

Alpha particles are larger and are quickly stopped by thin paper or a few centimetres of air. Alpha particles easily **ionise** atoms by knocking out the outer **electrons** from the atoms of the absorbing material. With each collision the alpha particle loses energy and so it does not travel far. It will not travel more than 5 cm in air.

Beta particles are very much smaller and they travel further at high speeds through other substances before losing their energy by colliding with the atoms of that substance. They ionise matter less well than alpha particles as they are so small. Beta particles can be stopped by a 5 mm thick sheet of aluminium.

Gamma rays are not charged particles like alpha and beta particles, but are high-energy waves belonging to the electromagnetic spectrum. They are very penetrating as they are poor ionisers and do not lose their energy quickly. Gamma radiation is never entirely stopped, but can be reduced to almost zero levels by a 10 cm thick sheet of lead or metres of concrete.

Figure 18.8

Selective absorption of radioactive emissions

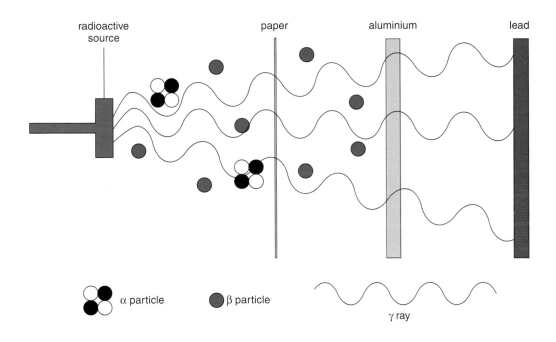

When radioactive substances need to be shielded they are placed in thick lead cans or stored behind several metres of concrete. This blocks almost all the radiation.

Gamma rays have very short wavelengths, shorter than those of X-rays. Both types of wave belong to the electromagnetic spectrum, both are penetrating and poor ionisers, but they are produced in different ways. X-rays are not radioactive emissions, as they do not come from the nucleus of the atom.

Figure 18.9
Radioactive material is transported in this thick metal container

Identifying emissions

The type of emission from a source can be found by measuring the radioactivity level with various absorbers placed between the source and the detector.

The results in Figure 18.10 show that paper does not reduce the count rate very much, so there are no alpha particles emitted. Aluminium reduces the count rate considerably – this suggests that beta particles are emitted. Lead reduces the count rate down to almost the level of background radiation (0.4 counts/second) so gamma rays are emitted.

Figure 18.10
Identifying radiation from an unknown source, X

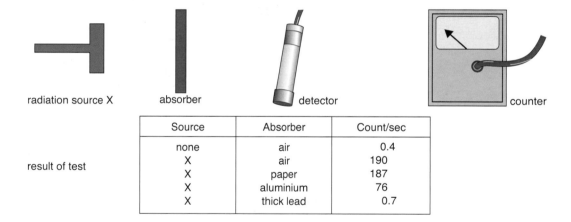

radiation source X   absorber   detector   counter

result of test

Source	Absorber	Count/sec
none	air	0.4
X	air	190
X	paper	187
X	aluminium	76
X	thick lead	0.7

Radiation and living cells

Particles and rays come out of the nucleus with great energy allowing them to pass through other substances. As the radiation is absorbed, it ionises molecules of that substance, causing the substance to change. This is why radioactivity can be dangerous to living cells, because their chemical structure, particularly their DNA, is altered and those cells do not work properly any longer. In certain cases, the cells are so changed that they become a threat to the living organism, affecting the normal cells and causing cancers (see section 15.3).

Radioactive sources are handled according to strict government guidelines. Great care is taken to prevent penetrating emissions, such as beta particles or gamma rays, from reaching the body from an external source. Protective clothing and special gloves stop beta and gamma radiation travelling through the skin and reaching the internal organs.

Radioactivity

Alpha particles are unlikely to penetrate the skin. However if a radioactive source is taken internally, then alpha particles are most dangerous as they are powerful ionisers of matter.

Uses of radiation

Harmful bacteria in food can be killed by exposing the fresh food to gamma radiation. This prevents the bacteria from multiplying and spoiling the food. Germs on hospital instruments and dressings, which could infect patients, are made harmless by gamma radiation. This method of sterilising surgical instruments is useful when boiling in water could damage the instrument.

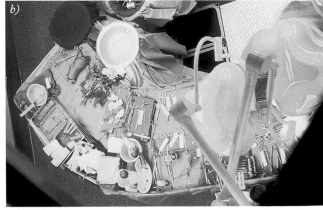

Figure 18.11
Some uses of radiation:
a) harmful bacteria in food can be killed by gamma rays,
b) operating equipment is sterilised by gamma rays

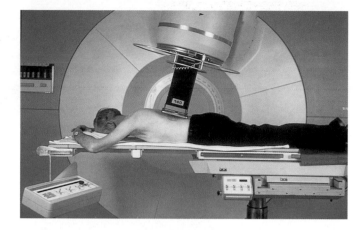

Radiation can be used to cure cancer. It may seem strange that radioactivity can cause cancer and cure it. In radiotherapy large doses of radioactivity are carefully given from an external source of gamma rays so the cancerous area receives the most radiation and nearby healthy tissue is less affected (see section 15.3).

Figure 18.12
Radiotherapy apparatus

For some conditions the radioactive isotope is given internally. For this method to work, the radioactive isotope must go to the particular part to be treated. The substance used is usually a beta emitter so that the effects are localised. Gamma rays would penetrate too far and could affect healthy tissue.

Figure 18.13
Iodine-131 localised in the thyroid gland

Radioactive sources are commonly used in medical tests. Injecting a person with iodine-131 can show if that person has an enlarged thyroid gland. Radioactive isotopes with fast decay rates are chosen for internal doses. This is to prevent long-term radiation affecting normal, living cells.

Industrial uses of radiation

If a **radioactive tracer** is put in a fluid, the path of the fluid can be followed even when it cannot be seen. Gamma rays from the tracer will penetrate pipes and soil to reach a detector. Such tracers are used to test liquid flow rates through pipes, the dispersion of effluents into larger quantities of water, the uptake of fertilisers by plants and leaks in underground pipelines (Figure 18.14).

The tracers are chosen with care not to damage the environment. The element is usually non-toxic, it is well diluted and its radioactivity decays quickly.

Radioactive isotopes can also be used industrially to control the quality of material. For example the thickness of an aluminium sheet can be controlled by measuring how much beta radiation passes through it. If the amount of radiation is too high, the sheet is too thin. The position of the rollers can then be automatically adjusted to produce the correct thickness (Figure 18.15).

Radioactivity

Figure 18.14
Using radioactive tracers to locate a leak in a pipe

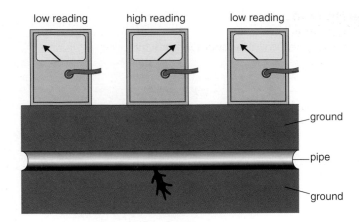

Figure 18.15
Using a beta source to control the thickness of an aluminium sheet

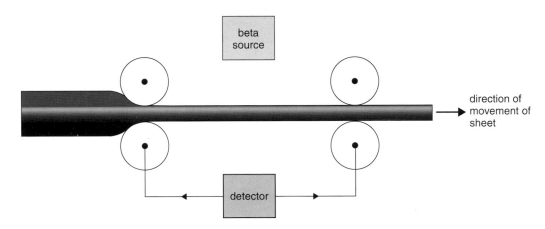

Summary

◆ **Radioactivity** is the random emission of particles and rays from an unstable **nucleus**. It can be detected by photographic film or a **Geiger–Müller tube**.

◆ The radiation from space, food , rocks, air and buildings is called **background radiation**.

◆ Radiation can damage living cells.

◆ Radiation has a number of medical and industrial uses.

Type of emission	Nature	Penetrating ability
alpha	large, + charged particle	5 cm in air; stopped by paper
beta	small, − charged particle	50 cm in air; stopped by 5 mm thickness of aluminium foil
gamma	high frequency electromagnetic wave	intensity falls with distance; blocked by 10 cm thickness of lead

Topic questions

1 Fill in the missing words in the sentences, using some of these words

> temperature   stable   different   chemical
> unstable   particle   random   atom   nucleus

Radiations come from an _____ nucleus. The activity is _____ and cannot be affected by _____ , pressure or _____ reactions. When a _____ leaves a nucleus, the _____ changes to that of a _____ atom.

2 Can you find the radioactive elements from these nonsense words?

MONOPULI   DIAMUR   RUINAUM
DROAN   RIMOUTH

3 Complete these sentences.
a) The activity of all radioactive materials _____ with time.
b) _____ particles travel a short distance in air and are easily _____ .

c) _____ rays travel from space and are a _____ source of radiation.
d) Radioactive _____ are used in medicine to treat _____ .

4 Put the following sources of radiation into a table, under the correct heading, Natural or Man-made:

building materials, cosmic rays, X-rays, plants, radon gas, nuclear waste.

5 A source was examined for the type of emissions it produced. Find the emissions from the results in table below.

no absorber	360 counts/sec
paper absorber	180 counts/sec
aluminium absorber	185 counts/sec
lead absorber	3 counts/sec

18.2   Atomic structure and radioactivity

Co-ordinated	Modular
DA 12.23	DA 12
SA 12.15	SA 18

Radioactivity is the result of changes in the nuclei of atoms. Only after the structure of the atom had been discovered could the various forms of radiation be explained.

The story of the atom – John Dalton to Ernest Rutherford

In the early part of the 19th century John Dalton had proposed that the smallest part of an element was a tiny solid indivisible particle called an atom (see section 7.1). In 1897, J J Thomson discovered the existence of a particle much smaller than an atom. This particle was called an electron. Thomson knew that because electrons have a negative charge the rest of the atom should have a positive charge if the atom was to be held together. In 1904 he proposed what has been called the 'plum-pudding' model of the atom. It was given this name because the model looked like a pudding, which represented a sphere of positively charged material, with bits of plum, representing the negatively charged electrons, scattered in the pudding.

In 1906, Ernest Rutherford discovered the existence of alpha particles. These particles had a double positive charge and so were thought to be the positive parts of the Thomson atom.

Ernest Rutherford investigated his idea for the structure of the atom with Hans Geiger and Ernest Marsden, at Manchester University between 1909–1911. They fired a thin beam of alpha particles, which they knew had a double postive charge, at very fine gold foil (Figure 18.17). Instead of passing straight through the foil, as would be expected if the atom was like the one proposed by Thomson, the alpha particles were scattered like the light from a torch.

blob of positive charge

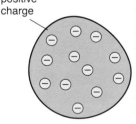

Figure 18.16
J J Thomson's 1904 'plum pudding' model of the atom

379

Figure 18.17
How they set up their experiment

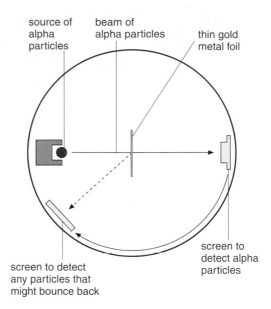

Figure 18.18
What the results showed

The results after many months, during which over 100 000 measurements were made, showed that whilst most alpha particles went straight through the foil, a few were deflected through quite wide angles and some even bounced back towards the source. These unexpected results came as a surprise.

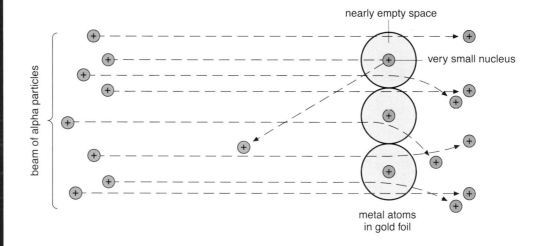

Rutherford decided that:

● The alpha particles which were not deflected had been unaffected by the atoms in the gold foil and were going through empty space. Because of this the atoms in the foil must contain a lot of empty space.

● The positively charged alpha particles that were deflected had been repelled by the positively charged matter within each gold atom. Because of this the parts of the atom containing the positive charges must be concentrated in a very small, dense part of the atom – the nucleus.

● Any alpha particle that had bounced back must have collided with the nucleus of one of the gold atoms. Because few alpha particles bounced back the nucleus must be very small.

● The electrons orbited around in the empty space surrounding the nucleus.

Rutherford concluded that atoms are made up of a dense positively charged nucleus around which was empty space in which orbited the negatively charged electrons. He also concluded that the amount of positive charge in the nucleus should equal the negative charge on the electrons. Rutherford's atom was the same size as Thomson's but contained mainly empty space.

Figure 18.19
The Rutherford model of the atom

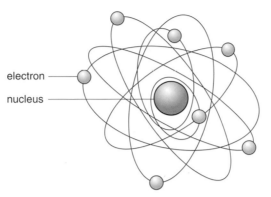

electron

nucleus

The Rutherford Model of an atom
A miniature solar system with the electrons moving like planets around the nucleus.

Rutherford presented his findings on the structure of the atom in 1911. The clear objectives and thorough methods of his experiments convinced many scientists that his idea for a nuclear atom was correct. Niels Bohr incorporated Rutherford's conclusions into a theoretical framework that explained how the electrons were able to orbit a central positive nucleus. Bohr used the light spectrum of hydrogen, the simplest atom, to confirm that electrons were able to orbit the nucleus without falling into it. The nuclear atom became accepted as the correct structure for the atom, and it is still the view we have today, with some modified details.

Did you know?

Joseph John Thomson, a British physicist (1856–1940) found the charge to mass ratio for the electron. He won the Nobel prize for physics in 1906.

Joseph John Thomson

Ernest Rutherford (1871–11937) was born in New Zealand of a Scottish father. He was considered the greatest experimental physicist of his generation and won the Nobel prize for chemistry in 1908.

Ernest Rutherford

381

Hans Geiger (1882–1945), a German nuclear physicist who took part in many of Rutherford's experiments, also developed a better instrument for detecting radiation, called the Geiger–Müller tube in 1928.

Hans Geiger

Ernest Marsden (1889–1970) a British physicist.

Ernest Marsden

Niels Bohr (1885–1962), a Danish theoretical physicist, the 'father of atomic theory'. He won the Nobel prize for physics in 1922.

Niels Bohr

Radioactivity and the atom

Atomic structure

The nucleus of an atom contains **nucleons**, which are closely packed **protons** and **neutrons**. The protons have a positive charge and the neutrons have no charge (see section 7.1). The electrons orbiting the nucleus each have a negative charge of the same size as that on a proton. The number of protons and electrons is the same, making a single atom electrically neutral. Atoms of the same elements which have different numbers of neutrons but the same number of protons are called **isotopes**.

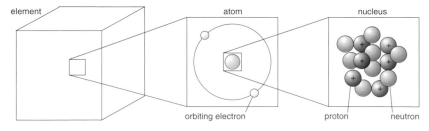

Figure 18.20
Atomic structure

The mass of a neutron is almost the same as that of a proton, and these masses are 2000 greater than that of an electron. As these are such tiny particles, it is easier to talk of their mass in units, which compare the mass of each sub-particle, to that of the simplest atom hydrogen.

On this scale:

sub-atomic particle	mass	charge
proton	1	+1
neutron	1	0
electron	negligible	−1

Because of the work on atomic structure it is now known that radioactivity occurs because of changes to the unstable nuclei of atoms. Atoms with unstable nuclei are called radioactive isotopes (or **radioisotopes** or **radionuclides**). Radiation is produced when an unstable nucleus emits a particle or ray (or both) in order to become more stable. Once a particle has been emitted by the nucleus, then the nucleus changes to that of another element. The number of protons in the new nucleus is different.

The forces which hold the nucleus together are very strong, so any particles that come from a nucleus have large amounts of energy – rather like a bullet from a gun.

Figure 18.21
Unstable nucleus emitting a particle and a ray

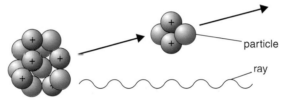

The rays and particles are emitted totally at random from a radioactive substance.

Types of emission

Alpha and beta particles

Alpha particles are the largest particles to be emitted from an unstable nucleus. As they pass through a substance, for example air, the large alpha particles collide with the air particles. Alpha particles consist of two protons and two neutrons. This is the same structure as the nucleus of helium.

Figure 18.22
The structure of alpha and beta particles

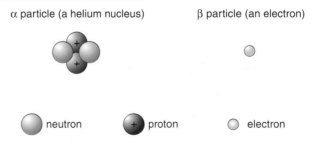

α particle (a helium nucleus)   β particle (an electron)

neutron   + proton   electron

Each alpha particle has a **mass number** of 4 and an **atomic number** of 2. The atomic number is the number of protons, and the mass number, or nucleon number, is the number of protons plus the number of neutrons (see section 7.1). When an alpha particle is emitted from an unstable nucleus, the atomic number of the original nucleus goes down by 2. This produces the nucleus of the element two places lower in the periodic table (section 11.2). The mass number decreases by 4. For example, the nucleus of a radium atom decays into radon and an alpha particle. This is represented by the following equation:

$$^{226}_{88}\text{Ra} \rightarrow {}^{222}_{86}\text{Rn} + {}^{4}_{2}\text{H}$$

383

Beta particles are fast moving electrons. As an electron is emitted from an unstable nucleus, a neutron changes to a proton. The emission of a beta particle produces a nucleus of an element one place higher in the periodic table (see section 11.2). For example, the nucleus of a radioactive carbon atom decays into nitrogen and an electron. This is represented by the following equation:

$$^{14}_{6}\text{C} \rightarrow \, ^{14}_{7}\text{N} + \, ^{0}_{-1}\text{e}$$

Gamma radiation

Gamma radiation is short wavelength electromagnetic radiation (see section 15.3). It is emitted when an unstable nucleus loses excess energy. The emission of gamma radiation causes no change to either the mass number or the atomic number of the nucleus.

Summary

◆ Radiation occurs because of changes in atoms with unstable nuclei.

◆ The work of Rutherford helped to develop the modern ideas about the structure of the atom.

◆ Alpha particles are helium nuclei – two protons and two neutrons.

◆ Beta particles are electrons that come from the nucleus. Each electron released changes a neutron into a proton.

◆ Gamma radiation is high energy, short wavelength electromagnetic waves.

Topic questions

1   Match the conclusion with the observed result for Rutherford's experiments on alpha scattering.

Conclusion	Observations
Atoms mainly empty space	Curved paths of deflected particles
Concentrated positive nucleus	Most particles passed straight through
Electrostatic repulsion laws obeyed	Some particles bounced back with no energy loss

2   What are the differences between the atomic models of Thomson and Rutherford?

3   Complete the gaps in the following table about sub atomic particles.

sub-atomic particle	mass	charge
	1	+1
neutron		
	negligible	

4   What is a radioisotope?

5   a) An isotope of radium is written as $^{226}_{88}$Ra. Find the proton and neutron numbers.
    b) An isotope of calcium (Ca) has 20 protons and 24 neutrons. Write this in symbol form.

6   Complete the gaps in the table about radioactive emissions.

type of emission	what it is	charge	absorbed by	causes ionisation
	two neutrons and two protons		_____ of air	very strong
beta particle			_____ of aluminium	
gamma ray		no charge	_____ of lead	

18.3   Half-life

Co-ordinated	Modular
DA 12.22	DA 12
SA n/a	SA n/a

The **half-life** of a radioisotope (radionuclide) is the time taken for half the unstable atoms to decay. That is the time it takes for the activity rate to fall to half its original value. It doesn't matter how much of the radioisotope there is to start with, the half-life is the same. Isotopes with a short half-life give off their radiation more quickly. Therefore they are more dangerous. Figure 18.23 below gives some data for the radioactive decay of an isotope with a half-life of two hours.

Figure 18.23

	Activity in counts/min	Time in hours		
	1800	0		
	1300	1		
(1800/2)	900	2	one half-life	2 hours
	650	3		
(900/2)	450	4	two half-lives	2 × 2 = 4 hrs
	320	5		
(450/2)	225	6	three half-lives	3 × 2 = 6 hrs
	160	7		

These data are shown as a graph in Figure 18.24.

Figure 18.24
The radioactive decay curve for a substance with a half-life of 2 hours

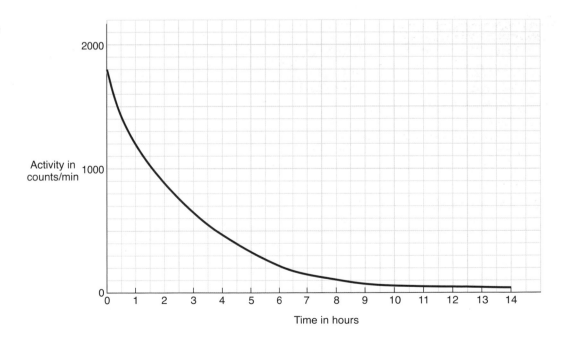

Sources which are very radioactive have a fast decay rate, which means they have a short half-life. Less active sources have a slow decay rate, and a long half-life.

Radioactive dating

The radioactive isotope, carbon-14, is a tiny part of all the carbon taken in by living organisms, it is mostly carbon-12. The proportion of carbon-14 to carbon-12 is constant during the lifetime of an organism. When the organism dies, this fraction decreases as the carbon-14 decays. By comparing the activity rate for the carbon of a dead sample, to that for living things, the age of the sample can be found. The half-life for carbon-14 is 5730 years. Items over 10 000 years old are not reliably dated by their activity as it is too small to detect and recent items have an activity rate similar to that for living things.

Another method measures the proportion of carbon-14 atoms to carbon-12 atoms directly. This method destroys less of the original sample and can date items up to 100 000 year of age, but it needs special equipment.

Figure 18.25
These human remains were analysed by carbon dating

Rocks can be dated by comparing the proportion of uranium and lead atoms in a sample. The half-life of uranium-238 is 4500 million years so this radioisotope can be used to date some of the Earth's oldest rocks.

The radioisotope potassium-40 decays to produce argon as the stable product. The proportion of potassium-40 and argon can be used to date igneous rocks in which the argon has been trapped.

Summary

◆ Very active sources decay very quickly.

◆ The **half-life** of a **radioisotope** is the time taken for half the unstable nuclei to decay to nuclei of different atoms, or for its activity rate to fall to half its original value. A short half-life indicates a very active source.

◆ Radioactive decay can be used to date materials.

Topic questions

1 A smoke detector uses a radioisotope to ionise the air inside it. This allows a small electric current to flow. This allows a small electric current to flow. In the presence of smoke, the current falls and sets off the alarm.
   a) Which is the best choice of radioisotope given in the table below?
   b) Explain your choice.

Radioisotope	Emission	Decay rate
cobalt-60	gamma	fast
americium	alpha	slow
iodine-131	beta	very fast
technetium-99	gamma	very fast

2 The table shows the rate of decay for a radioisotope. What is the half-life of this radioisotope. You may plot a graph.

Activity in counts/min	Time in days
8000	0
6100	10
4600	20
3500	30
2700	40
2000	50
1500	60
1200	70
870	80

3 The half life of carbon-14 is 5700 years. The activity of the carbon in a living sample is 15 counts/mm. An ancient axe-handle sample of the same mass gives an activity of 3.75 counts/mm. What age is this handle?

4 A radioisotope of lead has a half-life of 10.6 hours. A small sample of lead containing the isotope has a count of 6000 counts per minute. How long will it be before the count rate reaches 375 counts per minute?

18.4   Nuclear fission

Co-ordinated	Modular
DA 12.23	DA 12
SA n/a	SA n/a

Producing the heat energy in a nuclear reactor

Atoms, such as those of uranium-235, are unstable. If a slow moving neutron is captured by an atom of uranium-235 then the large nucleus breaks into two large parts and some neutrons. This splitting up releases a large amount of energy and is called **nuclear fission**. This energy is very many times greater than the energy released in a chemical reaction when bonds between two atoms are made. The fission of 1 kg of uranium-235 releases more energy than the burning of 2 000 000 kg of coal. It is the energy released from such a chain reaction that provides the heat energy in the core of a nuclear power station.

Figure 18.26
A neutron hitting the nucleus of an atom of uranium-235

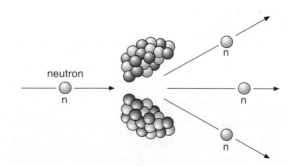

neutron

n

n

n

n

Radioactivity

The released neutrons may then go on to to be captured by more uranium atoms. This makes them split and fire off more neutrons which may be captured by other atoms and so on. This is called a **chain reaction**. New atoms formed by nuclear fission are radioactive.

Figure 18.27
The chain reaction

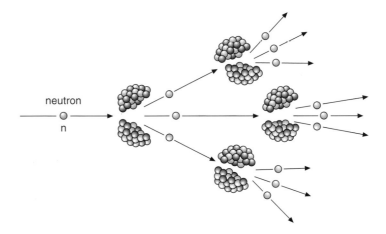

The products of nuclear fission are themselves highly radioactive. When the uranium atom splits, it usually does so in two unequal parts, with a few neutrons. The two larger parts carry more neutrons than their stable isotopes, so they emit particles rapidly. These products are shielded and cooled whilse the more active elements decay. Then they are sorted in protective containers before being reprocessed.

Summary

◆ In a nuclear power station the heat is produced by **nuclear fission**.

◆ During nuclear fission a vast amount of energy is released together with two smaller nuclei and several neutrons. The neutrons can split nuclei of other large atoms creating a **chain reaction**.

Topic questions

1 Describe what happens to an atom of uranium-235 when it captures a slowmoving neutron and how this leads to a chain reaction.

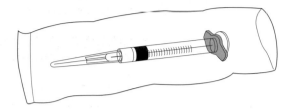

Examination questions

1 a) The different sources of radiation to which we are exposed are shown in the pie chart. Some sources are natural and some artificial.

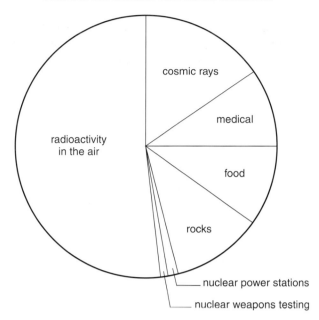

i) Name *one* natural source of radiation shown in the pie chart. *(1 mark)*

ii) Name *one* artificial source of radiation shown in the pie chart. *(1 mark)*

b) A radioactive source can give out three types of emission: alpha particles, beta particles, gamma radiation.
The diagram shows the paths taken by the radiation emitted by two sources, X and Y. What types of radiation are emitted by each of the sources? *(2 marks)*

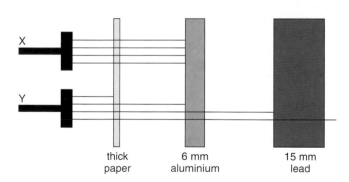

c) The diagram shows a disposable syringe sealed inside a plastic bag. After the bag has been sealed the syringe is sterilised using radiation. Explain why radiation can be used to sterilise the syringe. *(3 marks)*

2 a) The diagram shows the apparatus used by a teacher to investigate an alpha (α) source.

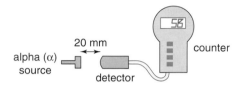

i) Which piece of apparatus could be used as a radiation dectector?

| Geiger-Müller Tube   Oscilloscope   Voltmeter |

(1 mark)

ii) Complete the following sentence.
When a piece of paper is placed between the detector and the alpha source the count rate will go _____ .

(1 mark)

b) Two sheets of steel were joined together by welding.

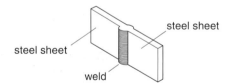

Radiation was used to check how well the welding had been done.

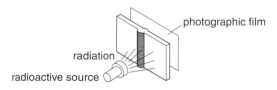

i) Which type of radiation should be used? Give a reason for your answer. *(2 marks)*

ii) The diagram shows the exposed photographic film.

Does the photographic film show that the weld was good or bad? Give a reason for your answer. *(2 marks)*

3 The diagram shows a film badge worn by people who work with radioactive materials. The badge has been opened. The badge is used to measure the amount of radiation to which the workers have been exposed.

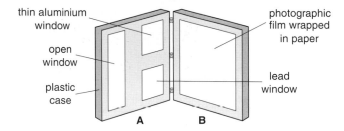

The detector is a piece of photographic film wrapped in paper inside part **B** of the badge. Part **A** has "windows" as shown.

a) Use words from the list to complete the sentences.

alpha   beta   gamma

When the badge is closed

i) _____ radiation and _____ radiation can pass through the open window and affect the film. *(1 mark)*

ii) _____ radiation and _____ radiation will pass through the thin aluminium window and affect the film. *(1 mark)*

iii) Most of the _____ radiation will pass through the lead window and affect the film. *(1 mark)*

b) Other detectors of radiation use a gas which is ionised by the radiation.

i) Explain what is meant by *ionised*. *(1 mark)*

ii) Explain why ionising radiation is dangerous to people who work with radioactive materials. *(2 marks)*

4 a) The table gives information about five radioactive isotopes.

Isotope	Type of radiation emitted	Half-life
Californium-241	alpha (α)	4 minutes
Cobalt-60	gamma (γ)	5 years
Hydrogen-e	beta (β)	12 years
Strontium-90	beta (β)	28 years
Technetium-99	gamma (γ)	6 hours

i) What is an alpha (α) particle? *(1 mark)*

ii) What is meant by the term half-life? *(1 mark)*

iii) Which **one** of the isotopes could be used as a tracer in medicine? Explain the reason for your choice. *(3 marks)*

b) The increased use of radioactive isotopes is leading to an increase in the amount of radioactive waste. One method for storing the waste is to seal it in containers which are then placed deep underground.

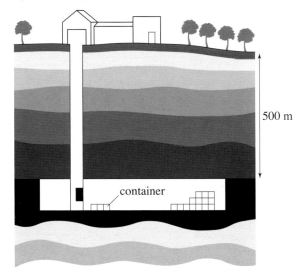

Some people may be worried about having such a storage site close to the area in which they live. Explain why. *(3 marks)*

5 a) The graph shows how a sample of barium-143, a radioactive *isotope* with a short *half-life*, decays with time.

i) What is meant by the term *isotope*? *(1 mark)*

ii) What is meant by the term *half-life*? *(1 mark)*

iii) Use the graph to find the half-life of barium-143. *(1 mark)*

b) Humans take in the radioactive isotope carbon-14 from their food. After their death, the proportion of carbon-14 in their bones can be used to tell how long it is since they died. Carbon-14 has a half-life of 5700 years.

i) A bone in a living human contains 80 units of carbon-14. An identical bone taken from a skeleton found in an ancient burial ground contains 5 units of carbon-14. Calculate the age of the skeleton. Show clearly how you work out your answer. *(2 marks)*

ii) Why is carbon-14 unsuitable for dating a skeleton believed to be about 150 years old? *(1 mark)*

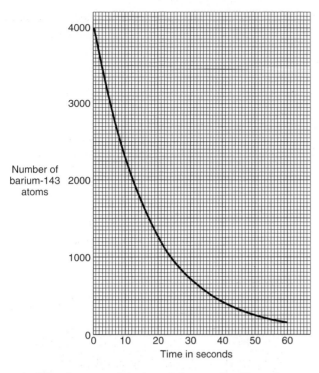

Number of barium-143 atoms

Time in seconds

c) The increased industrial use of radioactive materials is leading to increased amounts of radioactive waste. Some people suggest that radioactive liquid waste can be mixed with water and then safely dumped at sea. Do you agree with this suggestion? Explain the reason for your answer. *(3 marks)*

6 The isotope of sodium with a mass number of 24 is radioactive. The following data were obtained in an experiment to find the half-life of sodium-24.

Time in hours	Count rate in counts per minute
0	1600
10	1000
20	600
30	400
40	300
50	150
60	100

a) Draw a graph of the results and find the half-life for the isotope. On the graph show how you obtain the half-life. *(4 marks)*

b) Sodium-24 decays by beta emission. The G-M tube used in the experiment is shown in the diagram. Each beta particle which gets through the glass causes a tiny electric current to pass in the circuit connected to the counter.

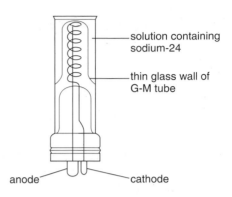

solution containing sodium-24

thin glass wall of G-M tube

anode   cathode

i) Why must the glass wall of the G-M tube be very thin?

ii) Why is this type of arrangement of no use if the radioactive decay is by alpha emission? *(1 mark)*

c) Sodium chloride solution is known as saline. It is the liquid used in 'drips' for seriously-ill patients. Radioactive sodium chloride, containing the isotope sodium-24, can be used as a tracer to follow the movement of sodium ions through living organisms. Give one advantage of using a sodium isotope with a half-life of a few hours compared to using an isotope with a half-life of:

i) five years *(1 mark)*

ii) five seconds. *(1 mark)*

Glossary

Absorption (digestion) The process in digestion whereby water or small soluble particles pass through the lining of the intestine into the bloodstream.

Absorption (plants) The process whereby plants take in water or dissolved mineral ions through their root hairs.

Acceleration Rate of change of velocity. Units are m/s².

Acid A substance that dissolves in water to give a solution with a pH of less than 7.

It is a substance that forms hydrogen ions, H^+(aq), when added to water.

Activation energy The minimum amount of energy needed by reactant particles before a reaction can occur.

Active transport The process by which cells take up substances against a concentration gradient. The process requires energy from respiration.

Activity rate/count rate The number of radioactive emissions in a certain time.

Adaptation A feature of an organism that helps it survive in a particular environment.

Addiction Being unable to do without a drug such as nicotine, alcohol or heroin.

Addition polymerisation A reaction in which unsaturated alkene molecules join to form saturated polymer molecules.

ADH (anti-diuretic hormone) A hormone released from the pituitary gland which targets the kidney tubules to ensure water is reabsorbed into the blood so reducing the amount of water in urine and controlling water balance.

Aerobic respiration Respiration that takes place in the presence of oxygen.

Air A mixture of gases made up of approximately 4/5 nitrogen and 1/5 oxygen.

Air resistance (drag) The force from the air that opposes movement.

Alkali A base (metal oxide or hydroxide) that dissolves in water to form a solution with a pH greater than 7.

It is a substance that forms hydroxide ions, OH^-(aq), when added to water.

Alkali metals The name given to the metals in Group 1 of the Periodic Table.

Alkanes A family of hydrocarbons with the general formula C_nH_{2n+2}. Methane (CH_4) is the simplest alkane. Alkanes have a single covalent bond between the atoms.

Alkenes A family of unsaturated hydrocarbons with the general formula C_nH_{2n}. Ethene (C_2H_4) is the simplest alkene. Alkenes have a double covalent bond between two carbon atoms. Alkenes decolourise bromine water.

Alleles One of the different forms of a particular gene. For example the allele for eye colour can be for blue or brown eyes.

Alloys A mixture of metals (or of carbon with a metal). Alloys can have properties different from the parent metal(s).

Alpha (radiation/particles) A type of radioactive emission with low penetrating power blocked by paper.

An alpha particle is made up of two neutrons and two protons (a helium nucleus).

Alternating current (a.c.) A current that changes direction as the supply voltage changes from + to −, or − to +.

Alveoli (singular: alveolus) A microscopic air sac in the lungs which acts as the surface for gaseous exchange.

Amino acids The breakdown products of the digestion of proteins, and the building blocks for making new proteins.

Ammeter An instrument used to measure the size of an electric current.

Ampere (amp) The unit of electric current.

Amplitude The maximum displacement of a wave from the equilibrium position.

Amylase An enzyme that digests starch into sugar.

Anaerobic respiration Respiration which takes place in the absence of oxygen.

Analogue signals Signals carried as a series of 'on' and 'off' pulses.

Anhydrous Crystals from which water has been removed.

Anions Atoms or groups of atoms that have gained electrons to become negatively-charged ions.

Anode The positively-charged electrode.

Antibodies A protein produced by some white cells which neutralises the effects of foreign cells.

Antitoxins Substances which neutralise a toxin (poison) produced by bacteria by combining with it.

Anus The opening at the end of the digestive system.

Aorta The main artery of the body which carries oxygenated blood away from the heart.

Artery A blood vessel that carries blood away from the heart.

Artificial satellite A satellite put into orbit from the Earth.

Artificial selection The process of deliberately selecting and breeding organisms with desired characteristics.

Asexual reproduction Reproduction that does not involve the formation and fusion of gametes. The offspring have identical genetic information to the parent.

Atmosphere The layer of gases around the Earth.

Atom The smallest part of an element that can exist. Atoms have a nucleus consisting of protons and neutrons around which are shells of electrons.

Atomic number The number of protons present in an atomic nucleus (and the number of electrons present in the neutral atom).

Atria (singular: atrium) The upper chambers of the heart which receive blood from the veins.

Background radiation The radioactivity that is always present in the environment.

Bacteria (singular, bacterium) A single celled organism consisting of cytoplasm and a membrane surrounded by a cell wall, with genes not organised to form a distinct nucleus.

Base An oxide or hydroxide of a metal.

Battery A number of electrical cells joined together.

Bauxite The main ore of aluminium containing aluminium oxide (Al_2O_3).

Beta (radiation/particles) A type of radioactive emission with moderate penetrating power blocked by thin sheets of aluminium.

Beta particles are high-energy electrons.

Big Bang theory A theory that considers that the Universe started from a gigantic explosion.

Bile A liquid produced by the gall bladder that breaks up fats into droplets.

Biodegradable Materials which can be broken down (decomposed) by bacteria.

Biomass The dry mass of living material in an ecosystem.

Black hole An object in space that is so dense and its gravitational field so strong that light and other forms of electromagnetic radiation cannot escape from it.

Bladder The sac which fills with urine from the kidney.

Blast furnace The industrial method used for extracting metals such as iron, from their ores.

Bonding The forces that hold atoms together.

Brain The part of the central nervous system which controls and co-ordinates most of the body's activities.

Braking distance The distance a vehicle travels before stopping after the brakes have been applied.

Breathing The action of passing air into and out of the lungs.

Bromine water A saturated solution of bromine dissolved in water. Used to distinguish between alkanes and alkenes.

Cancer A group of cells that are dividing very much more rapidly than is normal.

Capillaries Fine, thin-walled blood vessels which form a network for the exchange of substances with the tissue cells.

Carbon dioxide A gas formed during respiration and in the combustion of hydrocarbons. It turns clear limewater milky.

Catalyst A substance that changes the speed of a reaction but remains unchanged after the reaction.

Cathode The negatively-charged electrode.

Cations Atoms or groups of atoms that have lost electrons to become positively-charged ions.

Cell (biological) The smallest part of an animal or plant.

Cell (electrical) The single unit from which batteries are made.

Cellulose The material that makes up the cell wall of plant cells.

Cell division The processes of mitosis and meiosis.

Cell membrane The very thin membrane on the outside of a cell that controls the movement of substances in and out of the cell.

Cell wall The outer part of a plant cell that gives strength and shape to the plant cell.

Chain reaction A reaction in which a nucleus is split and neutrons released that can split other nuclei to produce a continuous chain of events.

Characteristic A feature of an organism that can be observed.

Charge A feature of atomic particles. Protons and electrons have a charge. Electrons have a negative charge and protons have a positive charge. Opposite charges attract; like charges repel.

Chlorophyll A green pigment found in the leaves and stems of plants. It traps light energy for use in photosynthesis.

Chloroplast That part of the plant cell containing chlorophyll.

Chromosome One of the thread-like structures found in the nucleus which contains genetic material. They are made of a single very long molecule of DNA. In humans there are 23 pairs of chromosomes in each body cell, but 23 single chromosomes in each gamete.

Ciliary muscles Muscles in the eye which control the thickness of the lens when focussing on near and distant objects.

Circuit breaker A device that uses the action of an electromagnet to switch off an electrical supply very rapidly.

Clone An organism produced asexually from one parent. The clone will be genetically identical to the parent.

Coke The substance used as the reducing agent in the blast furnace. It is almost pure carbon and is made by heating coal.

Combustion The burning of a substance in oxygen to release heat energy.

Comet An object made of ice and rock which orbits the Sun in a different plane to the planets.

Competition The interaction of organisms which are trying to obtain the same food or occupy the same space.

Compound A substance which contains two or more elements chemically joined together.

Compression (forces) The process of being squashed.

Compression (waves) The region in a sound wave where the vibrating particles of the medium particles are closer together than normal.

Concentration gradient This exists wherever there is a difference in the concentration of a substance in two areas.

Conduction (electrical) The transfer of electrical energy along a material by mobile electrons.

Conduction (heat) The transfer of heat energy along a material.

Conductor (electrical) A substance allowing electrical charge to pass through.

Conductor (heat) A substance allowing heat energy to pass through.

Connector neurone (relay neurone) Neurones found in the spinal cord which link sensory and motor neurones.

Constructive plate boundary A boundary where two tectonic plates are moving away from each other so that magma comes to the surface to form new rocks.

Consumer An organism which has to rely on food made by green plants (producers) or other animals.

Convection The transfer of heat energy in a liquid or a gas (fluid) caused by differences in density. Warmer, less dense fluids rise. Cooler, more dense fluids sink.

Core (Earth) The innermost part of the Earth.

Core (electromagnets) The central part of electromagnet around which current-carrying coils of wire are wound.

Cornea The transparent front part of the eye. Plays an important part in focussing light onto the retina.

Corrosion A reaction between a metal and substances in the atmosphere.

Cosmic ray Rays and particles from space reaching Earth.

Coulomb Unit of electric charge.

Covalent bond The bonding of atoms caused by the sharing of pairs of electrons in their outer electron shells.

Cracking A form of thermal decomposition in which large hydrocarbon molecules are broken down into smaller ones.

Crest The point of maximum displacement in a transverse wave.

Critical angle When a ray of light, travelling in a more dense medium, hits the boundary between the more dense medium and the less dense medium and only just emerges by refraction, the angle of incidence of the ray in the more dense medium is called the critical angle.

Crude oil A mixture of substances, most of which are hydrocarbons, formed by the anaerobic decomposition of marine organisms over a long period of time.

Crust The surface layer of the Earth.

Cryolite Used with bauxite in the extraction of aluminium by electrolysis. It lowers the melting point of bauxite and increases the electrical conductivity of the bauxite.

Current A flow of electrons, ions or electric charge.

Cycle (vibration) For a transverse wave one cycle is one trough and one crest. In a longitudinal wave one cycle is one compression and one rarefaction.

Cytoplasm All the material in a cell inside the membrane (apart from the nucleus), where chemical reactions take place under the control of enzymes.

Decay (atomic) The break up of unstable nuclei resulting in the production of radioactive emissions.

Decay (biological) The breaking down by decomposers of complex organic materials into simple ones.

Decelerate To slow down.

Decomposers Organisms which break down complex organic materials into simple ones during decay. Most are bacteria and fungi.

Deforestation The removal of trees from woodland and mountain sides. It often leads to flooding of rivers as the trees can no longer take up the rain that falls on them.

Denaturing The process by which enzymes are destroyed when heated above a temperature of about 40 °C

Denitrifying bacteria Bacteria which convert nitrates in the soil into nitrogen gas.

Deoxygenated blood Blood that is not rich in oxygen.

Deposition The laying down in water of a layer of rock fragments.

Destructive plate boundary A boundary where two tectonic plates are moving towards each other where rocks are squeezed together to form mountains or where rocks melt back to magma and are recycled.

Detrivores Organisms, such as termites, earthworms, fungi and bacteria that obtain energy and nutrients by feeding on dead organic matter. The decomposers are a class of detrivores.

Diabetes A disease caused by lack of insulin production from the pancreas. Blood sugar levels cannot be controlled properly.

Dialysis The treatment of kidney failure by taking blood from the patient and removing the urea and other waste products by diffusion.

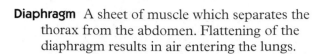

Glossary

Diaphragm A sheet of muscle which separates the thorax from the abdomen. Flattening of the diaphragm results in air entering the lungs.

Diffraction The spreading out of a wave as it passes through a narrow gap or moves past an object.

Diffusion The movement of particles resulting in a net movement from a region where they are at a high concentration to a region where they are at a lower concentration.

Digestion The process by which food is broken down into particles small enough to be absorbed into the blood.

Digital signals signals carried as continuous waves that vary in frequency and amplitude.

Dilate To get wider.

Diode An electrical device that only conducts electricity flowing in one direction.

Direct current (d.c.) Electric current that does not change direction.

Displacement reaction A reaction in which one metal displaces another metal.

DNA The chemical that carries the genetic information on the genes.

Dominant The allele that controls an observable characteristic (phenotype) in the offspring even when it is present on only one chromosome (heterozygous).

Drag The force from the air that opposes movement.

Dynamo (generator) Device supplying a voltage from the relative motion of a conductor with a magnetic field.

Earthing The linking of a low resistance wire to a metal object to provide an easy path to the Earth's surface for an electric current.

Ecosystem The organisms living and surviving in a particular place.

Eddy current Small electric current loops in objects in a changing magnetic field.

Effector neurone Nerve cell which carries impulses away from the spinal cord towards effectors.

Effectors Structures such as muscles or glands which carry out responses to stimuli.

Efficiency The ratio of useful energy transferred by device to total energy transferred to device.

Elastic potential energy The energy stored in an object when work has been done to change its shape.

Electric charge A quantity of electricity.

Electric current The flow of electrons or ions. The rate of transferring electric charge. Units are amperes (amps)

Electrical energy Energy transferred by a charge moving due to an electric force.

Electrode A negatively or positively charged conductor.

Electrolysis The process of splitting up a chemical compound using an electric current.

Electromagnetic induction The production of a voltage or current across a conductor in relative motion within a magnetic field.

Electromagnetic spectrum The range of frequencies and wavelengths of electromagnetic waves.

Electromagnetic waves Transverse waves that have a common speed in air or a vacuum; can travel through a vacuum.

Electrons Negatively-charged sub-atomic particles orbiting in shells around the atomic nucleus.

Electrostatic forces Forces due to stationary electric charge. Like charges repel, unlike charges attract.

Element A substance made up of atoms which contain the same number of protons so contain only one type of atom, and which cannot be broken down into anything simpler by chemical means.

Emulsifying Breaking down of a liquid into very fine droplets.

Endocrine gland A gland which discharges its products, called hormones, straight into the blood.

Endothermic reaction A reaction in which heat energy is transferred from the surroundings because more energy is needed to break the existing bonds in the reactants than is released when new bonds are made in the products.

Energy level diagrams Diagrams which show the energy content of the reactants and the products during a chemical reaction.

Environment The surroundings and conditions that affect the growth and behaviour of plants and animals.

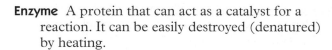

Enzyme A protein that can act as a catalyst for a reaction. It can be easily destroyed (denatured) by heating.

Equilibrium (reversible reactions) When the forward reaction proceeds at the same rate as the reverse reaction.

Erosion The wearing away of the Earth's surface.

Eutrophication A process caused when large amounts of nitrates and phosphates are discharged into rivers and streams. The nutrients cause the rapid growth of algae and water plants. The eventual death of the algae and plants soon leads to the rapid growth of aerobic bacteria. These decomposers soon use up all the available oxygen in the water. This in turn causes other animal life in the water to suffocate and die.

Evaporation The loss of the more energetic particles from the surface of a liquid.

Excretion The removal of chemical waste material made in the body or a cell.

Exhale To breathe out.

Exothermic reaction A reaction in which heat energy is transferred to the surroundings because more energy is given out making the new chemical bonds in the products than is taken in to break the existing bonds in the reactants.

Extinct A description of an organism no longer living today but which according to the fossil record has lived in the past.

Extrusive rocks Igneous rocks formed when lava cools on the surface of the Earth. Because the rock cools rapidly, the crystals are small.

Eye A sense organ that contains the receptors sensitive to light.

Faeces The indigestible food remaining once digestion has taken place.

Fatty acids The breakdown products of fats.

Fermentation The changing of glucose into ethanol (alcohol) and carbon dioxide by the action of enzymes in yeast.

Fertile The ability of a male or female to produce sex cells which are capable of producing viable offspring.

Fertilisation The fusion of an egg and a sperm.

Fertiliser A substance which can be natural or artificial applied to soil to improve the growth of plants.

Fertility drugs Drugs that stimulate the release of eggs from the ovaries.

Fetal imaging Using ultrasound waves to view a fetus.

Fetus The name given to an unborn child more than 8 weeks after conception.

Filtration Filtration helped by the high pressure of the blood in the capillaries in the glomerulus resulting in the first stage of the formation of urine.

Filtration (kidney) Filtration of water and ions from the bloodstream, to form urine.

Flammable Easily set on fire.

Fluid A liquid or a gas.

Focus The formation of a sharp image of near and distant objects by altering the shape of the lens.

Food chain A diagram which shows feeding relationships of some organisms in an ecosystem. All food chains start with producers which trap light energy.

Formula mass The mass in grams of one mole of a substance.

Fossil The remains or imprints of dead plants or animals trapped in sedimentary rocks when the rocks were formed. The remains or imprints may have been mineralised and turned into stone.

Fossil fuels The non-renewable energy resources: crude oil, natural gas and coal.

Fossilisation The process that produces fossils.

Fractional distillation A method of separating liquids whose boiling points are close together. The process used to separate the different substances in crude oil.

Free electron The electrons in metals that move around inside the metal and do not remain in orbit around a nucleus. The presence of these free electrons allows the metal to conduct electricity and heat.

Frequency The number of cycles (vibrations) per second. Units are Hertz (Hz).

Friction A force which opposes the movement of an object.

FSH (follicle-stimulating hormone) The hormone secreted by the pituitary gland that causes eggs to mature and stimulates the ovaries to produce oestrogen.

Fuse A wire fitted in plugs that is designed to melt if too large a current flows through it.

Fusion (atomic) The joining of small nuclei to form a large nucleus. The process transfers heat energy to the surroundings.

Fusion (biological) The process that occurs when the nucleus of a male gamete combines with the nucleus of a female gamete.

Galaxy A vast number of star systems held together by gravitational forces.

Gall bladder A small organ joined to the liver that stores bile.

Galvanising The process of covering iron or steel with a layer of zinc.

Gamete A sex cell.

Gamma (radiation) A type of radioactive emission with high penetrating power blocked by concrete/lead. Gamma radiation is part of the electromagnetic spectrum and has a very high frequency.

Gaseous exchange Occurs in the alveoli in the lungs when oxygen diffuses across the alveolar membrane from the lungs to the capillaries and carbon dioxide diffuses from the blood capillaries into the alveoli.

Geiger–Müller tube (GM tube) A detector of radioactive emissions.

Gene A unit of inheritance controlling one particular characteristic and made up of a length of chromosomal DNA.

Generator (dynamo) Device supplying a voltage from the relative motion of a conductor within a magnetic field.

Genetic Related to inheritance.

Genetic engineering The deliberate changing of the characteristics of an organism by manipulating chromosomal DNA.

Genotype The genetic make-up of an individual. The sum total of all the genes even if they are not shown in the individual.

Geostationary satellite A satellite which takes 24 hours to orbit the Earth.

Geothermal energy The energy produced in the Earth by natural heating process.

Giant structure Ionic compounds that have high melting points, usually dissolve in water and are good conductors of electricity when molten or in aqueous solution.

Gland A structure that releases hormones into the bloodstream.

Global warming An international problem caused partly by the increase in the amounts of carbon dioxide and methane in the atmosphere which results in an increase in the average temperature of the Earth.

Glucagon A hormone released by the pancreas that causes the liver to convert glycogen into glucose.

Glycerol A breakdown product from the digestion of fats and oils.

Glycogen The form in which excess glucose in the blood is stored in the liver and muscles.

Gravitational potential energy The energy stored in an object due to the vertical height through which it has been lifted.

Gravity (gravitational force) A force of attraction that acts between all objects.

Greenhouse effect The effect in the atmosphere of heat energy being absorbed by gases such as carbon dioxide and methane.

Group A vertical column of elements in the Periodic Table having similar chemical properties due to the atoms of the elements having the same number of electrons in their outer shells.

Guard cells Pairs of cells which surround the stomata on the surface of leaves which by means of osmosis open and close the stomata thus regulating the flow of gases into and out of the leaf.

Gullet *see oesophagus.*

Haematite A common ore of iron containing iron(III) oxide.

Haemoglobin The red pigment in the red blood cells which combines with and transports oxygen.

Half-life The time taken for half a given number of radioactive atoms to decay to different atoms.

Halide The compound formed when a halogen reacts with another element.

Halogens The name given to the elements in Group 7 in the Periodic Table.

Heart A double pump with the right side pumping blood at low pressure to the lungs to release carbon dioxide and collect oxygen and the left side pumping oxygenated blood at higher pressure around the body.

Herbicide A chemical used to destroy unwanted plants.

Herbivore An organism that feeds only on plants.

Hertz (Hz) The unit of frequency.

Heterozygous The inheriting of one dominant allele and one recessive allele for a particular characteristic.

Homeostasis The automatic control system by which the internal conditions of an organism are kept steady.

Homozygous The inheriting of two dominant or two recessive alleles for a particular characteristic.

Hormone A substance secreted by endocrine glands directly into the blood in one part of the body and carried in the blood plasma to a target organ. Plant hormones are called auxins.

Hydrocarbons Compounds containing only hydrogen and carbon.

Hydroelectric Electrical power generated by the flow of moving water.

Hydrogen The chemical element with the lowest density. Small amounts of it burn with a squeaky pop.

Hydrogen ion An ion (H^+) present in all acids.

Hydroxide ion An ion (OH^-) present in all alkalis.

Igneous rocks Rocks formed by magma rising upwards from the mantle, cooling and solidifying into a hard crystalline rock.

Image The picture formed on the retina of the eye.

Indicator A dye which changes colour when mixed with acidic, alkaline or neutral solutions.

Inert Unreactive.

Inhale To breathe in.

Insoluble A substance that will not dissolve in a liquid, usually water.

Insulator (electrical) A substance not allowing an electric current to flow and charges to move.

Insulator (heat) A substance not allowing a transfer of heat energy from a hot region to a cold region.

Insulin A hormone released by the pancreas which helps to control sugar level in the blood.

Insulin controls the conversion of excess glucose into glycogen which is then stored in the liver and muscle cells.

Intrusive rock Igneous rocks formed when lava cools beneath the surface of the Earth. Because the rock cools slowly, the crystals are large.

Ion An atom or group of atoms which have lost or gained electrons to become positively or negatively charged.

Ionic bond The electrostatic attraction between opposite charges responsible for holding metal and non-metal elements together in a compound. The ions are formed when the metal atoms transfer electrons to the non-metal atoms in order to achieve full outer electron shells.

Ionic equation Equation which shows the ions taking part in a reaction. In ionic equations the charges must balance as well as the number of ions/atoms involved.

Ionise To remove or add electrons to atoms or groups of atoms so giving them positive or negative charges.

Iris A ring of muscle which controls the amount of light entering the eye.

Isotope Atoms of the same element which contain different numbers of neutrons.

Joule The unit of energy.

Kidney An organ which removes excess water from the blood and excretes urine made from the urea produced in the liver.

Kilowatt 1000 watts.

Kilowatt hour A unit of electrical energy.

Kinetic energy The energy possessed by an object due to its motion.

Lactic acid A product of anaerobic respiration in very active human muscles which is a mild tissue poison (causes the muscles to hurt).

Large intestine The part of the digestive system where water is removed from indigestible food.

Lava Magma that has erupted through the Earth's crust.

Law of Conservation of Mass This states that during any chemical reaction matter (material) is neither created nor destroyed.

LDR (light dependant resistor) An electrical component the resistance of which decreases when light shines on it.

Lens A transparent structure within the eye that is flexible and helps light to form a sharp image on the retina during focussing.

LH (luteinising hormone) The hormone secreted by the pituitary gland that stimulates the release of an egg.

Light year The distance a light ray travels in one year.

Limewater A solution of calcium hydroxide that turns milky when carbon dioxide is passed through it.

Limiting factor The factor such as light intensity (brightness), light wavelength, water and carbon dioxide that limits the rate of photosynthesis at a given time.

Lipids Foods made up of fats and oils.

Lipase An enzyme that digests fats to fatty acids.

Lithosphere The outer shell of the Earth made from the crust and the upper part of the mantle.

Liver An organ where excess glucose in the blood is stored as glycogen, where bile is produced and where poisons such as alcohol are removed from the blood.

Longitudinal wave A wave in which the vibrations of the particles are in the same direction as the energy transferred along the wave.

Magma Molten rock below the Earth's crust.

Magnet An object that attracts magnetic materials such as iron, steel, nickel and cobalt.

Magnetic field The region around a magnet where a magnetic material experiences a magnetic force.

Mantle The layer of the Earth between the crust and the core.

Mass number The total number of protons and neutrons in an atomic nucleus.

Mass The amount of matter an object contains Units are kg.

Meiosis Cell division that leads to the production of gametes in which there has been some reassortment of genetic material so producing variation. It is a reduction division so each gamete has only half the number of chromosomes as the parent.

Metamorphic rocks Rocks formed from rocks which became buried deep underground and had their structure changed by high temperatures and or high pressures.

Migration The mass movement of organisms on a regular basis. Most migrations are connected with seasonal changes and enable organisms to maintain food supplies.

Milky Way The galaxy containing our solar system.

Mineral A solid element or compound found naturally in the Earth's crust.

Mitochondria (singular: mitochondrion) The parts of the cell in which aerobic respiration takes place producing cellular energy.

Mitosis Cell division that occurs during growth and asexual reproduction and involves each chromosome making an exact copy of itself, resulting in the formation of two daughter cells each with the same number of chromosomes as the parent.

Mixture Two or more substances which are usually easy to separate.

Molar volume The volume occupied by 1 mole of any gas. The molar volume of any gas at room temperature and atmospheric pressure is $24 \, dm^3$ ($24 \, 000 \, cm^3$).

Mole The mass in grams of 6×10^{23} particles of any substance. It is the relative atomic mass of an element or the relative formula mass of a substance expressed in grams.

Molecule A particle containing atoms of the same or different elements bonded together. The smallest part of an element or compound that can take part in a chemical reaction.

Monomers Small molecules which join together to form a long chain of molecules called a polymer.

Moon A natural satellite in orbit around a planet.

Motor effect The motion of a current-carrying conductor in a magnetic field

Motor neurones Neurones that carry electrical impulses from the brain or spinal cord to an effector.

Mucus A sticky fluid which traps dust or protects surfaces.

Mutation A change suddenly occurring in one or more of the genes or chromosomes or in the number of chromosomes.

Natural selection The process by which beneficial characteristics with greater survival value are selected and increase in proportion in the population. Natural selection leads to evolution.

Negative The charge on an electron.

Negative feedback An automatic control mechanism in which a change from the normal condition triggers off a response which restores the normal condition.

Nerve impulses Electrical signals which travel along nerve pathways made up of nerve cells (neurones).

Neurone A cell in the nervous system.

Neutral (charge) Having no overall charge.

Neutral (indicators) Having a pH of 7.

Neutralisation A reaction between an acid and a base or a carbonate.

Neutron A particle with no electrical charge found in the nucleus of most atoms. Its mass is similar to that of a proton.

Newton The unit of force (N).

Nicotine The addictive substance in tobacco.

Nitrates Chemicals containing NO_3 ions, frequently used in fertilisers to help plants synthesise proteins.

Nitrifying bacteria Bacteria which convert ammonium compounds in the soil into nitrates.

Noble gases The name given to the elements in Group 0 in the Periodic Table.

Non-metals Elements in the Periodic Table which usually have low melting points and boiling points, are poor conductors of electricity and heat, and as solids are brittle.

Non-renewable (finite) energy resources Energy resources that, once used, cannot be replaced.

Normal The line drawn at right angles to a surface.

Nuclear fission The breaking up of a large atomic nucleus to release energy.

Nucleon The protons and neutrons in the nucleus of an atom.

Nucleus (atom) The central part of an atom that contains positively-charged protons and uncharged neutrons.

Nucleus (cells) The part of a cell that contains the chromosomes which carry the genes controlling the cell's characteristics.

Nutrition The process by which organisms obtain their raw materials and absorb useful substances from it.

Oesophagus (gullet) The muscular tube which carries food from the mouth to the stomach.

Oestrogen A hormone produced by the ovaries which controls female sexual characteristics.

It inhibits the production of FSH and stimulates the release of LH.

Ohm The unit of electrical resistance.

Optic nerve A bundle of nerve cells which carries impulses from the eye to the brain.

Oral contraceptive Tablets, usually containing oestrogen, that inhibit the production of FSH so that no eggs mature.

Orbit The regular path taken by an object which passes around another object.

Ores Minerals or mixtures of minerals from which a metal can be extracted in economically viable amounts.

Organ A group of tissues working together to carry out a particular function.

Organ system A group of organs working together to carry out a particular function or group of related functions.

Organic Compounds of carbon found in large quantities in living and dead organisms.

Organism An individual plant or animal.

Osmosis The diffusion of water through a partially-permeable membrane – the water flowing from a region of high water concentration to a region of lower water concentration.

Oxidation A chemical reaction which involves the addition of oxygen.

A reaction involving the loss of electrons.

Oxygen The chemical element that is vital to life. It will relight a glowing spill.

Oxygenated blood Blood rich in oxygen.

Oxygen debt The oxygen needed to remove the lactic acid from the muscles produced as a result of muscles respiring anaerobically during vigorous exercise.

Oxyhaemoglobin The chemical formed when oxygen combines with haemoglobin.

Ozone layer The layer of gas in the upper atmosphere that reduces the amount of harmful ultraviolet radiation reaching the Earth's surface.

Palisade cells the cells in the upper part of green leaves which contain most chlorophyll and carry out most of the photosynthesis in the leaf.

Pancreas An organ of the digestive system that produces the hormone insulin and the enzyme lipase.

Parallel circuits Closed electrical circuits that provide several pathways for an electric current.

Partially permeable membrane Allows small molecules to pass through quickly but not large molecules.

Period A horizontal row of elements in the Periodic Table.

Periodic table The arrangement of the elements in order of increasing atomic number.

Pesticide A chemical designed to kill unwanted organisms.

pH A scale used to measure acidity and alkalinity.

pH scale A set of numbers from 1 to 14 used to measure the acidity or alkalinity of an aqueous solution.

Phenotype The way an individual appears as a result of the alleles it carries and the environment in which it has grown up.

Phloem A column of cells in a plant responsible for the transport of food made in photosynthesis to wherever it is needed.

Phosphates Chemicals containing PO_4 ions, frequently used as fertilisers to help plants photosynthesise and respire.

Photosynthesis The process in green plants which produces biomass (initially carbohydrates) and oxygen, and requires carbon dioxide and water as raw materials and chlorophyll to enable the plant to absorb light energy.

Pituitary gland A gland, found at the base of the brain, that secretes FSH and LH.

Planet A very large object which orbits the Sun.

Plasma The straw-coloured liquid part of the blood which transports cells and dissolved substances.

Platelets Cell fragments which help in forming blood clots at wounds.

Poles The parts of a magnet where the magnetic forces are strongest.

Pollution The introduction of harmful substances into an environment.

Polymer A long chain molecule made up of many smaller molecules called monomers.

Polymerisation A reaction in which small molecules join together to make larger molecules.

Population The numbers of one species of animal living in a particular area.

Positive The charge on a proton.

Potassium An element used by plants to help the action of the enzymes involved in photosynthesis and respiration.

Potential difference The voltage between two points a circuit.

Power The rate of transfer of energy. Units are watts.

Precipitate The formation of an insoluble solid during the reaction between two solutions.

Precipitation (chemical) The type of reaction in which a precipitate is formed.

Precipitation (weather) The name given to rain, hail and snow.

Predation The process by which one animal (**predator**) catches then eats another (**prey**).

Primary coil The input coil in a transformer.

Producer A green plant which photosynthesises to make its own food.

Products The new materials produced as a result of a chemical reaction.

Proteases Enzymes that digest proteins into amino acids.

Proton A positively-charged particle found in the nucleus of an atom. It has a mass similar to that of a neutron and the number of protons present decides which element is present.

Pulmonary artery The blood vessel that takes deoxygenated blood from the heart to the lungs.

Pulmonary vein The blood vessel that takes oxygenated blood from the lungs to the left atrium of the heart.

Pupil The gap surrounded by the iris through which light passes into the eye. Pupil size can be changed by dilation and constriction of the iris.

Putrefying bacteria These break down animal waste and produce ammonia.

P waves Longitudinal seismic waves which travel through solids and liquids.

Pyramids Diagrams which illustrate quantitatively the relationships between organisms in a food chain. Each organism is represented by a block in the pyramid. Pyramids can show number, biomass or energy relationships.

Radiation (heat transfer) A process by which heat is transferred.

Radiation (nuclear) The random emission of energy from an atomic nucleus as the result of the breakdown of unstable nuclei.

Radioactive (radiocarbon) dating The use of half-life to date ancient organic objects.

Radioactive decay The emission of particles or rays from an unstable atomic nucleus.

Radioactive emissions The particles and rays produced as the result of the breakdown of unstable nuclei.

Radioactive tracer A radioactive substance, usually with a relatively short half-life, that is passed into the body and used to detect, for example, the presence of cancers, tumours or the direction of blood flow. Tracers can be used in the treatment of cancers and tumours. Tracers can also be used to monitor the flow of liquids and gases in underground pipes.

Radioactivity The random emission of energy from an atomic nucleus as the result of the breakdown of unstable nuclei.

Radioisotope A radioactive isotope.

Radionuclides Materials which produce ionising radiation, such as X rays, gamma radiation, alpha particles and beta particles.

Random Spontaneous not regular.

Rarefaction The region in a sound wave where the vibrating particles of the medium particles are further apart than normal.

Reabsorption The way in which substances needed by the body are taken back into the blood from the tubules in the kidney.

Reactants The starting materials in a chemical reaction.

Reactivity series A list of metals arranged in order of their chemical reactivity. The most reactive metals are at the top of the list.

Receptors Special cells which are capable of detecting environmental changes.

Recessive The allele which must be present on both chromosomes to show an effect in the phenotype.

Red blood cells Cells that contain haemoglobin and whose function is to transport oxygen around the body.

Red giant A relatively cool giant star.

Red shift The effect on the spectrum of a galaxy being moved to the red end due to the galaxy moving away from us.

Reduction A chemical reaction which involves the loss of oxygen.

A reaction involving the addition of electrons.

Reflex action A rapid automatic response to a stimulus, during which nerve impulses are sent by receptors through the nervous system to effectors.

Reflex arc The route taken by a nerve impulse through the nervous system to bring about a reflex action.

Refraction The change in direction of a wave when it passes from one medium to another due to a change in speed when passing from one medium to another.

Relative atomic mass The average mass of an atom of an element on a scale on which the mass of a hydrogen atom = 1 or the mass of the ^{12}C isotope of carbon = 12. It takes into account the relative abundance of different isotopes with different mass numbers.

Relative molecular mass (relative formula mass) This is found by adding together the relative atomic masses of all the atoms in one molecule of the substance.

Relay neurone (connector neurone) Neurones found in the spinal cord which link sensory and motor neurones.

Renal To do with the kidney.

Renal artery The blood vessel that carries blood to the kidneys.

Renal vein The blood vessel that carries blood away from the kidneys.

Renewable energy resources Energy resources that will always be available.

Reproduction The formation of offspring.

Resistance A measurement describing the difficulty of electric current flow in a conductor. Units are ohms.

Resources Natural materials available for the use of organisms.

Respiration The process taking place in living cells transferring energy from food molecules (glucose) to cellular energy.

Respire The cellular process of obtaining energy from food.

Response The reaction of an organism to a stimulus.

Retina The light receptor surface at the back of the eye where light sensitive receptors convert light into nerve impulses.

Reversible reaction A reaction that can proceed in either direction depending on the reaction conditions. Reactants can be changed into products which in turn can be changed back into reactants.

Rib muscles The muscles between the ribs which contract to raise the rib cage for inhalation.

Root hairs Cells with a large surface area and thin cell wall that absorb water and mineral salts from the soil by osmosis, diffusion and active transport.

Rusting The corrosion of iron in the presence of air and water to form hydrated iron oxides.

Sacrificial protection Used to reduce the rusting of iron by attaching a more reactive metal such as magnesium or zinc.

Saliva A liquid containing the enzyme amylase produced in the salivary glands.

Salivary glands Glands with tubes emptying saliva into the mouth. Glands in the mouth that secrete saliva and the enzyme amylase.

Satellite An object which orbits a planet.

Saturated hydrocarbons Hydrocarbons in which the carbon atoms are all linked together with single C — C bonds.

Sclera The tough outer layer of the eye.

Secondary coil The output coil in a transformer.

Sedimentary rocks Rocks formed by deposition, burial and compression of weathered rock fragments or the shells of dead animals. They can also be formed by the precipitation of calcium carbonate usually in warm, shallow seas.

Seismic waves These are waves created in the Earth by vibrations due to earthquakes.

Selective breeding The process of deliberately breeding animals or plants according to desirable characteristics.

Sensory neurone A nerve cell which carries impulses from sense cells or organs to the spinal cord.

Series circuits Closed electrical networks giving only one pathway for an electric current.

Sexual reproduction This involves two parents who each produce sex cells that must join together. The offspring are genetically different from the parents and each other.

Skin A water-proof, germ-proof layer that contains receptors sensitive to touch, pressure and temperature and plays a part in temperature control.

Small intestine That part of the digestive system where the absorption of soluble foods into the blood occurs.

Smelting The process of getting a metal from its ore by heating the ore with carbon.

Solar system A system made up of the Sun, planets, moons, asteroids and comets.

Soluble Able to be dissolved (usually in water).

Species A group of organisms which look similar and that can breed together to produce fertile offspring.

Speed The distance an object travels in a unit of time. Units are m/s.

Star A source of light due to heat caused by nuclear fusion.

Stimulus A change in the environment of an organism which produces a response.

Stomach The part of the digestive system after the gullet where food is churned into a liquid mass.

Stomata (singular: stoma) The tiny openings in the surface of a leaf through which gases can pass by diffusion. The size of the openings is regulated by the guard cells.

Structural formula A way of displaying the formula of a compound so that the bonds and the approximate positions of the atoms are shown. Usually only used for organic compounds.

Subduction zone An area where two tectonic plates are coming together to cause some of the Earth's crust to be pushed back into the mantle where it gets recycled.

Substrate A liquid enzymes can work on.

Sun A star at the centre of a solar system.

Suspensory ligaments The muscles in an eye that hold the lens in place.

S waves Transverse seismic waves that can only travel through solids.

Synapse The gap between two neurones.

Synthesis The process in which elements are chemically combined to make a new compound.

Target organ The organ affected by the release of a hormone.

Tectonic plates The separate slow-moving adjacent sections of the Earth's lithosphere that move because of convection currents within the Earth's mantle caused by the natural radioactive processes within the Earth.

Temperature How hot or how cold an object is. Units are °C.

Terminal velocity The constant speed reached by a falling object when the forces acting on it (in the direction of its motion) are balanced.

Thermal decomposition The breaking down of a compound by the action of heat.

Thermistor An electrical component in which the resistance decreases when it gets warm.

Thermit process A method of joining two lengths of railway track together using the exothermic reaction between aluminium and iron(III) oxide.

Thinking distance The distance travelled by a car during the driver's reaction time.

Thorax The chest cavity.

Tissue A group of cells working together to carry out a particular function.

Tissue fluid A liquid formed from the blood plasma and carries soluble substances from the blood to the tissue cells.

Total internal reflection This takes place at the boundary of two materials when light travelling in the more dense material strikes the boundary at an angle of incidence greater than the critical angle.

Toxic Poisonous.

Toxins Poisons.

Trachea The tube which connects the throat and the lungs and through which air passes into the lungs.

Transformer A device that changes the size of an alternating voltage.

Transition elements (transition metals) The name given to the elements in the Periodic Table between Groups 2 and 3.

Transpiration The process by which water evaporates from the leaf through the stomata, creating a pull causing water to rise up the plant in the transpiration stream.

Transportation The removal of rocks broken down by weathering and erosion.

Transverse wave A wave in which the vibrations of the particles are at right angles to the direction of the energy transferred along the wave.

Trough The point of maximum displacement in a transverse wave in the opposite direction to a peak.

Turbine A device that turns a generator.

Turgor The pressure that the cytoplasm and vacuole of a cell exert on the cell wall.

Ultrasound Sound of too high a frequency to be heard by humans.

Ultraviolet radiation Electromagnetic radiation, produced by the Sun, against which the skin needs to be protected to avoid skin cancers developing.

Universal indicator An indicator used to measure the pH of a solution to show whether the solution is acidic, neutral or alkaline.

Universe Made up of innumerable galaxies.

Unsaturated hydrocarbons Hydrocarbons in which some of the carbon atoms are all linked together with C=C double bonds.

Urea The breakdown product of amino acids produced in the liver and excreted by the kidneys in urine.

Ureter The tube taking urine from the kidney to the bladder.

Urine The waste fluid produced in the kidneys that contains urea, excess water and salts.

Vaccine A liquid containing dead or weakened disease-producing microorganisms that causes the body to produce antibodies.

Vacuole A cavity in the cytoplasm which is surrounded by a membrane. The vacuole contains cell sap.

Variation The differences in characteristics between members of the same species.

Vein A blood vessel taking blood to the heart.

Velocity The speed of an object in a particular direction. Units are m/s.

Vena cava The blood vessel that carries blood from the body to the heart.

Ventilation Movement of air in and out of the lungs during breathing.

Ventricles The lower pumping chambers of the heart.

Vibration The movement needed to produce a wave.

Villi (singular: villus) The finger-like projections in the small intestine that provide a large, thin, moist surface and good blood supply through which the soluble products of digestion are rapidly absorbed.

Virus An organism that consists only of a protein coat surrounding a few genes.

Viscosity A measure of how runny a liquid is.

Volt The unit of voltage; joules per coulomb.

Voltage The electrical energy difference of a charge moved across two points in a circuit; energy transferred per unit charge.

Voltmeter An instrument used to measure voltage.

Watt The unit of power.

Wavelength The distance between adjacent crests in a wave equivalent to the distance taken by one complete cycle.

Waves Vibrations that transfer energy but not matter.

Wave speed The distance travelled by a wave in a second. Units are m/s.

Weathering The chemical, physical or biological action by which rocks are broken down into rock fragments.

Weight The force due to gravity on an object. Units are newtons.

White blood cells These cells are important in the defence against disease by ingesting bacteria, producing antibodies or producing antitoxins which neutralise the toxins produced by bacteria.

White dwarf A small very dense star.

Wilting A condition brought about by loss of water from cells in a plant. The cells cease to be turgid and support for cells and plants is reduced.

Work That which is done when a force moves an object a certain distance. Units are joules.

Xylem A column of dead cells in a plant that are responsible for the transport of water and mineral ions upwards in the plant.

Index

Note: Glossary entries are in bold.